MW00835411

Oxford Lecture Series in
Mathematics and its Applications 20

*Series Editors*
John Ball Dominic Welsh

OXFORD LECTURE SERIES
IN MATHEMATICS AND ITS APPLICATIONS

1. J. C. Baez (ed.): *Knots and quatum gravity*
2. I. Fonseca and W. Gangbo: *Degree theory in analysis and applications*
3. P. L. Lions: *Mathematical topics in fluid mechanics, Vol. 1: Incompressilble models*
4. J. E. Beasley (ed.): *Advances in linear and integer programming*
5. L. W. Beineke and R. J. Wilson (eds): *Graph connections: Relationships between graph theory and other areas of mathematics*
6. I. Anderson: *Combinatorial designs and tournaments*
7. G. David and S. W. Semmes: *Fractured fractals and broken dreams*
8. Oliver Pretzel: *Codes and algebraic curves*
9. M. Karpinski and W. Rytter: *Fast parallel algorithms for graph matching problems*
10. P. L. Lions: *Mathematical topics in fluid mechanics, Vol. 2: Compressible models*
11. W. T. Tutte: *Graph theory as I have known it*
12. Andrea Braides and Anneliese Defranceschi: *Homogenization of multiple integrals*
13. Thierry Cazenave and Alain Haraux: *An introduction to semilinear evolution equations*
14. J. Y. Chemin: *Perfect incompressible fluids*
15. Guiseppe Buttazzo, Mariano Giaquinta and Stefan Hildebrandt: *One-dimensional variational problems: an introduction*
16. Alexander I. Bobenko and Ruedi Seiler: *Discrete integrable geometry and physics*
17. Doina Cioranescu and Patrizia Donato: *An introduction to homogenization*
18. E. J. Janse van Rensberg: *The statistical mechanics of interactiong walks, polygons, animals and vesicles*
19. S. Kuksin: *Hamiltonian partial differential equations*
20. Alberto Bressan: *Hyperbolic systems of conservation laws: the one-dimensional Cauchy problem*

# Hyperbolic Systems of Conservation Laws

## The One-Dimensional Cauchy Problem

Alberto Bressan

*Professor at the International School for Advanced Studies (SISSA)*
*Trieste, Italy*

OXFORD
UNIVERSITY PRESS

OXFORD
UNIVERSITY PRESS

Great Clarendon Street, Oxford OX2 6DP

Oxford University Press is a department of the University of Oxford.
It furthers the University's objective of excellence in research, scholarship,
and education by publishing worldwide in

Oxford New York

Athens Auckland Bangkok Bogotá Buenos Aires Calcutta
Cape Town Chennai Dar es Salaam Delhi Florence Hong Kong Istanbul
Karachi Kuala Lumpur Madrid Melbourne Mexico City Mumbai
Nairobi Paris São Paulo Singapore Taipei Tokyo Toronto Warsaw

with associated companies in
Berlin Ibadan

Oxford is a registered trade mark of Oxford University Press
in the UK and in certain other countries

Published in the United States
by Oxford University Press Inc., New York

© Alberto Bressan 2000

The moral rights of the authors have been asserted

Database right Oxford University Press (maker)

First published 2000

All rights reserved. No part of this publication may be reproduced,
stored in a retrieval system, or transmitted, in any form or by any means,
without the prior permission in writing of Oxford University Press,
or as expressly permitted by law, or under terms agreed with the appropriate
reprographics rights organization. Enquiries concerning reproduction
outside the scope of the above should be sent to the Rights Department,
Oxford University Press, at the address above

You must not circulate this book in any other binding or cover
and you must impose this same condition on any acquirer

A catalogue record of this book is available from the British Library

Library of Congress Cataloging in Publication Data

Bressan, Alberto, 1956-
Hyperbolic systems of conservation laws: the one dimensional Cauchy problem /
Alberto Bressan.
(Oxford lecture series in mathematics and its applications; 20)
1. Conservation laws (Mathematics) 2. Cauchy problem. I. Title. II. Series.

QA377 .B74 2000 515′.353–dc21 00-040645

ISBN 0 19 850700 3 (Hbk)

Typeset by Newgen Imaging Systems (P) Ltd., Chennai, India
Printed in Great Britain
on acid-free paper by
Biddles Ltd,
Guildford & King's Lynn

*To my parents*

*Anna and Aldo*

# Preface

This book provides a self-contained introduction to the mathematical theory of hyperbolic systems of conservation laws, including the main recent advances on the uniqueness and stability of entropy weak solutions. It has evolved from a set of lecture notes used at SISSA during the years 1993–1999.

A particular feature of non-linear hyperbolic systems is the appearance of shock waves. Global solutions can be found only in the space of discontinuous functions. Because of this lack of regularity, most modern mathematical techniques are not applicable, and new methods must be devised specifically to deal with these problems.

Since its appearance in 1965, the global existence theorem of Glimm has provided the foundation for most of the subsequent research on the subject. In particular, the *wave interaction functional* introduced by Glimm remains the basic tool for controlling the total strength of waves. By Helly's theorem, this yields the compactness of the set of approximate solutions, and hence the existence result. Our discussion of weak solutions will remain within the framework adopted in Glimm's paper, namely,

(i) one space dimension,
(ii) strictly hyperbolic systems,
(iii) each characteristic field either genuinely non-linear or linearly degenerate,
(iv) Cauchy problem with small $BV$ data.

In addition to the basic existence theory, this book will cover the new results on uniqueness and $\mathbf{L}^1$ stability of entropy admissible weak solutions. In the main, these results on the well-posedness of the Cauchy problem, within a space of functions with bounded variation, have been obtained by the author together with several young collaborators at SISSA during the years 1992–1998. Thanks to a major contribution by Liu and Yang, we believe that this theory has now reached a mature stage and is suitable for presentation in a textbook.

While all previous monographs on the subject relied on the Glimm random approximation scheme, a particular feature of the present work is the systematic use of wave-front tracking approximations both in the construction of entropy weak solutions, as well as in the analysis of their qualitative properties. The two approaches are essentially equivalent, in the sense that both types of approximations converge to the same limit solutions and yield the same theoretical results. Moreover, the fundamental paper of Liu (1975) shows that is is possible to trace a discrete number of wave-fronts within an approximate solution constructed by the Glimm scheme. Still, we believe that a direct construction of front tracking approximations allows better control of approximate solutions and often leads to technically simpler proofs, compared with the Glimm scheme.

After a short introduction in Chapter 1, Chapter 2 presents a number of results in mathematical analysis for later use. This chapter may be skipped in a first reading, but readers may go back to its various sections whenever they are required in the remainder of the book.

The classical theory of semilinear and quasilinear systems is presented in Chapter 3. We first describe the method of characteristics for non-linear scalar equations. Solutions to semilinear hyperbolic systems are then obtained as fixed points of a suitable integral transformation. In the last section we show how to solve the quasilinear Cauchy problem in terms of a sequence of semilinear problems.

In Chapter 4 we begin the study of weak solutions, introducing the basic definitions and discussing various entropy admissibility conditions. The classical Rankine–Hugoniot equations are proved here in a quite general setting, which will be needed later for the study of uniqueness properties. The last section contains an analysis of the Volpert entropy conditions, which are valid in the scalar case.

Chapter 5 introduces the key concepts of genuinely non-linear and of linearly degenerate characteristic fields. After a discussion of elementary waves (centred rarefactions, shocks and contact discontinuities), we construct the standard self-similar solution of the Riemann problem, following Lax (1957). Two elementary examples are presented in the last section.

Chapter 6 is concerned with global solutions to a scalar conservation law. Among the several techniques available in the literature, we choose here the wave-front tracking algorithm of Dafermos (1972), to familiarize the reader with this type of construction in its simplest setting. After establishing the existence of $BV$ solutions, we prove the famous uniqueness theorem of Kruzhkov (1970), which is valid for general $\mathbf{L}^\infty$ solutions. This allows us to treat the scalar Cauchy problem for a large class of initial data, obtaining a contractive semigroup of generalized solutions in a domain of $\mathbf{L}^1$ functions.

The Cauchy problem for $n \times n$ systems of conservation laws is studied in Chapter 7. In the earlier sections we give a detailed description of the front tracking algorithm, with several illustrations. We then introduce the Glimm interaction functional, establishing the crucial a priori estimates of the total variation and proving the main theorem on the global existence of $\varepsilon$-approximate front tracking solutions. Letting $\varepsilon \to 0$ provides by a compactness argument the global existence of weak solutions.

In Chapter 8 we construct a Lyapunov-type functional, equivalent to the $\mathbf{L}^1$ distance, which is non-increasing along each couple of solutions. This argument shows that solutions constructed by front tracking approximations converge to a unique limit, depending Lipschitz-continuously on the initial data. Our presentation follows the recent paper by Bressan *et al.* (1999). The meaning of the various terms contained in the functional is explained in detail, with the help of several figures.

Chapter 9 contains various uniqueness results for entropy admissible weak solutions. Having already established the existence of a Lipschitz semigroup, a natural method is now available for proving uniqueness. That is, given any entropic solution of the Cauchy problem, it suffices to show that it coincides with the corresponding (unique) solution obtained as the limit of front tracking approximations. Various regularity conditions are introduced, which guarantee uniqueness, namely tame variation, tame oscillation, bounded variation along space-like segments. In all cases, the uniqueness is established within the same class of functions where an existence theorem has been proved.

In Chapter 10 we present various applications of the technique of front tracking, deriving a number of qualitative results for entropy weak solutions. In the first section we extend the Glimm wave interaction functional to general *BV* functions, proving a lower semicontinuity result w.r.t. convergence in $\mathbf{L}^1$. The second section provides an estimate of the decay of positive waves of genuinely non-linear families. For $n \times n$ systems, this provides an analogue of the classical decay property proved by Oleinik in the scalar case. The last section contains a description of the global structure of a general *BV* solution, always based on the analysis of front tracking approximations. In particular, we show that the solution is continuous outside a countable family of shock curves and a countable set of interaction points. Along each shock curve the solution satisfies a.e. the Rankine–Hugoniot equations and the Lax entropy conditions.

As mentioned above, we shall confine ourselves to the study of the one-dimensional Cauchy problem in the standard setting considered by Glimm. The results presented here can be extended in several ways. In particular, one can consider solutions with large total variation, initial boundary value problems, systems of balance laws with a forcing term on the right hand side, or hyperbolic systems where the assumptions of genuine non-linearity or linear degeneracy fail. Our present goal, however, is not to describe the most general theorems available to date. Rather, we prefer to present the key ideas and techniques in a basic setting. Various extensions and applications are discussed in the bibliographical notes in Chapter 11, which contains a survey of recent literature.

This book can be used for a course on conservation laws at graduate level. Care has been taken in particular to explain the more advanced material, helping the reader's understanding with several figures. A set of problems, of varying difficulty, is given at the end of each chapter. These exercises are designed to verify and expand a student's understanding of the concepts and techniques previously discussed.

The book also intends to provide research specialists with a standard reference for the method of front tracking and for the semigroup approach to the uniqueness problem. Until now, this material has been scattered in the literature, where many versions of the front tracking algorithm have appeared, together with different constructions of the Riemann semigroup. We believe that a systematic presentation of these results will be useful. Although they are not always as sharp as in the original research papers, the theorems collected here often have more polished and accessible proofs. This can provide a better starting point for future research in a fast developing area of mathematics.

It is a pleasure to acknowledge the contributions of several students, friends and collaborators at SISSA who helped me during the writing of this book. In particular, F. Ancona, P. Baiti, R. M. Colombo, G. Guerra, H. K. Jenssen, M. Lewicka, A. Marson, B. Piccoli, W. Shen and H. Zhao, who in various measures took part in the development of the new theory, suggested better lines of proofs, or carefully read the original manuscript pointing out misprints and inaccuracies. My warm thanks go to all of them, and my best wishes: may they succeed in rendering the present book obsolete by creating a still better theory in the future.

*Trieste,*
*January 2000*

A. B.

# Contents

# 1

# Introduction

A single conservation law in one space dimension is a first-order partial differential equation of the form

$$u_t + f(u)_x = 0. \tag{1.1}$$

Here $u$ is the *conserved quantity* while $f$ is the *flux*. Integrating (1.1) over the interval $[a, b]$ one obtains

$$\begin{aligned}
\frac{d}{dt}\int_a^b u(t,x)\,dx &= \int_a^b u_t(t,x)\,dx \\
&= -\int_a^b f(u(t,x))_x\,dx \\
&= f(u(t,a)) - f(u(t,b)) \\
&= [\text{inflow at } a] - [\text{outflow at } b].
\end{aligned} \tag{1.2}$$

In other words, the quantity $u$ is neither created nor destroyed: the total amount of $u$ contained inside any given interval $[a, b]$ can change only due to the flow of $u$ across the two endpoints.

Using the chain rule, (1.1) can be written in the quasilinear form

$$u_t + a(u)u_x = 0, \tag{1.3}$$

where $a = f'$ is the derivative of $f$. For smooth solutions, the two equations (1.1) and (1.3) are entirely equivalent. If $u$ has a jump, however, the left hand side of (1.3) will contain the product of a discontinuous function $a(u)$ with the distributional derivative $u_x$, which in this case contains a Dirac mass at the point of the jump. In general, such a product is not well defined. Hence (1.3) is meaningful only within a class of continuous functions. On the other hand, working with the equation in divergence form (1.1) allows us to consider discontinuous solutions as well, interpreted in a distributional sense. More precisely, a locally integrable function $u = u(t, x)$ is a *weak solution* of (1.1) provided that

$$\iint \{u\phi_t + f(u)\phi_x\}\,dx\,dt = 0 \tag{1.4}$$

for every differentiable function with compact support $\phi \in \mathcal{C}_c^1$.

**Example 1.1 (Traffic flow).** Let $u(t, x)$ be the density of cars on a highway at point $x$ at time $t$. For example, $u$ may be the number of cars per kilometre. In a first approximation, we shall assume that $u$ is continuous and that the speed $s$ of the cars depends only on their density, say

$$s = s(u), \quad \text{with } \frac{ds}{du} < 0.$$

Given any two points $a, b$ on the highway, the number of cars between $a$ and $b$ therefore varies according to the law

$$\begin{aligned}\frac{d}{dt}\int_a^b u(t,x)\,dx &= [\text{inflow at } x = a] - [\text{outflow at } x = b] \\ &= s(u(t,a)) \cdot u(t,a) - s(u(t,b)) \cdot u(t,b) \\ &= -\int_a^b [s(u)u]_x\,dx. \end{aligned} \tag{1.5}$$

Since (1.5) holds for all $a, b$, this leads to the conservation law

$$u_t + [s(u)u]_x = 0,$$

where $u$ is the conserved quantity and $f(u) = s(u)u$ is the flux function. In practice, one often takes

$$f(u) = a_1\left(\ln\frac{a_2}{u}\right)u \quad (0 < u \le a_2),$$

for suitable constants $a_1, a_2$.

This book will be mainly concerned with $n \times n$ system of conservation laws of the form

$$\begin{cases} \dfrac{\partial}{\partial t}u_1 + \dfrac{\partial}{\partial x}f_1(u_1, \ldots, u_n) = 0, \\ \quad\vdots \\ \dfrac{\partial}{\partial t}u_n + \dfrac{\partial}{\partial x}f_n(u_1, \ldots, u_n) = 0. \end{cases} \tag{1.6}$$

For simplicity, this will still be written in the form (1.1), but bearing in mind that now $u = (u_1, \ldots, u_n)$ is a vector in $\mathbb{R}^n$ and that $f = (f_1, \ldots, f_n)$ is a map from $\mathbb{R}^n$ into itself. If $A(u) \doteq Df(u)$ is the $n \times n$ Jacobian matrix of the map $f$ at the point $u$, the system (1.6) can be written in the quasilinear form

$$u_t + A(u)u_x = 0. \tag{1.7}$$

We say that this system is *strictly hyperbolic* if every matrix $A(u)$ has $n$ real, distinct eigenvalues, say $\lambda_1(u) < \cdots < \lambda_n(u)$.

**Example 1.2 (Gas dynamics).** The Euler equations for a compressible, non-viscous gas in Lagrangian coordinates take the form of a system of three conservation laws:

$$\begin{cases} v_t - u_x = 0 & \text{(conservation of mass)}, \\ u_t + p_x = 0 & \text{(conservation of momentum)}, \\ \left(e + \dfrac{u^2}{2}\right)_t + (pu)_x = 0 & \text{(conservation of energy)}. \end{cases} \tag{1.8}$$

Here $\rho$ is the density, $v = \rho^{-1}$ the specific volume, $u$ the velocity, $e$ the internal energy and $p$ the pressure. The system is closed by an additional equation $p = p(e, v)$, called the equation of state, depending on the particular gas under consideration.

**Example 1.3.** The *p-system* consists of the two equations

$$v_t - u_x = 0, \qquad u_t + p(v)_x = 0, \tag{1.9}$$

where $p' < 0$, $p'' > 0$. Regarding $v$ as the specific volume, $u$ as the velocity, and taking $p(v) = kv^{-\gamma}$, this system provides a simple model for isentropic gas dynamics. The choice $\gamma \in [1, 3]$ is appropriate for most gases; in particular $\gamma \approx 1.4$ for air. In the region where $v > 0$, the system is strictly hyperbolic. Indeed, the Jacobian matrix is

$$A \doteq Df = \begin{pmatrix} 0 & -1 \\ p'(v) & 0 \end{pmatrix}$$

with real distinct eigenvalues

$$\lambda_1 = -\sqrt{-p'(v)}, \qquad \lambda_2 = \sqrt{-p'(v)}.$$

Integrating the first equation in (1.9) we obtain a scalar function $w = w(t, x)$ such that $w_t = u$, $w_x = v$. Inserting this in the second equation yields the non-linear second-order equation

$$w_{tt} + p(w_x)_x = 0. \tag{1.10}$$

In a neighbourhood of a given state $v_0$, one can approximate $p$ by a linear function $p(v) \approx p(v_0) - c^2(v - v_0)$. In this case (1.10) reduces to the familiar wave equation

$$w_{tt} - c^2 w_{xx} = 0. \tag{1.11}$$

A basic feature of non-linear systems of the form (1.1) is that, even for smooth initial data, the solution of the Cauchy problem may develop discontinuities in finite time. To achieve a global existence result, it is thus essential to work within a class of discontinuous functions, interpreting the equation in (1.1) in its distributional sense (1.4).

**Example 1.4.** Consider the scalar conservation law (inviscid Burgers' equation)

$$u_t + \left(\frac{u^2}{2}\right)_x = 0 \tag{1.12}$$

with initial condition

$$u(0, x) = \bar{u}(x) = \frac{1}{1 + x^2}.$$

For $t > 0$ small the solution can be found by the method of characteristics. Indeed, if $u$ is smooth, (1.12) is equivalent to

$$u_t + uu_x = 0. \tag{1.13}$$

By (1.13) the directional derivative of the function $u = u(t, x)$ along the vector $(1, u)$ vanishes. Therefore, $u$ must be constant along the characteristic lines in the $t$–$x$ plane:

$$t \mapsto (t, x + t\bar{u}(x)) = \left(t, x + \frac{t}{1+x^2}\right).$$

For $t < T \doteq 8/\sqrt{27}$, these lines do not intersect (Fig. 1.1). The solution to our Cauchy problem is thus given implicitly by

$$u\left(t, x + \frac{t}{1+x^2}\right) = \frac{1}{1+x^2}\,. \tag{1.14}$$

On the other hand, when $t > T$, the characteristic lines start to intersect. Indeed, in this case the map

$$x \mapsto x + \frac{t}{1+x^2}$$

is not one-to-one and (1.14) no longer defines a single valued solution of our Cauchy problem.

An alternative point of view is the following (Fig. 1.2). As time increases, points on the graph of $u(t, \cdot)$ move horizontally with speed $u$, equal to their distance from the $x$-axis. This determines a change in the profile of the solution. As $t$ approaches the critical time $T \doteq 8/\sqrt{27}$, one has

$$\lim_{t \to T-} \left\{ \inf_{x \in \mathbb{R}} u_x(t, x) \right\} = -\infty,$$

and no classical solution exists beyond time $T$. The solution can be prolonged for all times $t \geq 0$ only within a class of discontinuous functions.

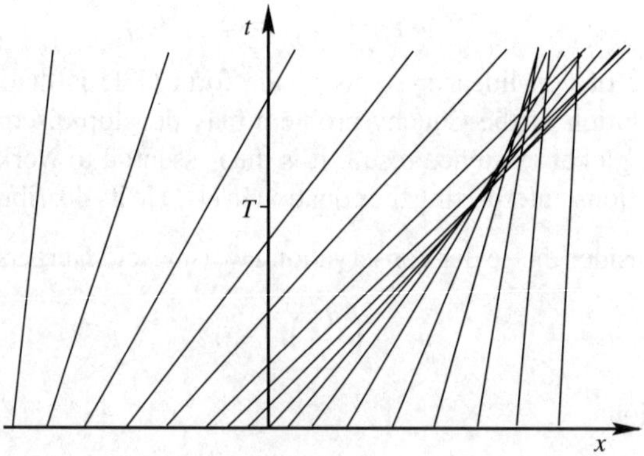


**Figure 1.1**

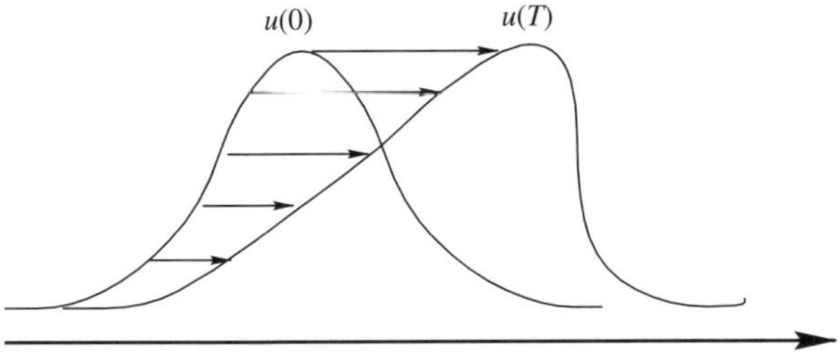


**Figure 1.2**

A major part of the theory of conservation laws is thus concerned with discontinuous solutions. In particular, the following topics are of interest:

1. Local and global existence of solutions to the Cauchy problem, with various initial data: smooth, piecewise constant, locally integrable, with small total variation, etc.
2. Admissibility conditions for weak solutions, characterizing the physically relevant ones.
3. Uniqueness of the 'admissible' solutions and their continuous dependence on the initial data.
4. Regularity of solutions. Local and global structure of discontinuous solutions.
5. Asymptotic shape of a globally defined solution, as $t \to \infty$. Asymptotic blow-up patterns, for solutions whose $\mathbf{L}^\infty$ or $BV$ norms explode in finite time.
6. Relations between the solutions of a hyperbolic system of conservation laws and the solutions of various approximating systems, modelling more complex physical phenomena. In particular, vanishing viscosity approximations, relaxations, kinetic models.
7. Initial boundary value problems, with appropriate boundary conditions. Balance laws, with the presence of source terms.
8. Problems of controllability and optimal control for systems of conservation laws. Variational structure of the flow generated by conservation laws.
9. Numerical algorithms for the construction of approximate solutions. Estimates of the difference between an approximate solution and the exact one.

In this monograph we shall not attempt to cover all of the above topics. Rather, we concentrate on the Cauchy problem for a class of strictly hyperbolic systems of conservation laws, studying the global existence, uniqueness, stability and regularity of weak solutions. Many of our results will be obtained by the analysis of piecewise constant approximate solutions, constructed by a wave-front tracking algorithm.

# 2
# Mathematical preliminaries

This chapter presents various background material which will be used in later chapters.

Section 2.1 recalls the implicit function theorem, while in Section 2.2 we discuss the continuous dependence on parameters of the eigenvalues and eigenvectors of a hyperbolic matrix. Section 2.3 reviews the most basic definitions in the theory of distributions. The concept of distributional derivative given here will become essential in what follows, when we consider discontinuous solutions to a conservation law.

Sections 2.4 and 2.5 are concerned with functions having bounded variation. We first consider maps defined on the real line and prove Helly's compactness theorem. This will provide the key ingredient for the proof of existence of weak solutions to systems of conservation laws given in Chapter 7. We then consider *BV* functions of two independent variables, introducing the concepts of approximate continuity and approximate jump discontinuity. Having stated the fundamental theorem of Volpert (1967) on the structure of a multidimensional *BV* function, we then derive some particular consequences of this result. These will be used in Chapter 10 for the analysis of the global structure of *BV* solutions.

Banach's contraction mapping theorem is proved in Section 2.6. Since weak solutions to conservation laws cannot be obtained as fixed points of a continuous transformation, this existence theorem is not much use here. Its main applications will be confined to Chapter 3, in connection with broad solutions to semilinear hyperbolic systems.

In Section 2.7 we prove the theorem of Rademacher on the a.e. differentiability of Lipschitz functions. Section 2.8 provides an error estimate for an approximate trajectory of a Lipschitz-continuous semigroup. Although quite elementary, this estimate turns out to be very useful in the analysis of uniqueness and stability of weak solutions, given in Chapter 9. In Section 2.9 we derive some a priori bounds on the size of a function of several variables, based on information concerning some of its Taylor coefficients at the origin. These estimates will be repeatedly used in Chapter 7 to obtain uniform bounds on the *BV* norm of approximate solutions, and in Chapter 8 to estimate the distance between two front tracking approximations.

A basic theorem on the weak convergence of Radon measures is stated in Section 2.10. In the same section we also prove some additional properties of weak convergence. These will play a key role in Chapter 10, where we establish the lower semicontinuity of the Glimm functionals and analyse the global structure of entropy weak solutions with small total variation. Finally, Section 2.11 contains a brief review of the basic existence and uniqueness results for solutions of ordinary differential equations.

## 2.1 Implicit function theorems

In the following, $\mathcal{C}^k$ denotes the space of functions which are $k$ times continuously differentiable, with uniformly bounded derivatives up to order $k$. We write $F \in \mathcal{C}^k_c$ to indicate that the function $F$ also has compact support. If $U \subseteq \mathbb{R}^p$, $V \subseteq \mathbb{R}^m$ are open sets and $F = (F_1, \ldots, F_m)(u_1, \ldots, u_p, v_1, \ldots, v_m)$ is a $\mathcal{C}^1$ mapping from $U \times V$ into $\mathbb{R}^m$, we denote by $D_u F(u, v)$ and $D_v F(u, v)$ respectively the $m \times p$ Jacobian matrix of partial derivatives $(\partial F_i / \partial u_j)$ and the $m \times m$ matrix $(\partial F_i / \partial v_j)$, computed at the point $(u, v)$.

**Theorem 2.1.** *Let $U \subseteq \mathbb{R}^p$, $V \subseteq \mathbb{R}^m$ be open sets and let $F : U \times V \mapsto \mathbb{R}^m$ be $k$ times continuously differentiable, with $k \geq 1$. If $F(\bar{u}, \bar{v}) = 0$ and the Jacobian matrix $D_v F(\bar{u}, \bar{v})$ is invertible, then there exists a neighbourhood $\mathcal{N}$ of $\bar{u}$ and a $\mathcal{C}^k$ mapping $\varphi : \mathcal{N} \mapsto V$ such that*

$$\varphi(\bar{u}) = \bar{v}, \quad F(u, \varphi(u)) = 0 \qquad \text{for all } u \in \mathcal{N}. \tag{2.1}$$

*If the $k$-th derivatives of $F$ are Lipschitz continuous, then the same is true of the $k$-th derivatives of $\varphi$. The derivative of $\varphi$ at the point $\bar{u}$ is the $m \times p$ Jacobian matrix*

$$D\varphi(\bar{u}) = -[D_v F(\bar{u}, \bar{v})]^{-1} \cdot D_u F(\bar{u}, \bar{v}). \tag{2.2}$$

For a proof, see for example Diéudonne (1960).

We shall often use a parametrized version of the implicit function theorem, assuming that the function $F$ depends also on a parameter $\omega$, defined on a neighbourhood of a compact set $K$. In this case, the size of the neighbourhood $\mathcal{N}$ where (2.1) holds can be chosen uniformly w.r.t. $\omega \in K$.

**Theorem 2.2.** *Let $U \subseteq \mathbb{R}^p$, $V \subseteq \mathbb{R}^m$, $\Omega \subseteq \mathbb{R}^m$ be open sets. Let $(u, v, \omega) \mapsto F(u, v, \omega)$ be a $\mathcal{C}^k$ map from $U \times V \times \Omega$ into $\mathbb{R}^m$, with $k \geq 1$. Let $\omega \mapsto (\bar{u}_\omega, \bar{v}_\omega)$ be a $\mathcal{C}^k$ map from $\Omega$ into $U \times V$ such that $F(\bar{u}_\omega, \bar{v}_\omega; \omega) = 0$ for every $\omega$. If the Jacobian matrix $D_v F(\bar{u}_\omega, \bar{v}_\omega; \omega)$ is invertible for every $\omega$ in a compact set $K \subset \Omega$, then there exists $\delta > 0$ and a $\mathcal{C}^k$ function $(u, \omega) \mapsto \varphi(u, \omega)$ such that*

$$\varphi(\bar{u}_\omega, \omega) = \bar{v}_\omega, \quad F\big(u, \varphi(u, \omega); \omega\big) = 0 \qquad \text{whenever } \omega \in K, \ |u - \bar{u}_\omega| \leq \delta. \tag{2.3}$$

*If the $k$-th derivatives of $F$ are Lipschitz continuous, then the same is true for the $k$-th derivatives of $\varphi$.*

## 2.2 Some linear algebra

Let $A$ be an $n \times n$ matrix with $n$ real distinct eigenvalues, say $\lambda_1 < \lambda_2 < \cdots < \lambda_n$. Then $A$ admits $n$ linearly independent right eigenvectors $r_1, \ldots, r_n$ and $n$ linearly independent left eigenvectors $l_1, \ldots, l_n$. These eigenvectors can be uniquely determined, except for the orientation, by imposing the relations

$$|r_i| = 1, \qquad l_i \cdot r_j = \delta_{ij} \doteq \begin{cases} 1 & \text{if } i = j, \\ 0 & \text{if } i \neq j. \end{cases} \tag{2.4}$$

We then have the identities

$$l_i A = \lambda_i l_i, \quad A r_i = \lambda_i r_i \qquad i \in \{1, \dots, n\}, \tag{2.5}$$

$$v = \sum_{i=1}^{n} (l_i \cdot v) r_i \quad v \in \mathbb{R}^n. \tag{2.6}$$

A useful consequence of (2.6) is the following.

**Proposition 2.1.** *A vector $v \in \mathbb{R}^n$ is parallel to $r_i$ iff the inner products $l_j \cdot v$ vanish for all $j \neq i$.*

The next result shows that, if the matrix $A$ is a smooth function of a parameter $\eta$, then the corresponding eigenvalues and eigenvectors also depend smoothly on $\eta$.

**Proposition 2.2.** *Assume that the entries of the matrix $A = (a_{ij})$ are $C^k$ functions of a parameter $\eta \in \mathbb{R}^m$, with $k \geq 1$. When $\eta = \bar{\eta}$, let $A(\bar{\eta})$ have real distinct eigenvalues $\lambda_1(\bar{\eta}) < \cdots < \lambda_n(\bar{\eta})$. Then there exists $\delta > 0$ such that, for $|\eta - \bar{\eta}| < \delta$, the matrix $A(\eta)$ has distinct eigenvalues $\lambda_1(\eta) < \cdots < \lambda_n(\eta)$. Moreover, one can choose bases of left and right eigenvectors $l_i(\eta), r_i(\eta)$ satisfying (2.4) such that the maps $\eta \mapsto \lambda_i(\eta)$, $\eta \mapsto l_i(\eta)$, $\eta \mapsto r_i(\eta)$ are all $C^k$.*

*Proof.* Consider the polynomial

$$P(\eta, \lambda) \doteq \det[\lambda I - A(\eta)].$$

For each $i \in \{1, \dots, n\}$ we have

$$P(\bar{\eta}, \lambda_i(\bar{\eta})) = 0, \qquad \frac{\partial P}{\partial \lambda}(\bar{\eta}, \lambda_i(\bar{\eta})) \neq 0,$$

because $\lambda_i(\bar{\eta})$ is an isolated zero of $P(\bar{\eta}, \cdot)$. Hence the implicit function theorem yields the local existence of a $C^k$ map $\eta \mapsto \lambda_i(\eta)$ such that $P(\eta, \lambda_i(\eta)) = 0$.

Let $\{r_1(\bar{\eta}), \dots, r_n(\bar{\eta})\}$ be a basis of eigenvectors of $A(\bar{\eta})$ such that, at $\eta = \bar{\eta}$, one has

$$A(\eta) r_i(\eta) = \lambda_i(\eta) r_i(\eta), \quad |r_i(\eta)| = 1 \qquad \text{for all } i \in \{1, \dots, n\}. \tag{2.7}$$

We claim that there exists $C^k$ functions $\eta \mapsto r_i(\eta)$ such that (2.7) holds in a neighbourhood $\mathcal{N}$ of $\bar{\eta}$. Indeed, for every $i$, at $\eta = \bar{\eta}$ the matrix $A(\eta) - \lambda_i(\eta) I$ has rank $n - 1$. Hence, $n - 1$ of its row vectors are linearly independent, say

$$v_j(\bar{\eta}) = (a_{j1}(\bar{\eta}), \dots, a_{jj}(\bar{\eta}) - \lambda_i(\bar{\eta}), \dots, a_{jn}(\bar{\eta})), \quad j \in \{1, \dots, n\}, \quad j \neq j^*$$

for some index $j^*$. For $\eta$ in a neighbourhood of $\bar{\eta}$, the vector $r_i = (r_{i1}, \dots, r_{in})(\eta)$ is then implicitly defined by the system of $n$ equations

$$r_i \cdot r_i = \sum_{\ell=1}^{n} r_{i\ell} r_{i\ell} = 1, \qquad v_j(\eta) \cdot r_i = \sum_{\ell=1}^{n} a_{j\ell}(\eta) r_{i\ell} - \lambda_i(\eta) r_{ij} = 0 \quad j \neq j^*. \tag{2.8}$$

Since $r_i(\bar{\eta})$ is perpendicular to each $v_j(\bar{\eta})$ with $j \neq j^*$, the $n$ vectors $r_i(\bar{\eta})$, $v_j(\bar{\eta})$ in (2.8) are linearly independent. Therefore, the implicit function theorem can be applied, yielding the existence of $\mathcal{C}^k$ functions $r_1, \dots, r_n$ such that (2.7) holds, in a neighbourhood of $\bar{\eta}$.

In turn, the second equation in (2.4) uniquely determines a basis of left eigenvectors $l_1, \dots, l_n$, which are $\mathcal{C}^k$ functions of $\eta$. □

## 2.3 Distributions

When one studies solutions of partial differential equations within a class of locally integrable, possibly discontinuous functions, the differentiations must be interpreted in a generalized sense. We review here a few basic facts about distributions, which will be used in later chapters.

Let $\Omega$ be an open subset of $\mathbb{R}^m$. For every compact set $K \subset \Omega$, by $\mathcal{D}_K(\Omega)$ we denote the set of all $\mathcal{C}^\infty$ functions from $\Omega$ into $\mathbb{R}$ which vanish outside $K$. The set of all $\mathcal{C}^\infty$ functions $\phi : \Omega \mapsto \mathbb{R}$ with compact support contained in $\Omega$ is written $\mathcal{D}(\Omega)$.

A *multi-index* is an ordered $m$-tuple of non-negative integers $\alpha = (\alpha_1, \dots, \alpha_m)$. To every multi-index $\alpha$ we associate the differential operator

$$D^\alpha = \left(\frac{\partial}{\partial x_1}\right)^{\alpha_1} \cdots \left(\frac{\partial}{\partial x_m}\right)^{\alpha_m},$$

with order $|\alpha| = \alpha_1 + \cdots + \alpha_m$. For each $N \geq 0$, we define the norm

$$\|\phi\|_N = \max\left\{ |D^\alpha \phi(x)|;\ x \in \Omega,\ |\alpha| \leq N \right\}.$$

With the above notation, the space $\mathcal{D}'(\Omega; \mathbb{R}^n)$ of (vector-valued) *distributions* on $\Omega$ is the set of all linear mappings $\Lambda : \mathcal{D}(\Omega) \mapsto \mathbb{R}^n$ with the following property. For every compact set $K \subset \Omega$, there exists a non-negative integer $N$ and a constant $C$ such that

$$|\Lambda(\phi)| \leq C\, \|\phi\|_N \quad \text{for all } \phi \in \mathcal{D}_K(\Omega). \tag{2.9}$$

If there exists some integer $N$ for which (2.9) holds for all $K$ (possibly with different constants $C$), the smallest such $N$ is called the *order* of $\Lambda$. If $\Lambda$ has order 0, then it can be extended by continuity to the set of all continuous functions $\varphi : \Omega \mapsto \mathbb{R}$ with compact support. In this case, $\Lambda$ can be identified with a vector measure on $\Omega$.

The *characteristic function* of a set $K$ is defined as

$$\chi_K(x) = \begin{cases} 1 & \text{if } x \in K, \\ 0 & \text{if } x \notin K. \end{cases}$$

We say that a function $f : \Omega \mapsto \mathbb{R}^n$ is *locally integrable*, and write $f \in \mathbf{L}^1_{\text{loc}}$, if, for every compact set $K \subset \Omega$, the product $f \cdot \chi_K$ of $f$ with the characteristic function of $K$ is integrable. A sequence of functions $(f_\nu)_{\nu \geq 1}$ converges to $f$ in $\mathbf{L}^1_{\text{loc}}$ if the sequence $f_\nu \cdot \chi_K$ converges to $f \cdot \chi_K$ in $\mathbf{L}^1$ for every compact set $K$.

Every $f \in \mathbf{L}^1_{\text{loc}}(\Omega; \mathbb{R}^n)$ determines a distribution of order 0, defined as

$$\Lambda_f(\phi) = \int_\Omega f(x)\phi(x)\,dx \quad \text{for all } \phi \in \mathcal{D}(\Omega). \tag{2.10}$$

If $\alpha$ is a multi-index and $\Lambda \in \mathcal{D}'(\Omega)$, then the derivative $D^\alpha \Lambda$ is the distribution defined as

$$(D^\alpha \Lambda)(\phi) = (-1)^{|\alpha|} \Lambda(D^\alpha \phi) \quad \text{for all } \phi \in \mathcal{D}(\Omega). \tag{2.11}$$

If a function $f$ is $N$ times continuously differentiable and $|\alpha| \leq N$, then with the above notation one has $D^\alpha \Lambda_f = \Lambda_{D^\alpha f}$.

**Example 2.1.** A case that will be frequently encountered in applications is the following (Fig. 2.1). The open set $\Omega$ is contained in the plane $\mathbb{R}^2$ with coordinates $t, x$. Inside $\Omega$ we are given a $\mathcal{C}^1$ curve

$$\{(t, x);\ x = \gamma(t)\ a < t < b\}$$

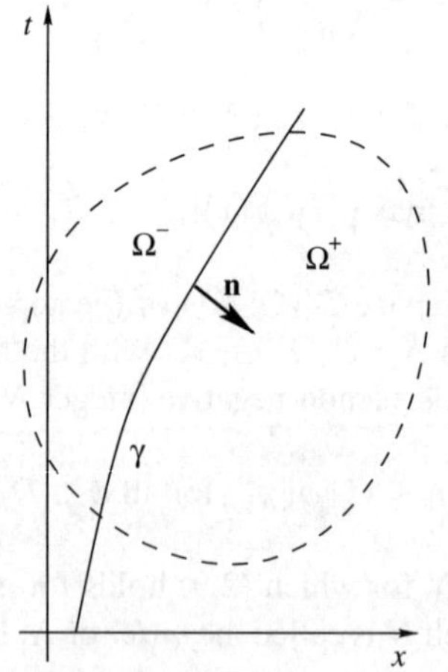


**Figure 2.1**

and two functions $g = g(t,x)$, $h = h(t,x)$ which are continuously differentiable for $x \neq \gamma(t)$ but possibly discontinuous on $\gamma$. We wish to compute the distribution $\Lambda$ associated with the sum of distributional derivatives

$$D_t g + D_x h.$$

At each point $(t, \gamma(t))$ consider the jumps

$$\Delta g(t) = \lim_{x\to\gamma(t)+} g(t,x) - \lim_{x\to\gamma(t)-} g(t,x),$$
$$\Delta h(t) = \lim_{x\to\gamma(t)+} h(t,x) - \lim_{x\to\gamma(t)-} h(t,x),$$

assuming that the above limits exist for all $a < t < b$. Given any $\mathcal{C}^1$ function $\phi = \phi(t,x)$ with compact support contained inside $\Omega$, we apply the divergence theorem to the vector field $\mathbf{v} \doteq (\phi g,\ \phi h)$ on the domains

$$\Omega^+ \doteq \{(t,x) \in \Omega, x > \gamma(t)\}, \qquad \Omega^- \doteq \{(t,x) \in \Omega, x < \gamma(t)\}.$$

Let $d\sigma$ be the differential of the arc-length along the curve $\gamma$ and denote the derivative w.r.t. time by an upper dot. Moreover, let $\mathbf{n}$ be the outer unit normal to the boundaries of $\Omega^-, \Omega^+$. By assumption, $\phi = 0$ on $\partial\Omega$; hence the only portion of these boundaries where $\phi$ does not vanish is the line $x = \gamma(t)$. On this line, an elementary calculation yields $\mathbf{n} \cdot d\sigma = \pm(-\dot\gamma, 1)\, dt$. Therefore

$$\begin{aligned}\Lambda(\phi) &\doteq -\iint_\Omega \{g\phi_t + h\phi_x\}\, dx\, dt \\ &= \iint_\Omega \{g_t + h_x\}\phi\, dx\, dt + \int_a^b \{\Delta h\,(t,\gamma(t)) - \dot\gamma(t)\Delta g\,(t,\gamma(t))\}\phi\,(t,\gamma(t))\, dt.\end{aligned} \tag{2.12}$$

## 2.4 Functions with bounded variation

Consider a (possibly unbounded) interval $J \subseteq \mathbb{R}$ and a map $u : J \mapsto \mathbb{R}^n$. The *total variation* of $u$ is then defined as

$$\text{Tot. Var.}\{u\} \doteq \sup\left\{\sum_{j=1}^N |u(x_j) - u(x_{j-1})|\right\}, \tag{2.13}$$

where the supremum is taken over all $N \geq 1$ and all $(N+1)$-tuples of points $x_j \in J$ such that $x_0 < x_1 < \cdots < x_N$. If the right hand side of (2.13) is bounded, we say that $u$ has bounded variation, and write $u \in BV$. Some elementary properties of $BV$ functions are collected in the following lemmas.

**Lemma 2.1.** *Let $u : ]a,b[ \mapsto \mathbb{R}^n$ have bounded variation. Then, for every $x \in ]a,b[$, the left and right limits*

$$u(x-) \doteq \lim_{y \to x-} u(y), \qquad u(x+) \doteq \lim_{y \to x+} u(y)$$

*are well defined. Moreover, u has at most countably many points of discontinuity.*

*Proof.* Let $x \in ]a,b[$ be given and consider any strictly increasing sequence of points $x_\nu$ tending to $x$. Since

$$\sum_{\nu \geq 1} |u(x_\nu) - u(x_{\nu-1})| \leq \text{Tot. Var.}\{u\} < \infty,$$

the sequence $u(x_\nu)$ is Cauchy and converges to some limit $u^-$. Observing that any two such sequences $x_\nu \to x$, $x'_\nu \to x$ can be combined into a unique non-decreasing sequence, it is clear that the above limit is independent of the choice of the $x_\nu$. This proves the existence of the left limit $u(x-)$. The case of $u(x+)$ is entirely similar.

To prove the last statement, for each $\nu \geq 1$ we observe that the number of points contained in the set

$$A_\nu \doteq \left\{x \in ]a,b[\,;\ |u(x-) - u(x)| + |u(x+) - u(x)| > 1/\nu\right\}$$

cannot be bigger than $\nu \cdot \text{Tot. Var.}\{u\}$. Hence the set of points where $u$ is discontinuous, being contained in the union of all $A_\nu$, $\nu \geq 1$, is at most countable. □

**Remark 2.1.** By the above lemma, if $u$ has bounded variation, we can redefine the value of $u$ at each jump point by setting $u(x) \doteq u(x+)$. In particular, if we are only interested in the $\mathbf{L}^1$-equivalence class of a *BV* function $u$, by possibly changing the values of $u$ at countably many points we can assume that $u$ is right continuous.

**Remark 2.2.** If $u : \mathbb{R} \mapsto \mathbb{R}^n$ has bounded variation, the same arguments used in the proof of Lemma 2.1 show that the limits $u(-\infty)$, $u(\infty)$ are well defined.

**Lemma 2.2.** *Let $u : \mathbb{R} \mapsto \mathbb{R}^n$ be right continuous with bounded variation. Then, for every $\varepsilon > 0$, there exists a piecewise constant function $v$ such that*

$$\text{Tot. Var.}\{v\} \leq \text{Tot.Var.}\{u\}, \qquad \|v - u\|_{\mathbf{L}^\infty} \leq \varepsilon. \tag{2.14}$$

*If, in addition,*

$$\int_{-\infty}^{0} |u(x) - u(-\infty)|\,dx + \int_{0}^{\infty} |u(x) - u(\infty)|\,dx < \infty,$$

*then one can find $v$ with the additional property*

$$\|u - v\|_{\mathbf{L}^1} < \varepsilon. \tag{2.15}$$

*Proof.* Define the scalar function

$$U(x) \doteq \sup \left\{ \sum_{j=1}^{N} |u(x_j) - u(x_{j-1})|;\ N \geq 1,\ x_0 < x_1 < \cdots < x_n = x \right\},$$

measuring the total variation of $u$ on the interval $]-\infty,\ x]$. Observe that $U$ is a right continuous, non-decreasing function which satisfies

$$\begin{aligned} U(-\infty) = 0, \qquad & U(\infty) = \text{Tot. Var.}\{u\}, \\ |u(y) - u(x)| \leq U(y) - U(x) \quad & \text{for all } x < y. \end{aligned} \tag{2.16}$$

Given $\varepsilon > 0$, let $N$ be the largest integer $<$ Tot. Var. $\{u\}$ and consider the points

$$x_0 \doteq -\infty, \quad x_N \doteq \infty, \quad x_j \doteq \min\{x\,;\, U(x) \geq j\varepsilon\}, \quad j = 1, \ldots, N-1.$$

Defining

$$v(x) \doteq u(x_j) \quad \text{if } x \in [x_j,\ x_{j+1}[\,,$$

by (2.16) the two estimates in (2.14) are both satisfied.

To prove (2.15) we observe that, under the additional assumption of the lemma, one can find $\rho$ large enough so that

$$\int_{-\infty}^{-\rho} |u(x) - u(-\infty)|\, dx + \int_{\rho}^{\infty} |u(x) - u(\infty)|\, dx < \frac{\varepsilon}{2}.$$

We can now construct a piecewise constant function $\tilde{v}$ such that

$$\|\tilde{v} - u\|_{\mathbf{L}^\infty} < \frac{\varepsilon}{2\rho}.$$

Defining

$$v(x) = \begin{cases} \tilde{v}(x) & \text{if } x \in [-\rho,\ \rho[\,, \\ u(-\infty) & \text{if } x < -\rho, \\ u(\infty) & \text{if } x \geq \rho, \end{cases}$$

one achieves the additional estimate (2.15). □

**Lemma 2.3.** *If $u : \mathbb{R} \mapsto \mathbb{R}^n$ has bounded variation, for every $\varepsilon > 0$ one has*

$$\frac{1}{\varepsilon} \cdot \int_{-\infty}^{\infty} |u(x+\varepsilon) - u(x)|\, dx \leq \text{Tot. Var.}\{u\}. \tag{2.17}$$

*Proof.* It is not restrictive to assume that $u$ is right continuous. As in the proof of Lemma 2.2, define the non-decreasing scalar function $U(x)$ as the total variation of $u$ on $]-\infty, x]$. By (2.16) we then have

$$\begin{aligned}\int_{-\infty}^{\infty} |u(x+\varepsilon) - u(x)|\, dx &\le \int_{-\infty}^{\infty} \big[U(x+\varepsilon) - U(x)\big]\, dx \\ &= \text{meas}\big\{(x,y) \in \mathbb{R}^2\,;\, U(x) < y < U(x+\varepsilon)\big\} \\ &= \int_{U(-\infty)}^{U(\infty)} \text{meas}\big\{x\,;\, U(x) < y < U(x+\varepsilon)\big\}\, dy \\ &= \int_0^{\text{Tot.Var.}\{u\}} \varepsilon\, dy \\ &= \varepsilon \cdot \text{Tot. Var. } \{u\}.\end{aligned}$$

This establishes (2.17). □

Bounded sets of $BV$ functions have a compactness property, stated in the following theorem, which will provide the key ingredient in the existence proof for weak solutions to systems of conservation laws.

**Theorem 2.3 (Helly).** *Consider a sequence of functions* $u_\nu : \mathbb{R} \mapsto \mathbb{R}^n$ *such that*

$$\text{Tot. Var. } \{u_\nu\} \le C, \quad |u_\nu(x)| \le M \quad \textit{for all } \nu, x, \tag{2.18}$$

*for some constants* $C, M$. *Then there exists a function* $u$ *and a subsequence* $u_\mu$ *such that*

$$\lim_{\mu\to\infty} u_\mu(x) = u(x) \quad \textit{for every } x \in \mathbb{R}, \tag{2.19}$$

$$\text{Tot. Var. } \{u\} \le C, \qquad |u(x)| \le M \quad \textit{for all } x. \tag{2.20}$$

*Proof.* 1. For every $\nu \ge 1$, let

$$U_\nu(x) \doteq \sup\left\{\sum_{j=1}^{N} |u_\nu(x_j) - u_\nu(x_{j-1})|;\ N \ge 1,\ x_0 < x_1 < \cdots < x_n = x\right\}$$

be the total variation of $u_\nu$ on $]-\infty, x]$. Observe that each $U_\nu$ is non-decreasing and satisfies

$$0 \le U_\nu(x) \le C, \quad |u_\nu(y) - u_\nu(x)| \le U_\nu(p_2) - U_\nu(p_1) \qquad \text{for all } p_1 \le x \le y \le p_2. \tag{2.21}$$

2. By a diagonal procedure, we construct a subsequence $U_{\nu'}$ whose limit exists at every rational point:

$$\lim_{\nu'\to\infty} U_{\nu'}(x) = U(x) \quad x \in \mathbb{Q}.$$

Because of (2.21), the function $U$ maps $\mathbb{Q}$ into $[0, C]$ and is non-decreasing. For each $n \geq 1$, consider the set of jump points

$$J_n \doteq \left\{ x \in \mathbb{R};\ \lim_{y \to x+} U(y) - \lim_{y \to x-} U(y) \geq \frac{1}{n} \right\}$$

where, of course, the variable $y$ ranges over $\mathbb{Q}$. By the properties of $U$, the set $J_n$ can contain at most $Cn$ points. Therefore, the set $J$ of points $x \in \mathbb{R}$ where the right and left limits of $U$ are distinct is at most countable, and indeed

$$J = \bigcup_{n \geq 1} J_n.$$

3. We now choose a further subsequence, say $u_\mu$, such that the limit

$$u(x) \doteq \lim_{\mu \to \infty} u_\mu(x) \tag{2.22}$$

exists for each $x$ in the countable set $J \cup \mathbb{Q}$. We claim that, for this subsequence, the limit (2.22) exists for every $x \in \mathbb{R}$ as well. Indeed, assume $x \notin J$. Then for each $n \geq 1$, since $x \notin J_n$, there exist rational points $p_1 < x < p_2$ such that $U(p_2) - U(p_1) < 2/n$. Using (2.21) and the fact that $u_\mu(p_1) \to u(p_1)$, we obtain

$$\begin{aligned}
\limsup_{h,k \to \infty} |u_h(x) - u_k(x)| &\leq \limsup_{h \to \infty} |u_h(x) - u(p_1)| + \limsup_{k \to \infty} |u_k(x) - u(p_1)| \\
&= 2 \cdot \limsup_{\mu \to \infty} |u_\mu(x) - u_\mu(p_1)| \leq 2 \cdot \limsup_{\mu \to \infty} \left(U_\mu(p_2) - U_\mu(p_1)\right) \\
&= 2\left(U(p_2) - U(p_1)\right) < \frac{4}{n}.
\end{aligned}$$

Since $n$ was arbitrary, our claim is proved. This establishes the first part of the theorem.

4. For any given points $x_0 < x_1 < \cdots < x_N$, we now have

$$\begin{aligned}
\sum_{j=1}^{N} |u(x_j) - u(x_{j-1})| &= \lim_{\mu \to \infty} \left( \sum_{j=1}^{N} |u_\mu(x_j) - u_\mu(x_{j-1})| \right) \\
&\leq \limsup_{\mu \to \infty} \left(\text{Tot. Var.}\{u_\mu\}\right) \leq C.
\end{aligned}$$

This proves the first inequality in (2.20). The second is obvious. □

**Theorem 2.4.** *Consider a sequence of functions* $u_\nu : [0, \infty[ \times \mathbb{R} \mapsto \mathbb{R}^n$ *with the following properties:*

$$\text{Tot. Var.}\{u_\nu(t, \cdot)\} \leq C, \quad |u_\nu(t, x)| \leq M \qquad \textit{for all } t, x, \tag{2.23}$$

$$\int_{-\infty}^{\infty} |u_\nu(t, x) - u_\nu(s, x)|\, dx \leq L|t - s| \quad \textit{for all } t, s \geq 0, \tag{2.24}$$

*for some constants C, M, L. Then there exists a subsequence $u_\mu$ which converges to some function u in $\mathbf{L}^1_{\text{loc}}([0,\infty)\times\mathbb{R};\ \mathbb{R}^n)$. This limit function satisfies*

$$\int_{-\infty}^{\infty} |u(t,x)-u(s,x)|\,dx \le L|t-s| \quad \textit{for all } t,s\ge 0. \tag{2.25}$$

*The point values of the limit function u can be uniquely determined by requiring that*

$$u(t,x)=u(t,x+)\doteq \lim_{y\to x+} u(t,y) \quad \textit{for all } t,x. \tag{2.26}$$

*In this case, one has*

$$\text{Tot. Var.}\,\{u(t,\cdot)\}\le C, \quad |u(t,x)|\le M \qquad \textit{for all } t,x. \tag{2.27}$$

*Proof.* Using Theorem 2.3 we construct a subsequence $\{u_\mu\}$ such that $u_\mu(t,\cdot)\to u(t,\cdot)$ pointwise and hence also in $\mathbf{L}^1_{\text{loc}}(\mathbb{R};\ \mathbb{R}^n)$, at each rational time $t\ge 0$. This limit function clearly satisfies (2.25) and (2.27), restricted to $t,s\in\mathbb{Q}$. By continuity, it can thus be uniquely extended to a map $t\mapsto u(t,\cdot)$ from $[0,\infty[$ into $\mathbf{L}^1_{\text{loc}}(\mathbb{R};\ \mathbb{R}^n)$, satisfying (2.25). More precisely, for each $t\ge 0$ we consider a sequence of rational times $t_m\to t$ and define

$$u(t,\cdot)\doteq \lim_{m\to\infty} u(t_m,\cdot).$$

Because of (2.25), this limit exists and does not depend on the choice of the sequence. Observing that

$$\text{Tot. Var.}\,\{u(t_m,\cdot)\}\le C, \quad |u(t_m,x)|\le M \qquad \text{for all } m\ge 1,\ x\in\mathbb{R},$$

by possibly modifying the limit function $u(t,\cdot)$ on a set of measure zero we achieve the bounds (2.27).

Finally, we define

$$u^\varepsilon(t,x)\doteq \frac{1}{\varepsilon}\int_x^{x+\varepsilon} u(t,x)\,dx.$$

Observe that each $u^\varepsilon$ is uniformly Lipschitz continuous w.r.t. both variables $t,x$. Indeed

$$|u^\varepsilon(t,x)-u^\varepsilon(s,x)|\le \frac{1}{\varepsilon}\int_x^{x+\varepsilon} |u(t,x)-u(s,x)|\,dx \le \frac{L}{\varepsilon}\cdot|t-s|,$$

$$|u^\varepsilon(t,x)-u^\varepsilon(t,x+h)|\le \frac{1}{\varepsilon}\left(\int_x^{x+h} + \int_{x+\varepsilon}^{x+\varepsilon+h}\right)|u(t,y)|\,dy \le \frac{2M}{\varepsilon}\cdot h.$$

Moreover, for every $t,x$ we have

$$\tilde u(t,x)\doteq \lim_{\varepsilon\to 0+} u^\varepsilon(t,x)=u(t,x+). \tag{2.28}$$

The function $\tilde{u}$, being the pointwise limit of continuous functions, is Borel-measurable. For each $t \geq 0$ the identity $\tilde{u}(t, x) = u(t, x)$ holds at all but countably many points $x$. By replacing $u$ with $\tilde{u}$, all requirements in (2.25)–(2.27) are clearly satisfied. □

## 2.5 *BV* functions of two variables

In the following, we denote $y = (y_1, \ldots, y_m)$ as the variable in $\mathbb{R}^m$. We say that a (vector-valued) function $u = u(y)$ has *locally bounded variation* if its distributional derivatives $D_{y_i} u$, $i = 1, \ldots, m$, are measures. By definition this is the case if, for every compact set $K \subset \mathbb{R}^2$, there exists a constant $C_K$ such that

$$\left| \int u \cdot \frac{\partial \phi}{\partial y_i} \, dy \right| \leq C_K \|\phi\|_{C^0} \tag{2.29}$$

for each $i$ and every $\phi \in C_c^1$ with support contained in $K$.

Functions of bounded variation possess much better regularity properties than arbitrary measurable functions. Some basic results in this direction will be presented below.

**Definition 2.1.** We say that a function $u$ has an *approximate jump discontinuity* at the point $\bar{y}$ if there exists vectors $u^+ \neq u^-$ and a unit normal vector $\mathbf{n} \in \mathbb{R}^m$ such that, setting

$$U(y) \doteq \begin{cases} u^- & \text{if } y \cdot \mathbf{n} < 0, \\ u^+ & \text{if } y \cdot \mathbf{n} > 0, \end{cases} \tag{2.30}$$

the following holds:

$$\lim_{r \to 0+} \frac{1}{r^m} \int_{|y| < r} |u(\bar{y} + y) - U(y)| \, dy = 0. \tag{2.31}$$

Moreover, we say that $u$ is *approximately continuous* at the point $\bar{y}$ if the above relations hold with $u^+ = u^-$ (and $\mathbf{n}$ arbitrary).

Observe that the above definitions depend only on the $\mathbf{L}^1$ equivalence class of $u$. Indeed, the limit (2.31) is unaffected if the values of $u$ are changed on a set $\mathcal{N} \subset \mathbb{R}^m$ of Lebesgue measure zero. The standard example of an approximate jump point is the following.

**Example 2.2.** Let $f_1, f_2 : \mathbb{R}^m \mapsto \mathbb{R}^n$ be continuous. Let $g : \mathbb{R}^m \mapsto \mathbb{R}$ be continuously differentiable. Consider the function

$$u(y) \doteq \begin{cases} f_1(y) & \text{if } g(y) \leq 0, \\ f_2(y) & \text{if } g(y) > 0. \end{cases}$$

At a point $\bar{y}$ where $g(\bar{y}) = 0$, call $u^- \doteq f_1(\bar{y})$, $u^+ \doteq f_2(\bar{y})$. If $u^+ = u^-$, then $u$ is continuous at $\bar{y}$, and hence also approximately continuous. On the other hand, if $u^+ \neq u^-$ and $\nabla g(\bar{y}) \neq 0$, then $u$ has an approximate jump at $\bar{y}$. Indeed, the limit (2.31) holds by choosing $\mathbf{n}$ as the unit vector in the direction of $\nabla g(\bar{y})$.

The following theorem provides a useful description of the structure of *BV* functions of two independent variables. A proof, valid also for arbitrary space dimensions, can be found in Evans and Gariepy (1992) or Ziemer (1989).

**Theorem 2.5.** *Let $\Omega$ be an open subset of $\mathbb{R}^2$ and let $u : \Omega \mapsto \mathbb{R}^n$ be a BV function. Then there exists a set $\widetilde{\mathcal{N}} \subset \Omega$ whose one-dimensional Hausdorff measure is zero and such that, at each point $y \notin \widetilde{\mathcal{N}}$, the function u either is approximately continuous or has an approximate jump discontinuity.*

A class of *BV* functions of two variables, particularly important for applications to conservation laws, is now considered.

**Theorem 2.6.** *Let $u : ]a, b[\times\mathbb{R} \mapsto \mathbb{R}^n$ satisfy*

$$\text{Tot. Var.}\{u(t,\cdot)\} \leq M \quad t \in ]a, b[\,, \tag{2.32}$$

$$\int_{-\infty}^{\infty} |u(t,x) - u(s,x)|\, dx \leq L|t-s| \quad s, t \in ]a, b[\,, \tag{2.33}$$

*for some constants L, M. Then u is a BV function of the two variables t, x. Moreover, there exists a set $\mathcal{N} \subset ]a, b[$ of measure zero such that, for every $(\tau, \xi) \in ]a, b[\times \mathbb{R}$ with $\tau \notin \mathcal{N}$, calling*

$$u^+ \doteq \lim_{x\to\xi+} u(\tau, x), \qquad u^- \doteq \lim_{x\to\xi-} u(\tau, x), \tag{2.34}$$

*the following holds. There exists a finite speed $\lambda \in \mathbb{R}$ such that the function*

$$U(t,x) \doteq \begin{cases} u^- & \text{if } x < \lambda t, \\ u^+ & \text{if } x > \lambda t \end{cases} \tag{2.35}$$

*satisfies*

$$\lim_{r\to 0+} \frac{1}{r^2} \int_{-r}^{r} \int_{-\lambda^* r}^{\lambda^* r} |u(\tau + t,\ \xi + x) - U(t,x)|\, dx\, dt = 0, \tag{2.36}$$

$$\lim_{r\to 0+} \frac{1}{r} \int_{-\lambda^* r}^{\lambda^* r} |u(\tau + r,\ \xi + x) - U(r,x)|\, dx = 0, \tag{2.37}$$

*for every $\lambda^* > 0$.*

*Proof.* To show that the distributional derivatives $D_t u$, $D_x u$ are measures, let $\phi \in \mathcal{C}_c^1$ be any function with compact support contained in the strip $]a, b[\,\times\mathbb{R}$. We then have

$$\begin{aligned} \left| \iint u\phi_t\, dx\, dt \right| &= \left| \lim_{h\to 0} \iint u(t,x) \cdot \frac{\phi(t+h,x) - \phi(t,x)}{h}\, dx\, dt \right| \\ &= \left| \lim_{h\to 0} \iint \frac{u(t,x) - u(t-h,x)}{h} \cdot \phi(t,x)\, dx\, dt \right| \end{aligned}$$

$$\leq \int_a^b \left\{ \limsup_{h\to 0} \frac{1}{h} \int_{-\infty}^{\infty} |u(t,x) - u(t-h,x)|\, dx \right\} \cdot \|\phi\|_{\mathcal{C}^0}\, dt$$

$$\leq (b-a)L \cdot \|\phi\|_{\mathcal{C}^0}.$$

Similarly,

$$\left| \iint u\phi_x\, dx\, dt \right| = \left| \lim_{h\to 0} \iint u(t,x) \cdot \frac{\phi(t,x+h) - \phi(t,x)}{h}\, dx\, dt \right|$$

$$= \left| \lim_{h\to 0} \iint \frac{u(t,x) - u(t,x-h)}{h} \cdot \phi(t,x)\, dx\, dt \right|$$

$$\leq \int_a^b \left\{ \limsup_{h\to 0} \frac{1}{h} \int |u(t,x) - u(t,x-h)|\, dx \right\} \cdot \|\phi\|_{\mathcal{C}^0}\, dt$$

$$\leq (b-a)M \cdot \|\phi\|_{\mathcal{C}^0}.$$

By the two previous estimates, $u$ is a *BV* function.

We can now apply Theorem 2.5 to the case where the variable $(y_1, y_2) \doteq (t, x)$ ranges in $\mathbb{R}^2$. This yields the existence of a set $\widetilde{\mathcal{N}} \subset ]a,b[\, \times \mathbb{R}$ of one-dimensional Hausdorff measure zero, such that $u$ either is approximately continuous or has a jump discontinuity at every point $(\tau, \xi) \notin \widetilde{\mathcal{N}}$. Calling

$$\mathcal{N} \doteq \left\{ t;\ (t,x) \in \widetilde{\mathcal{N}} \text{ for some } x \in \mathbb{R} \right\}$$

the projection of $\widetilde{\mathcal{N}}$ on the $t$-axis, it is clear that $\mathcal{N}$ has measure zero. Calling $y = (t,x)$ the variable in $\mathbb{R}^2$, at every point $\bar{y} = (\tau, \xi)$ with $\tau \notin \mathcal{N}$ the relations (2.30)–(2.31) hold for some states $u^-, u^+$ and some unit normal $\mathbf{n}$. In particular, (2.31) implies that (2.36) must hold for every $\lambda^* > 0$. If $u^+ = u^-$, we can trivially define $U$ as in (2.35), choosing $\lambda \doteq 0$. In the case $u^+ \neq u^-$, we claim that $\mathbf{n}$ in (2.30) is not parallel to the $t$-axis. Indeed, assume that (2.36) holds with

$$U(t,x) \doteq \begin{cases} u^- & \text{if } t < 0, \\ u^+ & \text{if } t > 0. \end{cases} \tag{2.38}$$

By (2.33) the map $t \mapsto u(t, \cdot)$ is Lipschitz continuous w.r.t. the $\mathbf{L}^1$ distance. We thus have the estimate

$$E^* \doteq \limsup_{r\to 0+} \frac{1}{r^2} \int_0^r \int_{-\lambda^* r}^{\lambda^* r} |u(\tau+h, \xi+x) - u(\tau-h, \xi+x)|\, dx\, dh$$

$$\leq \limsup_{r\to 0+} \frac{1}{r^2} \int_0^r 2Lh\, dh$$

$$= L. \tag{2.39}$$

On the other hand, by (2.36) it follows that

$$
\begin{aligned}
E^* &\geq \liminf_{r\to 0+} \frac{1}{r^2} \int_0^r \int_{-\lambda^* r}^{\lambda^* r} |U(h,x) - U(-h,x)|\,dx\,dh \\
&\quad - \limsup_{r\to 0} \frac{1}{r^2} \int_{-r}^{r} \int_{-\lambda^* r}^{\lambda^* r} |u(\tau+h, \xi+x) - U(h,x)|\,dx\,dh \\
&= 2|u^+ - u^-| \cdot \lambda^*.
\end{aligned} \tag{2.40}
$$

Since $\lambda^*$ can be arbitrarily large, from (2.39) and (2.40) we obtain a contradiction. Hence the function $U$ has the form (2.35) for some $\lambda \in \mathbb{R}$ and some states $u^-, u^+$.

To show that the right and left states $u^+$, $u^-$ are precisely given by (2.34), we define

$$
v(t,x) \doteq u(\tau + t,\ \xi + x) - U(t,x),
$$

$$
E \doteq |u^+ - u(\tau, \xi+)| + |u^- - u(\tau, \xi-)|.
$$

Observe that

$$
\lim_{r\to 0+} \frac{1}{r} \int_{\xi - r}^{\xi + r} |v(0,x)|\ dx = E. \tag{2.41}
$$

Moreover, the number $L' \doteq L + |\lambda|\,|u^+ - u^-|$ provides a Lipschitz constant for $v$, namely

$$
\int_{\mathbb{R}} |v(t,x) - v(s,x)|\,dx \leq L'|t-s| \quad \text{for all } s, t. \tag{2.42}
$$

By (2.36) and (2.41)–(2.42) it now follows that

$$
\begin{aligned}
0 &= \limsup_{r\to 0+} \frac{1}{r^2} \int_{-r}^{r}\int_{-r}^{r} |v(t,x)|\,dx\,dt \\
&\geq \limsup_{r\to 0+} \frac{1}{r^2} \int_0^{(E/L')r} \left( \int_{-r}^{r} |v(0,x)|\,dx - L'(t-\tau) \right) dt \\
&= \frac{E^2}{2L'}.
\end{aligned}
$$

Hence $E = 0$, proving (2.34).

The proof of (2.37) relies again on the Lipschitz continuity of $v$. If (2.37) failed, we could choose a constant $\delta < L'$ such that

$$
0 < \delta \leq \limsup_{r\to 0+} \frac{1}{r} \int_{-\lambda^* r}^{\lambda^* r} |u(\tau + r, \xi + x) - U(r,x)|\,dx.
$$

From (2.36) and (2.42) it then follows that

$$0 = \limsup_{r\to 0+} \frac{1}{r^2} \int_{-r}^{r} \int_{-\lambda^* r}^{\lambda^* r} |v(t,x)|\, dx\, dt$$

$$\geq \limsup_{r\to 0+} \frac{1}{r} \int_{r-(\delta/L')r}^{r} \frac{1}{r}\left( \int_{-\lambda^* r}^{\lambda^* r} |v(r,x)|\, dx - L'(r-t)\right) dt$$

$$\geq \frac{\delta^2}{2L'},$$

giving a contradiction. Hence (2.37) must hold. □

**Example 2.3.** Consider the scalar function (Fig. 2.2)

$$u(t,x) \doteq \begin{cases} 1 & \text{if } 0 < x < \min\{t^2, 1\}, \\ 0 & \text{otherwise.} \end{cases}$$

Then $u$ satisfies the assumptions of Theorem 2.6 and hence is a *BV* function of the two variables $t, x$. Observe that $u$ is continuous (hence also approximately continuous) at all points outside the two curves

$$\gamma_1 \doteq \{(t,x);\ x = 0\} \qquad \gamma_2 \doteq \{(t,x);\ x = \min\{t^2, 1\}\}.$$

Moreover, $u$ has an approximate jump discontinuity at all points of the curves $\gamma_1$, $\gamma_2$ except at the origin, where $u$ is approximately continuous, and at the points $P = (1, 1)$, $Q = (-1, 1)$. In this case, the set of irregular points is $\widetilde{\mathcal{N}} = \{P, Q\}$, which has one-dimensional Hausdorff measure zero.

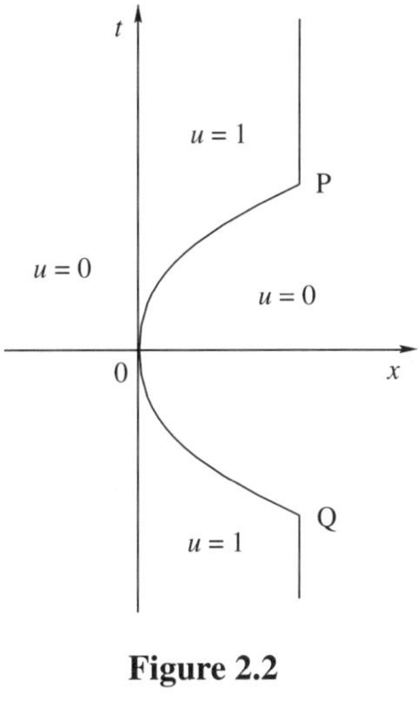


**Figure 2.2**

## 2.6 Fixed points of contractive maps

Let $\Lambda$ be a metric space, $X$ a Banach space. Given a map $\Phi : \Lambda \times X \mapsto X$, for each value of the parameter $\lambda \in \Lambda$ we seek a fixed point of the mapping $x \to \Phi(\lambda, x)$. If $\Phi$ is a strict contraction w.r.t. $x$, the existence of a unique solution is provided by a classical fixed point theorem.

**Theorem 2.7.** *Let $\Lambda$ be a metric space, $X$ a Banach space, and let $\Phi : \Lambda \times X \mapsto X$ be a mapping such that*

$$\|\Phi(\lambda, x) - \Phi(\lambda, y)\| \leq \kappa \|x - y\| \quad \textit{for all } \lambda, x, y, \tag{2.43}$$

*for some constant $\kappa < 1$. Then the following hold:*

*(i) For every $\lambda \in \Lambda$, there exists a unique $x(\lambda) \in X$ such that*

$$x(\lambda) = \Phi(\lambda, x(\lambda)). \tag{2.44}$$

*(ii) For every $\lambda \in \Lambda$, $y \in X$ one has*

$$\begin{aligned} \|x(\lambda) - \Phi(\lambda, y)\| &\leq \frac{\kappa}{1-\kappa} \|y - \Phi(\lambda, y)\|, \\ \|y - x(\lambda)\| &\leq \frac{1}{1-\kappa} \|y - \Phi(\lambda, y)\|. \end{aligned} \tag{2.45}$$

*(iii) For a fixed $\lambda_0 \in \Lambda$, if $D \subseteq X$ is a set whose closure satisfies*

$$\overline{D} \supseteq \Phi(\lambda_0, X) = \{\Phi(\lambda_0, x); x \in X\} \tag{2.46}$$

*and if*

$$\lim_{\lambda \to \lambda_0} \Phi(\lambda, y) = \Phi(\lambda_0, y) \qquad \textit{for all } y \in D, \tag{2.47}$$

*then*

$$\lim_{\lambda \to \lambda_0} x(\lambda) = x(\lambda_0). \tag{2.48}$$

*Proof.* Fix any point $y \in X$. For each $\lambda$, consider the sequence

$$y_0 = y, \qquad y_1 = \Phi(\lambda, y_0), \quad \ldots, \quad y_{\nu+1} = \Phi(\lambda, y_\nu), \quad \ldots.$$

By induction, for every $\nu \geq 0$ one checks that

$$\|y_{\nu+1} - y_\nu\| \leq \kappa^\nu \|y_1 - y_0\| = \kappa^\nu \|\Phi(\lambda, y) - y\|. \tag{2.49}$$

Since $\kappa < 1$, the sequence $y_\nu$ is Cauchy and converges to some limit point, which we call $x(\lambda)$. Since $\Phi$ is continuous w.r.t. the second variable, we now have

$$x(\lambda) = \lim_{\nu \to \infty} y_\nu = \lim_{\nu \to \infty} \Phi(\lambda, y_{\nu-1}) = \Phi(\lambda, \lim_{\nu \to \infty} y_{\nu-1}) = \Phi(\lambda, x(\lambda)),$$

and hence (2.44) holds. The uniqueness of $x(\lambda)$ is proved by observing that, if $x_1 = \Phi(\lambda, x_1)$ and $x_2 = \Phi(\lambda, x_2)$, then (2.43) implies

$$\|x_1 - x_2\| = \|\Phi(\lambda, x_1) - \Phi(\lambda, x_2)\| \leq \kappa \|x_1 - x_2\|.$$

Hence $x_1 = x_2$.

Next, observe that (2.49) implies

$$\|y_{\nu+1} - \Phi(\lambda, y)\| \leq \sum_{j=1}^{\nu} \|y_{j+1} - y_j\| \leq \sum_{j=1}^{\nu} \kappa^j \|y - \Phi(\lambda, y)\|$$
$$\leq \frac{\kappa}{1-\kappa} \|y - \Phi(\lambda, y)\| .$$

Letting $\nu \to \infty$ we obtain the first inequality in (2.45). The second follows immediately.

To prove the last statement, observe that (2.46) and the uniform Lipschitz continuity of the functions $y \mapsto \Phi(\lambda, y)$ imply

$$\lim_{\lambda \to \lambda_0} \Phi(\lambda, y) = \Phi(\lambda_0, y) \quad \text{for all } y \in \overline{D}. \tag{2.50}$$

Indeed, for every $y \in \overline{D}$, $\varepsilon > 0$, we can choose $\tilde{y} \in D$ such that $\|\tilde{y} - y\| \leq \varepsilon$. Hence

$$\limsup_{\lambda \to \lambda_0} \|\Phi(\lambda, y) - \Phi(\lambda_0, y)\| \leq \limsup_{\lambda \to \lambda_0} \Big\{ \|\Phi(\lambda, y) - \Phi(\lambda, \tilde{y})\|$$
$$+ \|\Phi(\lambda, \tilde{y}) - \Phi(\lambda_0, \tilde{y})\| + \|\Phi(\lambda_0, \tilde{y}) - \Phi(\lambda_0, y)\| \Big\}$$
$$\leq \kappa\varepsilon + 0 + \kappa\varepsilon.$$

Since $\varepsilon > 0$ was arbitrary, (2.50) holds. Using (2.50) and the second inequality in (2.45) with $y = x(\lambda_0) = \Phi(\lambda_0, x(\lambda_0)) \in \overline{D}$, we now obtain

$$\limsup_{\lambda \to \lambda_0} \|x(\lambda_0) - x(\lambda)\| \leq \limsup_{\lambda \to \lambda_0} \frac{1}{1-\kappa} \|\Phi(\lambda_0, x(\lambda_0)) - \Phi(\lambda, x(\lambda_0))\| = 0,$$

completing the proof. □

## 2.7 Differentiability of Lipschitz continuous functions

Let $f : \mathbb{R}^n \mapsto \mathbb{R}^m$. We say that $f$ is *differentiable* at a point $x_0$ if there exists a linear map $Df(x_0) : \mathbb{R}^n \mapsto \mathbb{R}^m$ such that

$$\lim_{w \to 0} \frac{f(x_0 + w) - f(x_0) - \big[Df(x_0)\big]w}{|w|} = 0.$$

We recall that $f$ is *Lipschitz continuous* if $|f(x) - f(y)| \leq L|x - y|$ for some constant $L$ and all $x, y$ in the domain of $f$. It is well known that a Lipschitz continuous function of a single real variable is absolutely continuous, and hence differentiable a.e. The following theorem extends this result to functions of several variables.

**Theorem 2.8 (Rademacher).** *Let $f : \mathbb{R}^n \mapsto \mathbb{R}^m$ be locally Lipschitz continuous. Then $f$ is differentiable a.e.*

*Proof.* It suffices to prove the theorem for $m = 1$, assuming that $f$ is Lipschitz continuous.

1. Let $v \in \mathbb{R}^n$ be any vector with unit length. Then for each $y \in \mathbb{R}^n$ the map $t \mapsto f(y+tv)$ of a single real variable is Lipschitz continuous, and hence differentiable a.e. Therefore, the directional derivative of $f$ along $v$, written $D_v f(x)$, exists a.e. for $x \in \mathbb{R}^n$. In particular, letting $v$ vary among the standard basis of $\mathbb{R}^n$, we see that the vector

$$\text{grad } f(x) \doteq \left(\frac{\partial f}{\partial x_1}(x), \dots, \frac{\partial f}{\partial x_n}(x)\right)$$

is well defined a.e. for $x \in \mathbb{R}^n$.

2. For every unit vector $v$, the equality

$$D_v f(x) = v \cdot \text{grad } f(x) \tag{2.51}$$

holds a.e. in $\mathbb{R}^n$. Indeed, let $\phi \in \mathcal{C}_c^1(\mathbb{R}^n)$. Then

$$\int_{\mathbb{R}^n} \left[\frac{f(x+\varepsilon v) - f(x)}{\varepsilon}\right] \phi(x)\, dx = -\int_{\mathbb{R}^n} f(x) \left[\frac{\phi(x) - \phi(x-\varepsilon v)}{\varepsilon}\right] dx. \tag{2.52}$$

Letting $\varepsilon \to 0+$ in (2.52), by the dominated convergence theorem we obtain

$$\begin{aligned}
\int_{\mathbb{R}^n} D_v f(x)\phi(x)\, dx &= -\int_{\mathbb{R}^n} f(x) D_v\phi(x)\, dx \\
&= -\sum_{i=1}^n v_i \int_{\mathbb{R}^n} f(x) \frac{\partial \phi}{\partial x_i}(x)\, dx \\
&= \sum_{i=1}^n v_i \int_{\mathbb{R}^n} \frac{\partial f}{\partial x_i}(x)\phi(x)\, dx \\
&= \int_{\mathbb{R}^n} (v \cdot \text{grad } f(x))\, \phi(x)\, dx.
\end{aligned}$$

Since the above equality holds for every $\phi \in \mathcal{C}_c^1$, the equality (2.51) follows.

3. Now choose a countable dense subset $\{v_k\}_{k \geq 1}$ of the sphere of unit vectors in $\mathbb{R}^n$. Define

$$A_k \doteq \{x \in \mathbb{R}^n;\ D_{v_k} f(x) = v_k \cdot \text{grad } f(x)\}, \qquad A \doteq \bigcap_{k=1}^{\infty} A_k.$$

By the previous step, one has

$$\text{meas}(\mathbb{R}^n \setminus A) \leq \sum_{k=1}^{\infty} \text{meas}(\mathbb{R}^n \setminus A_k) = 0. \tag{2.53}$$

We claim that $f$ is differentiable at every point of $A$. To prove this claim, we define

$$Q(x, v, t) \doteq \frac{f(x+tv) - f(x)}{t} - v \cdot \text{grad}\, f(x).$$

If $v'$ is any unit vector, calling $L$ the Lipschitz constant of $f$ we have

$$\begin{aligned} |Q(x, v, t) - Q(x, v', t)| &\leq \left| \frac{f(x+tv) - f(x+tv')}{t} \right| + |(v - v') \cdot \text{grad}\, f(x)| \\ &\leq L|v - v'| + |\text{grad}\, f(x)|\,|v - v'| \\ &\leq (1 + \sqrt{n})L|v - v'|. \end{aligned} \tag{2.54}$$

Now fix $\varepsilon > 0$ and choose $N$ so large that, for every unit vector $v$, one can find $k \in \{1, \ldots, N\}$ such that

$$|v - v_k| < \frac{\varepsilon}{2(1+\sqrt{n})L}. \tag{2.55}$$

Since

$$\lim_{t \to 0+} Q(x, v_k, t) = 0 \quad (k = 1, \ldots, N),$$

there exists $\delta > 0$ such that

$$|Q(x, v_k, t)| < \frac{\varepsilon}{2} \qquad \text{for all } t \in\, ]0, \delta], \quad k = 1, \ldots, N. \tag{2.56}$$

Now let $w \in \mathbb{R}^n \setminus \{0\}$, $|w| \leq \delta$. Set $v \doteq w/|w|$ so that $w = tv$ with $t \in\, ]0, \delta]$. Choose $k$ so that (2.55) holds. Together, the bounds (2.54)–(2.56) yield

$$\begin{aligned} \left| \frac{f(x+w) - f(x) - w \cdot \text{grad} f(x)}{|w|} \right| &= \left| \frac{f(x+tv) - f(x)}{t} - v \cdot \text{grad} f(x) \right| \\ &= |Q(x, v, t)| \\ &\leq |Q(x, v, t) - Q(x, v_k, t)| + |Q(x, v_k, t)| \\ &< \frac{\varepsilon}{2} + \frac{\varepsilon}{2}. \end{aligned}$$

Since $\varepsilon > 0$ was arbitrary, by (2.53) the theorem is proved. □

## 2.8 Approximation of Lipschitz flows

Let $\mathcal{D}$ be a closed subset of a Banach space $E$ and consider a continuous flow $S : \mathcal{D} \times [0, \infty[ \mapsto \mathcal{D}$ with the properties:

(i) $S_0 u = u, \;\; S_s S_t u = S_{s+t} u,$

(ii) $\|S_t u - S_s v\| \leq L \cdot \|u - v\| + L' \cdot |t - s|.$

Observe that (i) is the semigroup property, while (ii) states that the flow is globally Lipschitz continuous w.r.t. time and the initial data.

Given a Lipschitz-continuous map $w : [0, T] \mapsto \mathcal{D}$, we wish to estimate the difference between $w$ and the trajectory of the semigroup $S$ starting at $w(0)$.

**Theorem 2.9.** *Let $S : \mathcal{D} \times [0, \infty[ \mapsto \mathcal{D}$ be a continuous flow satisfying the properties (i)–(ii). For every Lipschitz-continuous map $w : [0, T] \mapsto \mathcal{D}$ one then has the estimate*

$$\|w(T) - S_T w(0)\| \leq L \cdot \int_0^T \left\{ \liminf_{h \to 0+} \frac{\|w(t+h) - S_h w(t)\|}{h} \right\} dt. \tag{2.57}$$

**Remark 2.3.** The integrand in (2.57) can be regarded as the instantaneous error rate for $w$ at time $t$. Since the flow is uniformly Lipschitz continuous, during the time interval $[t, T]$ this error is amplified at most by a factor $L$ (see Fig. 2.3).

*Proof.* As a preliminary, observe that the integrand

$$\phi(t) \doteq \liminf_{h \to 0+} \frac{\|w(t+h) - S_h w(t)\|}{h}$$

in (2.57) is bounded and measurable. Indeed, for every $h > 0$, the function

$$\phi_h(t) \doteq \frac{\|w(t+h) - S_h w(t)\|}{h} \tag{2.58}$$

is continuous. By the continuity of the maps $h \mapsto \phi_h(t)$, we have

$$\phi(t) = \lim_{\varepsilon \to 0+} \inf_{h \in \mathbf{Q} \cap ]0, \varepsilon]} \phi_h(t),$$

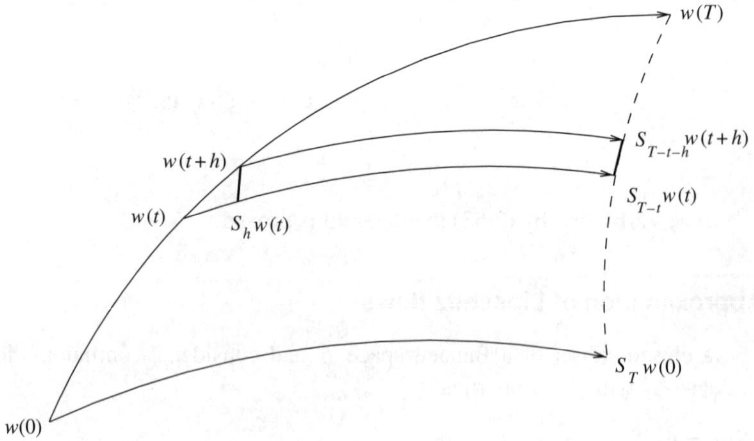


**Figure 2.3**

where the infimum is taken only over rational values of $h$. Hence the function $\phi$ is Borel measurable. Its boundedness is a consequence of the uniform Lipschitz continuity of $w$ and of all trajectories of the semigroup $S$.

Next, define

$$\psi(t) \doteq \|S_{T-t}w(t) - S_T w(0)\|\,, \tag{2.59}$$

$$x(t) \doteq \psi(t) - L\int_0^t \phi(s)\,ds. \tag{2.60}$$

We need to show that $x(T) \le 0$. Since $x$ is Lipschitz continuous and $x(0) = 0$, one has

$$x(T) = \int_0^T \dot{x}(s)\,ds, \tag{2.61}$$

where the upper dot denotes a differentiation w.r.t. time. By the theorems of Rademacher and of Lebesgue, there exists a null set $\mathcal{N} \subset [0, T]$ such that, for each $s \notin \mathcal{N}$, both functions $x$, $\psi$ are differentiable at time $s$ while $\phi$ is approximately continuous at $s$. For $s \notin \mathcal{N}$ we thus have

$$\dot{x}(s) = \lim_{h\to 0}\frac{x(s+h) - x(s)}{h} = \lim_{h\to 0}\frac{\psi(s+h) - \psi(s)}{h} - L\phi(s). \tag{2.62}$$

By the definition of $\psi$ and the properties of $S$ it follows

$$\begin{aligned}
\psi(s+h) - \psi(s) &= \|S_{T-s-h}w(s+h) - S_T w(0)\| - \|S_{T-s}w(s) - S_T w(0)\| \\
&\le \|S_{T-s-h}w(s+h) - S_{T-s}w(s)\| \\
&\le \|S_{T-s-h}w(s+h) - S_{T-s-h}S_h w(s)\| \\
&\le L \cdot \|w(s+h) - S_h w(s)\|\,.
\end{aligned}$$

Therefore

$$\lim_{h\to 0+}\frac{\psi(s+h) - \psi(s)}{h} \le L \cdot \liminf_{h\to 0+}\frac{\|w(s+h) - S_h w(s)\|}{h} = L\cdot\phi(s).$$

By (2.62) this implies

$$\dot{x}(s) \le 0 \quad \text{for all } s \notin \mathcal{N}.$$

Hence $x(T) \le 0$, proving the theorem. □

## 2.9 Taylor estimates

In this section we derive some a priori bounds on the size of a function $\Psi$ of several variables, in a neighbourhood of the origin. The assumption that $\Psi$ vanishes on certain sets, such as the coordinate axes or the diagonals, implies that some of the Taylor coefficients of $\Psi$ at the origin are equal to zero. By writing a Taylor approximation and using this additional information, we shall obtain various estimates that are useful for future applications.

Here, as in later chapters, the Landau notation $\mathcal{O}(1)$ is used to indicate a function whose absolute value remains uniformly bounded. For example,

$$\Psi(\sigma, \sigma', \omega) = \mathcal{O}(1) \cdot |\sigma|\,|\sigma'|$$

means that, for some constant $C$, the following holds:

$$\big|\Psi(\sigma, \sigma', \omega)\big| \le C \cdot |\sigma|\,|\sigma'|.$$

This notation is useful in order to avoid the introduction of a large number of auxiliary constants $C_1, C_2, C_3, \ldots$.

**Lemma 2.4.** *Let* $\Psi : \mathbb{R}^m \mapsto \mathbb{R}^n$ *be differentiable with Lipschitz-continuous derivative. If*

$$\Psi(0) = 0, \qquad \frac{\partial \Psi}{\partial x}(0) = 0, \tag{2.63}$$

*then one has*

$$\Psi(x) = \mathcal{O}(1) \cdot |x|^2. \tag{2.64}$$

**Lemma 2.5.** *Let* $\Psi : \mathbb{R}^p \times \mathbb{R}^q \mapsto \mathbb{R}^n$ *be twice differentiable, with Lipschitz-continuous second derivatives. If*

$$\Psi(x, 0) = \Psi(0, y) = 0 \quad \textit{for all } x \in \mathbb{R}^p, \quad y \in \mathbb{R}^q, \tag{2.65}$$

*then*

$$\Psi(x, y) = \mathcal{O}(1) \cdot |x|\,|y| \tag{2.66}$$

*for all* $x, y$ *in the neighbourhood of the origin. If in addition*

$$\frac{\partial^2 \Psi}{\partial x\, \partial y}(0, 0) = 0, \tag{2.67}$$

*then one has the sharper estimate*

$$\Psi(x, y) = \mathcal{O}(1) \cdot |x|\,|y|\,\left(|x| + |y|\right). \tag{2.68}$$

*Proof.* Observe that (2.65) implies

$$\begin{aligned} \Psi(x, y) &= \int_0^1 \left( \frac{\partial \Psi}{\partial y}(x, \theta y) \right) \cdot y \, d\theta \\ &= \int_0^1 \int_0^1 \left( \frac{\partial^2 \Psi}{\partial x\, \partial y}(\theta' x, \theta y) \right) \cdot (x \otimes y) \, d\theta' d\theta \\ &= \mathcal{O}(1) \cdot |x|\,|y|. \end{aligned} \tag{2.69}$$

This gives (2.66). If (2.67) also holds, then the Lipschitz continuity of the second derivative implies

$$\left|\frac{\partial^2\Psi}{\partial x\partial y}(x,y)\right| = \mathcal{O}(1)\cdot(|x|+|y|)\,. \tag{2.70}$$

Using the bound (2.70) inside the double integral in (2.69), we obtain (2.68). □

**Lemma 2.6.** *Let $\Psi = \Psi(\tilde{q}, q^*, \sigma)$ be a $\mathcal{C}^2$ mapping from $\mathbb{R}^{n-1}\times\mathbb{R}\times\mathbb{R}$ into $\mathbb{R}$, with Lipschitz-continuous second derivatives. Assume that, for all $\tilde{q}\in\mathbb{R}^{n-1}$, $s, q^*, \sigma\in\mathbb{R}$ the following holds:*

$$\Psi(\tilde{q}, q^*, 0) = \Psi(0, s, -s) = \Psi(0,0,\sigma) = 0. \tag{2.71}$$

*Then one has the estimate*

$$\Psi(\tilde{q}, q^*, \sigma) = \mathcal{O}(1)\cdot(|\tilde{q}|\,|\sigma| + |q^*|\,|\sigma|\,|q^*+\sigma|). \tag{2.72}$$

*If, in addition, for all $q^*, \sigma$*

$$\Psi(0, q^*, \sigma) = 0, \tag{2.73}$$

*then one has*

$$\Psi(\tilde{q}, q^*, \sigma) = \mathcal{O}(1)\cdot|\tilde{q}|\,|\sigma|. \tag{2.74}$$

*Proof.* To derive (2.72) we write

$$\Psi(\tilde{q}, q^*, \sigma) = \Psi(0, q^*, \sigma) + [\Psi(\tilde{q}, q^*, \sigma) - \Psi(0, q^*, \sigma)]. \tag{2.75}$$

By (2.71), for all $\tilde{q}, q^*$ one has

$$\frac{\partial\Psi}{\partial\tilde{q}}(\tilde{q}, q^*, 0) = 0.$$

Hence, by the Lipschitz continuity of the first derivative,

$$\frac{\partial\Psi}{\partial\tilde{q}}(\tilde{q}, q^*, \sigma) = \mathcal{O}(1)\cdot|\sigma|.$$

Using this last estimate we obtain

$$\begin{aligned}\Psi(\tilde{q}, q^*, \sigma) - \Psi(0, q^*, \sigma) &= \int_0^1 \left(\frac{\partial\Psi}{\partial\tilde{q}}(\theta\tilde{q}, q^*, \sigma)\right)\cdot\tilde{q}\,d\theta \\ &= \mathcal{O}(1)\cdot|\tilde{q}|\,|\sigma|.\end{aligned} \tag{2.76}$$

To establish (2.72) it thus remains to prove

$$\Psi(0, q^*, \sigma) = \mathcal{O}(1) \cdot |q^*|\,|\sigma|\,|q^* + \sigma|. \tag{2.77}$$

Observe that (2.71) implies

$$\frac{\partial^2 \Psi}{\partial q^*\,\partial \sigma}(0, 0, 0) = 0.$$

In the case where $q^*$ and $\sigma$ have the same sign, one has $|q^* + \sigma| = |q^*| + |\sigma|$. Therefore (2.77) follows from Lemma 2.5, being equivalent to the estimate (2.68).

To see what happens when $q^*$ and $\sigma$ have opposite signs, to fix the ideas assume that $q^* < 0 < \sigma$, $|q^*| < \sigma$, the other cases being entirely similar. We consider two possibilities: if $|q^*| < \sigma/2$, then

$$|q^* + \sigma| \geq \frac{|\sigma|}{2} \geq \frac{|q^*| + |\sigma|}{3}$$

and (2.77) again follows from (2.68) in Lemma 2.5. In the remaining case where $\sigma/2 \leq |q^*| \leq \sigma$, we write

$$\begin{aligned}
\Psi(0, q^*, \sigma) &= \Psi(0, -\sigma, \sigma) + \int_{-\sigma}^{q^*} \frac{\partial \Psi}{\partial q^*}(0, s, \sigma)\, ds \\
&= 0 + \int_{-\sigma}^{q^*} \int_{0}^{\sigma} \frac{\partial^2 \Psi}{\partial q^*\,\partial \sigma}(0, s, s')\, ds\, ds' \\
&= \int_{-\sigma}^{q^*} \int_{0}^{\sigma} \mathcal{O}(1) \cdot (|s| + |s'|)\, ds\, ds' \\
&= \mathcal{O}(1) \cdot |\sigma + q^*|\,|\sigma|\,(|\sigma| + |\sigma|)\,.
\end{aligned}$$

This yields (2.77) again, completing the proof of (2.72).

If the additional assumption (2.73) holds, then the estimate (2.74) is an immediate consequence of (2.75) and (2.76). □

**Lemma 2.7.** *Let $\Psi = \Psi(\tilde{q}, q^*, \sigma)$ be a $\mathcal{C}^1$ mapping from $\mathbb{R}^{n-1} \times \mathbb{R} \times \mathbb{R}$ into $\mathbb{R}$, with Lipschitz-continuous derivatives.*

*(a) If $\Psi$ satisfies the assumptions*

$$\frac{\partial \Psi}{\partial q^*}(0, 0, 0) = \frac{\partial \Psi}{\partial \sigma}(0, 0, 0) = \frac{1}{2}, \qquad \Psi(0, s, -s) = 0 \quad \textit{for all } s, \tag{2.78}$$

*then one has the estimate*

$$\Psi(\tilde{q}, q^*, \sigma) = \frac{q^* + \sigma}{2} + \mathcal{O}(1) \cdot (|\tilde{q}| + |q^* + \sigma|(|q^*| + |\sigma|)). \tag{2.79}$$

(b) *If instead* $\Psi$ *satisfies the assumptions*

$$\frac{\partial \Psi}{\partial q^*}(0,0,0) = \frac{1}{2}, \qquad \Psi(0,0,\sigma) = 0 \quad \textit{for all } \sigma, \tag{2.80}$$

*then one has the estimate*

$$\Psi(\tilde{q}, q^*, \sigma) = \frac{q^*}{2} + \mathcal{O}(1) \cdot (|\tilde{q}| + |q^*|(|q^*| + |\sigma|)). \tag{2.81}$$

*Proof.* We write $\Psi$ in the form

$$\Psi(\tilde{q}, q^*, \sigma) = [\Psi(\tilde{q}, q^*, \sigma) - \Psi(0, q^*, \sigma)] + \Psi(0, q^*, \sigma).$$

For the first term on the right hand side one has the easy bound

$$\Psi(\tilde{q}, q^*, \sigma) - \Psi(0, q^*, \sigma) = \mathcal{O}(1) \cdot |\tilde{q}|. \tag{2.82}$$

To estimate the second term, in case (a) we define $s \doteq (q^* - \sigma)/2$ and compute

$$\begin{aligned} \Psi(0, q^*, \sigma) &= \Psi(0, s, -s) + \int_0^{(q^*+\sigma)/2} \left[\frac{d}{d\theta}\Psi(0, s+\theta, -s+\theta)\right] d\theta \\ &= \int_0^{(q^*+\sigma)/2} \Big[1 + \mathcal{O}(1) \cdot \big(|q^*| + |\sigma|\big)\Big] d\theta \\ &= \frac{q^* + \sigma}{2} + \mathcal{O}(1) \cdot |q^* + \sigma|\,(|q^*| + |\sigma|). \end{aligned} \tag{2.83}$$

Combining (2.82) with (2.83) one obtains (2.79).

In case (b), we write

$$\begin{aligned} \Psi(0, q^*, \sigma) &= \int_0^{q^*} \left[\frac{d}{d\theta}\Psi(0, \theta, \sigma)\right] d\theta \\ &= \int_0^{q^*} \left[\frac{1}{2} + \mathcal{O}(1) \cdot (|q^*| + |\sigma|)\right] d\theta \\ &= \frac{q^*}{2} + \mathcal{O}(1) \cdot |q^*|\, \big(|q^*| + |\sigma|\big). \end{aligned} \tag{2.84}$$

Combining (2.82) with (2.84) one obtains (2.81). □

**Remark 2.4.** In the above lemmas, let us now assume that the functions $\Psi$ and their derivatives depend continuously on an additional parameter $\omega$ and satisfy the given assumptions for all values of $\omega$. Then all of the above estimates are still valid for quantities $\mathcal{O}(1)$ which remain uniformly bounded as $\omega$ ranges on compact sets.

## 2.10 Weak convergence of measures

In this section we review some basic facts on Radon measures and weak convergence.

Let $\Omega \subseteq \mathbb{R}^m$ be an open set and let $\mu$ be a signed Borel measure on $\Omega$. By the Hahn decomposition theorem, there exist unique positive measures $\mu^+, \mu^-$ which are mutually singular and such that $\mu = \mu^+ - \mu^-$. By *mutually singular* we mean that there exist $\Omega^+, \Omega^- \subseteq \Omega$ such that

$$\Omega^+ \cup \Omega^- = \Omega, \qquad \Omega^+ \cap \Omega^- = \emptyset,$$
$$\mu^-(E) = 0 \quad \text{for every } E \subseteq \Omega^+, \qquad \mu^+(E) = 0 \quad \text{for every } E \subseteq \Omega^-.$$

In connection with the above decomposition, we define the positive measure

$$|\mu| \doteq \mu^+ + \mu^-.$$

Clearly, the measures $\mu, \mu^+, \mu^-$ are absolutely continuous w.r.t. $|\mu|$.

Now let $\mu$ be a positive Borel measure. Given a Borel subset $E \subseteq \Omega$, we say that $\mu$ is *outer regular* on $E$ if

$$\mu(E) = \inf\{\mu(U);\ U \supset E, U \text{open}\}$$

and *inner regular* on $E$ if

$$\mu(E) = \sup\{\mu(K);\ K \subset E, K \text{compact}\}.$$

A *positive Radon measure* on $\Omega$ is a positive Borel measure which is finite on compact sets, outer regular on all Borel sets and inner regular on all open sets. A signed Borel measure $\mu$ is a *Radon measure* if its positive and negative parts are both Radon measures.

Call $\mathcal{C}_c(\Omega)$ the normed space of all continuous functions $f : \Omega \mapsto \mathbb{R}$ having compact support, with norm $\|u\| = \sup_{x\in\Omega} |u(x)|$. Its completion is the Banach space $\mathcal{C}_0(\Omega)$ of all continuous functions $f$ having the following property: for every $\varepsilon > 0$ there exists a compact set $K \subset \Omega$ such that $|f(x)| < \varepsilon$ for all $x \notin K$.

The Riesz representation theorem establishes an isometric isomorphism between the space $\mathcal{M}(\Omega)$ of bounded Radon measures on $\Omega$, with norm $\|\mu\| \doteq |\mu|(\Omega)$, and the dual space $\mathcal{C}_0(\Omega)^*$, whose elements are continuous linear functionals on $\mathcal{C}_0(\Omega)$. That is, for every $\mu \in \mathcal{M}(\Omega)$, the map

$$f \mapsto \int_\Omega f\, d\mu$$

is a bounded linear functional. Moreover,

$$\sup_{f\in\mathcal{C}_0(\Omega),\ \|f\|\le 1} \left| \int_\Omega f\, d\mu \right| = |\mu|(\Omega).$$

Let $\mu$ be a Radon measure on $\Omega$. A point $P \in \Omega$ is called an *atom* if $\mu(\{P\}) \neq 0$. Since $\Omega$ is a countable union of compact sets and $|\mu|(K) < \infty$ for every compact set

$K$, it is clear that $\mu$ can have at most countably many atoms. Let $A \doteq \{a_1, a_2, a_3, \ldots\}$ be this set of atoms. If $\mu(E) = 0$ for every set $E$ with $E \cap A = \emptyset$, we say that the measure $\mu$ is *purely atomic*. On the other hand, we say that $\mu$ is *continuous* if it has no atoms.

A special case is particularly important for our applications. Let $u : \mathbb{R} \mapsto \mathbb{R}$ be a right continuous function with bounded variation. Then its distributional derivative $\mu \doteq Du$ is a Radon measure, characterized by the property

$$\mu\,(]a, b]) = u(b) - u(a) \quad \text{for every } a < b.$$

In this case, the atoms of $\mu$ are precisely the points where $u$ has a jump. Indeed, for every $x \in \mathbb{R}$ one has

$$\mu\,(\{x\}) = u(x) - u(x-).$$

**Definition 2.2.** We say that a sequence of Radon measures $\mu_\nu \in \mathcal{M}(\Omega)$ *converges weakly to* $\mu \in \mathcal{M}(\Omega)$, and write $\mu_\nu \rightharpoonup \mu$, if

$$\lim_{\nu\to\infty} \int_\Omega f\, d\mu_\nu = \int_\Omega f\, d\mu \quad \text{for all } f \in \mathcal{C}_c(\Omega).$$

**Theorem 2.10.** *Let $\{\mu_j\}_{j\geq 1}$ be a bounded sequence in $\mathcal{M}(\Omega)$. Then there exists a subsequence $\{\mu_k\}_{k\geq 1}$ and a measure $\mu \in \mathcal{M}(\Omega)$ such that $\mu_k \rightharpoonup \mu$.*

For a proof, see Billingsley (1968) or Yosida (1980).

Some useful properties of weak convergence are proved in the following lemma.

**Lemma 2.8.** *Let $\mu_\nu \rightharpoonup \mu$ in $\mathcal{M}(\Omega)$.*

*(i) If all measures $\mu_\nu$ are positive, then*

$$\limsup_{\nu\to\infty} \mu_\nu(K) \leq \mu(K) \quad \textit{for every compact set } K \subset \Omega, \tag{2.85}$$

$$\liminf_{\nu\to\infty} \mu_\nu(U) \geq \mu(U) \quad \textit{for every open set } K \subset \Omega. \tag{2.86}$$

*(ii) Assume $|\mu_\nu| \rightharpoonup \tilde{\mu}$, and let $E \subset \Omega$ be a Borel set with compact closure whose boundary satisfies $\tilde{\mu}(\partial E) = 0$. Then*

$$\lim_{\nu\to\infty} \mu_\nu(E) = \mu(E). \tag{2.87}$$

*Proof.* Let $\varepsilon > 0$ be given. Given a compact set $K \subset \Omega$, by outer regularity there exists an open set $U \supset K$ with

$$\mu(U) - \mu(K) < \varepsilon. \tag{2.88}$$

Consider a function $f \in \mathcal{C}_c(\Omega)$ with support contained inside $U$, such that

$$\begin{cases} f(x) \in [0, 1] & \text{for all } x \in \Omega, \\ f(x) = 1 & \text{if } x \in K. \end{cases} \tag{2.89}$$

We then have

$$\limsup_{\nu\to\infty} \mu_\nu(K) \le \limsup_{\nu\to\infty} \int_\Omega f\,d\mu_\nu = \int_\Omega f\,d\mu \le \mu(U) < \mu(K)+\varepsilon.$$

Since $\varepsilon$ was arbitrary, this yields (2.85).

Similarly, given any open set $U$ we can find a compact set $K \subset U$ such that (2.88) holds. Consider again a function $f \in \mathcal{C}_c(\Omega)$ with support contained inside $U$, satisfying (2.89). We then have

$$\liminf_{\nu\to\infty} \mu_\nu(U) \ge \liminf_{\nu\to\infty} \int_\Omega f\,d\mu_\nu = \int_\Omega f\,d\mu \ge \mu(K) \ge \mu(U)-\varepsilon.$$

Since $\varepsilon$ was arbitrary, this yields (2.86).

Finally, given a set $E$ with $\tilde{\mu}(\partial E) = 0$, for any $\varepsilon > 0$ we can find a compact neighbourhood $K$ of $\partial E$ such that $\tilde{\mu}(K) < \varepsilon$. Let $f \in \mathcal{C}_c(\Omega)$ be a function such that

$$\begin{cases} f(x) \in [0,1] & \text{for all } x \in \Omega, \\ f(x) = 1 & \text{if } x \in E \setminus K, \\ f(x) = 0 & \text{if } x \notin E \cup K. \end{cases}$$

Calling $\chi_E$ the characteristic function of $E$, we now have

$$\begin{aligned}
&\limsup_{\nu\to\infty} |\mu_\nu(E) - \mu(E)| \\
&\quad = \limsup_{\nu\to\infty} \left| \int (\chi_E - f)\,d\mu_\nu + \int f\,(d\mu_\nu - d\mu) + \int (f-\chi_E)\,d\mu \right| \\
&\quad \le \limsup_{\nu\to\infty} \int |\chi_E - f|\,d|\mu_\nu| \\
&\qquad + \limsup_{\nu\to\infty} \left| \int f\,d\mu_\nu - \int f\,d\mu \right| + \int |f-\chi_E|\,d|\mu| \le \limsup_{\nu\to\infty} |\mu_\nu|(K) + |\mu|(K) \\
&\quad \le \tilde{\mu}(K) + \tilde{\mu}(K) \le 2\varepsilon.
\end{aligned}$$

Since $\varepsilon$ was arbitrary, this yields (2.87). □

## 2.11 Basic ODE theory

Let $\Omega$ be an open set in $\mathbb{R} \times \mathbb{R}^n$. Given a function $g : \Omega \mapsto \mathbb{R}^n$, by a (Carathéodory) solution of the differential equation

$$\dot{x} = g(t,x) \tag{2.90}$$

we mean an absolutely continuous function $x(\cdot)$ which satisfies (2.90) a.e. Equivalently, we require that

$$x(t) = x(t_0) + \int_{t_0}^{t} g\,(s, x(s))\,ds$$

for every $t_0, t$ in the interval where $x(\cdot)$ is defined.

To provide the existence and uniqueness of the solution to the Cauchy problem, the standard theory requires $g$ to be locally bounded, measurable in $t$ and Lipschitz continuous w.r.t. the variable $x$. More precisely, consider the following conditions:

(a) For every fixed $x \in \mathbb{R}^n$, the function $t \mapsto g(t, x)$ defined on the section $\Omega_x \doteq \{t\,;\ (t, x) \in \Omega\}$ is measurable.

(b) For every compact set $K \subset \Omega$ there exists constants $C_K$, $L_K$ such that

$$|g(t,x)| \le C_K, \quad |g(t,x) - g(t,y)| \le L_K|x-y| \qquad \text{for all } (t,x), (t,y) \in K. \tag{2.91}$$

**Theorem 2.12.** *Let $g : \Omega \mapsto \mathbb{R}^n$ satisfy the assumptions (a)–(b), and consider the Cauchy problem*

$$\dot{x} = g(t,x), \qquad x(t_0) = x_0, \tag{2.92}$$

*for some $(t_0, x_0) \in \Omega$. Then*

*(i) There exists $\delta > 0$ such that (2.92) has a solution $x(\cdot)$ defined on $[t_0,\ t_0 + \delta]$.*

*(ii) If, in addition, $g$ is defined on the whole set $[t_0,\ T] \times \mathbb{R}^n$ and there exists constants $C, L$ such that*

$$|g(t,x)| \le C, \quad |g(t,x) - g(t,y)| \le L|x-y| \qquad \textit{for all } t, x, y, \tag{2.93}$$

*then the problem (2.92) has a unique solution defined on $[t_0,\ T]$. Moreover, the map $x_0 \mapsto x(\cdot)$ is continuous from $\mathbb{R}^n$ into $\mathcal{C}^0\,([t_0,\ T];\ \mathbb{R}^n)$.*

*Proof.* 1. The proof of (ii) will be given first. In view of applying Theorem 2.7, we have to represent the solution of the Cauchy problem as the fixed point of a suitable contractive transformation. Since the initial condition $x_0 \in \mathbb{R}^n$ here plays the role of a parameter, we set $\Lambda \doteq \mathbb{R}^n$. Moreover, we let $X$ be the Banach space of all continuous functions from $[t_0,\ T]$ into $\mathbb{R}^n$ with the weighted norm

$$\|x(\cdot)\|_\dagger \doteq \max_{t_0 \le t \le T} e^{-2Lt}\, |x(t)|\,, \tag{2.94}$$

which is equivalent to the usual $\mathcal{C}^0$ norm. Finally, we define the map $\Phi : \Lambda \times X \mapsto X$ by setting

$$\Phi(x_0,\ w)(t) \doteq x_0 + \int_{t_0}^{t} g\,(s,\ w(s))\, ds \quad t \in [t_0,\ T]. \tag{2.95}$$

2. To prove that $\Phi$ is well defined, for any $w \in X$ consider the sequence of piecewise constant functions $\{w_\nu\}_{\nu \ge 1}$, with

$$w_\nu(t) = w\left(t_0 + \frac{j}{\nu}\right) \quad \text{if } t_0 + \frac{j}{\nu} \le t < t_0 + \frac{j+1}{\nu}.$$

By (a), all maps $t \mapsto g(t, w_\nu(t))$ are measurable. Moreover, for each $t$ the second assumption in (2.93) implies

$$\lim_{\nu\to\infty} |g(t, w(t)) - g(t, w_\nu(t))| \le \lim_{\nu\to\infty} L\,|w(t) - w_\nu(t)| = 0.$$

Hence the function $t \mapsto g(t, w(t))$ is measurable, being the limit of a sequence of measurable maps. Since $|g(s, w(s))| \le C$ for all $s$, the integral in (2.95) is well defined and depends continuously on $t$. Therefore, the map $\Phi$ is well defined and takes values inside $X$.

3. The continuous dependence of $\Phi$ on $x_0$ is obvious. To study its dependence on $w(\cdot)$, assume that $w, w' \in X$, $\|w - w'\|_\dagger = \delta$. Recalling (2.94) we have

$$|w(s) - w'(s)| \le \delta e^{2Ls},$$

$$\begin{aligned} e^{-2Lt}|\Phi(x_0, w)(t) - \Phi(x_0, w')(t)| &= e^{-2Lt}\left|\int_{t_0}^{t} \big\{g(s, w(s)) - g(s, w'(s))\big\}\,ds\right| \\ &\le e^{-2Lt}\int_{t_0}^{t} L|w(s) - w'(s)|\,ds \\ &\le e^{-2Lt}\int_{t_0}^{t} L\delta e^{2Ls}\,ds \\ &< \delta/2 \end{aligned}$$

for all $t \in [t_0, T]$. Therefore

$$\|\Phi(x_0, w) - \Phi(x_0, w')\|_\dagger \le \tfrac{1}{2}\|w - w'\|_\dagger.$$

4. We can now apply Theorem 2.7, obtaining the existence of a continuous mapping $x_0 \mapsto x(\cdot)$ such that $x = \Phi(x_0, x)$. By the definition of $\Phi$, for all initial data $x_0$ the corresponding function $x(\cdot)$ thus provides the unique Carathéodory solution to the Cauchy problem (2.92).

5. Finally, we prove (i), disregarding the assumption (2.93). Choose $\varepsilon > 0$ small enough so that the cylinder

$$K \doteq \{(t, x);\ t \in [t_0,\ t_0 + \varepsilon],\ |x - x_0| \le \varepsilon\}$$

is entirely contained inside $\Omega$. Then consider a smooth scalar function $\phi : \mathbb{R} \times \mathbb{R}^n \mapsto [0, 1]$ such that $\phi \equiv 1$ on $K$ while $\phi \equiv 0$ outside some larger compact set $K'$, with $K \subset K' \subset \Omega$. Fix any $T > t_0$ and observe that the function

$$g^*(t, x) \doteq \begin{cases} \phi(t, x)g(t, x) & \text{if } (t, x) \in \Omega, \\ 0 & \text{if } (t, x) \notin \Omega \end{cases}$$

satisfies (i)–(ii) together with the additional conditions (2.93), because it vanishes outside the compact set $K'$. By part (ii) of the theorem, there exists a solution to the Cauchy problem

$$\dot{x} = g^*(t, x), \qquad x(t_0) = x_0$$

defined on $[t_0, T]$. For some $\delta > 0$ small enough, the point $(t, x(t))$ remains inside $K$ as $t \in [t_0, t_0 + \delta]$. Since $g$ and $g^*$ coincide on $K$, the function $x(\cdot)$ thus provides a solution to the original Cauchy problem on the interval $[t_0,\ t_0 + \delta]$. □

The next lemma provides a useful tool for estimating the distance between two solutions of a differential equation. It represents the main ingredient in several uniqueness proofs, and in this respect it can replace the original version of Gronwall's lemma, which has more mathematical content and also a longer proof.

**Lemma 2.9.** *Let $z(\cdot)$ be an absolutely continuous non-negative function such that*

$$z(t_0) \le \gamma, \quad \dot{z}(t) \le \alpha(t)z(t) + \beta(t) \qquad \text{for a.e. } t \in [t_0, T], \tag{2.96}$$

*for some integrable functions $\alpha$, $\beta$ and some constant $\gamma \ge 0$. Then $z$ satisfies the estimate*

$$z(t) \le \gamma \exp\left(\int_{t_0}^{t} \alpha(s)\, ds\right) + \int_{t_0}^{t} \beta(s) \exp\left(\int_{s}^{t} \alpha(r)\, dr\right) ds \quad \text{for all } t \in [t_0, T]. \tag{2.97}$$

*Proof.* Observe that the right hand side of (2.97) is precisely the solution of the Cauchy problem

$$Z(t_0) = \gamma, \qquad \dot{Z}(t) = \alpha(t)Z(t) + \beta(t).$$

Consider the absolutely continuous function

$$\psi(t) \doteq \exp\left(-\int_{t_0}^{t} \alpha(r)\, dr\right)\left[z(t) - \int_{t_0}^{t} \beta(s) \exp\left(\int_{s}^{t} \alpha(r)\, dr\right) ds\right].$$

Using (2.96), a direct computation shows that $\dot{\psi}(t) \le 0$ for almost every $t$. Therefore

$$\psi(t) \le \psi(t_0) = z(t_0) \le \gamma \quad t \in [t_0,\ T]. \tag{2.98}$$

Multiplying (2.98) by $\exp(\int_{t_0}^{t} \alpha(r)dr)$, from the definition of $\psi$ we obtain (2.97). □

Applying the above lemma, one obtains a useful estimate on the distance between two solutions of (2.90).

**Lemma 2.10.** *Let the function $g = g(t, x)$ be measurable in $t$ and Lipschitz continuous in $x$ so that*

$$|g(t, x) - g(t, y)| \leq L|x - y| \quad \text{for all } t, x, y.$$

*Let $x_1, x_2$ be two solutions of the differential equation (2.90), defined on a common interval $[t_0, T]$. Then*

$$|x_1(t) - x_2(t)| \leq e^{L(t-t_0)}\,|x_1(t_0) - x_2(t_0)| \qquad t \in [t_0, T]. \tag{2.99}$$

*In particular, if $x_1(t_0) = x_2(t_0)$ then the two solutions coincide.*

*Proof.* Indeed, the absolutely continuous function $z(t) \doteq |x_1(t) - x_2(t)|$ satisfies

$$\dot{z}(t) \leq |\dot{x}_1(t) - \dot{x}_2(t)| \leq Lz(t).$$

Applying Lemma 2.9 with $\alpha = L$, $\beta = 0$, $\gamma = |x_1(t_0) - x_2(t_0)|$ we obtain (2.99). □

## Problems

(1) Let $\Phi : \Lambda \times X \mapsto X$ be as in Theorem 2.7. Let $x_0 \in X$ and let $(\lambda_\nu)_{\nu \geq 1}$ be a sequence in $\Lambda$, converging to $\lambda_0$. Prove that the sequence defined inductively by $x_{\nu+1} = \Phi(\lambda_\nu, x_\nu)$ converges to the unique point $x_0$, such that $\Phi(\lambda_0, x_0) = x_0$.

(2) Let $u : \mathbb{R} \mapsto \mathbb{R}^n$ have bounded variation. Consider a non-negative scalar function $\phi$ such that $\int \phi(y)\,dy = 1$ and define the convolution

$$(\phi * u)(x) = \int_{-\infty}^{\infty} u(x - y)\phi(y)\,dy.$$

Prove that Tot. Var. $(\phi * u) \leq$ Tot. Var. $(u)$.

(3) Let $u : \mathbb{R} \mapsto \mathbb{R}^n$ have bounded variation. Prove that, in addition to Lemma 2.3, the following holds:

$$\text{Tot. Var. } \{u\} = \sup_{\varepsilon > 0} \frac{1}{\varepsilon} \int_{-\infty}^{\infty} |u(x + \varepsilon) - u(x)|\,dx = \lim_{\varepsilon \to 0} \frac{1}{\varepsilon} \int_{-\infty}^{\infty} |u(x + \varepsilon) - u(x)|\,dx.$$

(4) Let $u : \mathbb{R} \mapsto \mathbb{R}^n$ be right continuous. Show that $u \in BV$ iff

$$\sup \left\{ \int_{-\infty}^{\infty} u(x) \cdot \phi_x(x)\,dx;\ \phi \in \mathcal{C}^1,\ \|\phi(x)\|_{\mathbf{L}^\infty} \leq 1 \right\} < \infty. \tag{2.100}$$

Conversely, assume that a function $u \in \mathbf{L}^1_{\text{loc}}(\mathbb{R}; \mathbb{R}^n)$ satisfies (2.100). Show that $u$ coincides a.e. with a right continuous $BV$ function.

(5) For each $\bar{x} \in \mathbb{R}$, denote by $t \mapsto S_t\bar{x}$ the unique strictly increasing solution of the Cauchy problem

$$\dot{x} = \sqrt{|x|}, \qquad x(0) = \bar{x}.$$

Show that $S$ is a continuous semigroup on $\mathbb{R}$, i.e. the map $(t, \bar{x}) \mapsto S_t\bar{x}$ is well defined and continuous on $[0, \infty[\, \times \mathbb{R}$, and satisfies $S_0\bar{x} = \bar{x}$, $S_s S_t \bar{x} = S_{s+t}\bar{x}$.

In connection with the null solution $w(t) \equiv 0$, for any $\tau > 0$ compute the two quantities

$$|w(\tau) - S_\tau w(0)|\,, \qquad \int_0^\tau \left\{ \liminf_{h \to 0+} \frac{|w(t+h) - S_h w(t)|}{h} \right\} dt.$$

Compare this result with the statement of Theorem 2.9.

(6) Let $\Psi : \mathbb{R}^m \mapsto \mathbb{R}^n$ be a $\mathcal{C}^2$ mapping with Lipschitz-continuous second derivatives. Assume that, for every $x_1, \ldots, x_m$, the following holds:

$$\Psi(x_1, 0, \ldots, 0) = \Psi(0, x_2, 0, \ldots, 0) = \cdots = \Psi(0, \ldots, 0, x_m) = 0.$$

Prove that, in a neighbourhood of the origin, one has the estimate

$$\Psi(x_1, \ldots, x_m) = \mathcal{O}(1) \cdot \sum_{i \neq j} |x_i x_j|.$$

Hint: observe that $f(x_1, \ldots, x_n) = \big[f(x_1, \ldots, x_n) - f(x_1, \ldots, x_{n-1}, 0)\big] + f(x_1, \ldots, x_{n-1}, 0)$. Use induction on $n$.

(7) Under the same assumptions as in Theorem 2.9, let $v : [0, T] \mapsto \mathcal{D}$ be a piecewise Lipschitz-continuous map, with jumps at the times $0 < t_1 < \cdots < t_m < T$. Prove the estimate

$$\|v(T) - S_T v(0)\| \leq L \cdot \int_0^T \left\{ \liminf_{h \to 0+} \frac{\|v(t+h) - S_h v(t)\|}{h} \right\} dt + L \cdot \sum_{i=1}^m \|v(t_i+) - v(t_i-)\|\,. \tag{2.101}$$

# 3

# Semilinear and quasilinear systems

A *quasilinear system* of $n$ equations in one space variable takes the form

$$u_t + A(t, x, u)u_x = h(t, x, u). \tag{3.1}$$

Here $A$ is an $n \times n$ matrix while $h, u$ take values in $\mathbb{R}^n$. If $A = A(t, x)$ is independent of $u$, we say that the system is *semilinear.*

Throughout the following, we shall assume that the system is *strictly hyperbolic*, i.e. each $n \times n$ matrix $A(t, x, u)$ has $n$ real distinct eigenvalues $\lambda_1 < \cdots < \lambda_n$. We can then select bases of left and right eigenvectors $\{l_1, \ldots, l_n\}$ and $\{r_1, \ldots, r_n\}$ such that

$$l_i A = \lambda_i l_i, \qquad Ar_i = \lambda_i r_i \tag{3.2}$$

at every point $(t, x, u)$. These eigenvectors can be normalized according to

$$|r_i| \equiv 1, \qquad l_j \cdot r_i = \begin{cases} 1 & \text{if } i = j, \\ 0 & \text{if } i \neq j. \end{cases} \tag{3.3}$$

We observe that (3.2)–(3.3) imply

$$v = \sum_{i=1}^{n} (l_i \cdot v)r_i, \qquad Av = \sum_{i=1}^{n} \lambda_i (l_i \cdot v)r_i \tag{3.4}$$

for every $v \in \mathbb{R}^n$. From the implicit function theorem it follows that, if the entries of the matrix $A$ are Lipschitz continuous, or $k$-times continuously differentiable as functions of $(t, x, u)$, then the same is true of the functions $\lambda_i,\ l_i,\ r_i,\ i = 1, \ldots, n$.

By a *classical solution* of (3.1) we mean a continuously differentiable function $u = u(t, x)$ which satisfies (3.1) at every point of its domain. In later sections, we shall also consider *broad solutions*, corresponding to fixed points of a suitable integral functional. For the initial value problem, the existence of such solutions can be proved by an application of the contraction mapping theorem, in a suitable Banach space.

A subsequent analysis will show that these solutions have the same regularity as the data of the problem, i.e. the functions $A, h$ and the initial value of $u$ at time $t = 0$. In particular, if all of these data are continuously differentiable, then a classical solution does exist.

The last sections of this chapter are concerned with the general quasilinear system (3.1). Clearly, a function $u = u(t, x)$ is a solution if it satisfies the semilinear system

$$u_t + \hat{A}(t, x)u_x = h(t, x, u),$$

with $\hat{A}(t,x) = A(t,x,u(t,x))$. Solutions of (3.1) will be constructed by recursively solving the sequence of semilinear problems

$$\overset{(\nu)}{u_t} + A\big(t,x,\overset{(\nu-1)}{u}(t,x)\big)\,\overset{(\nu)}{u_x} = h\big(t,x,\overset{(\nu)}{u}\big),$$

and proving the existence and regularity of the limit as $\nu \to \infty$.

## 3.1 Explicit solutions

The homogeneous scalar equation with constant coefficients has the form

$$u_t + \lambda u_x = 0. \tag{3.5}$$

Given initial data $u(0,x) = \bar{u} \in \mathcal{C}^1$, one can easily check that the travelling wave

$$u(t,x) = \bar{u}(x - \lambda t) \tag{3.6}$$

provides a classical solution to the corresponding Cauchy problem.

In the case where the initial condition $\bar{u}$, instead of $\mathcal{C}^1$, is only locally integrable, the function $u$ defined by (3.6) can still be interpreted as a 'solution' to (3.5) in various ways:

- As a *distributional solution*, since for every $\phi \in \mathcal{C}^1_c$ one has

$$\iint \{u\phi_t + \lambda u \phi_x\}\, dx\, dt = 0.$$

- As a *limit solution*, meaning that there exists a sequence of classical solutions $u^\nu$ to (3.5) such that $u^\nu \to u$ in $\mathbf{L}^1_{\mathrm{loc}}$ as $\nu \to \infty$.
- As a *broad solution*, meaning that the function $u$ is absolutely continuous (in fact, constant) along every line of the form $x = x_0 + \lambda t$. The directional derivative of $u = u(t,x)$ along the vector $\mathbf{v} = (1,\lambda)$ is everywhere defined and satisfies $D_{\mathbf{v}} u \equiv 0$.

Next, consider the initial value problem for the $n \times n$ homogeneous system with constant coefficients

$$u_t + A u_x = 0, \qquad u(0,x) = \bar{u}(x). \tag{3.7}$$

Here $A$ is an $n \times n$ hyperbolic matrix, with eigenvalues $\lambda_1 < \cdots < \lambda_n$ and eigenvectors $r_i$, $l_i$ normalized as in (3.3). Call $u_i \doteq l_i \cdot u$ the coordinates of a vector $u \in \mathbb{R}^n$ w.r.t. the basis of right eigenvectors $\{r_1, \ldots, r_n\}$. Multiplying (3.7) on the left by $l_1, \ldots, l_n$ we obtain

$$(u_i)_t + \lambda_i (u_i)_x = (l_i u)_t + \lambda_i (l_i u)_x = l_i u_t + l_i A u_x = 0,$$
$$u_i(0,x) = l_i \bar{u}(x) \doteq \bar{u}_i(x).$$

Therefore, the system (3.7) decouples into $n$ scalar Cauchy problems, which can be solved separately as in (3.6). An easy computation now shows that the function

$$u(t,x) = \sum_{i=1}^{n} \bar{u}_i(x - \lambda_i t) r_i = \sum_{i=1}^{n} (l_i \cdot \bar{u}(x - \lambda_i t)) r_i \tag{3.8}$$

provides a solution to (3.7). Indeed, by (3.4),

$$u_t(t,x) = \sum_{i=1}^{n} -\lambda_i (l_i \cdot \bar{u}_x(x - \lambda_i t)) r_i = -A u_x(t,x).$$

Observe that in the scalar case (3.6) the initial profile is shifted with constant speed $\lambda$. In the case of an $n \times n$ system, the initial profile is decomposed as a sum of $n$ waves, each travelling with one of the characteristic speeds $\lambda_1, \ldots, \lambda_n$.

**Example 3.1.** Given two states $u^-, u^+ \in \mathbb{R}^n$, a Cauchy problem with initial data having the special form

$$\bar{u}(x) = \begin{cases} u^- & \text{if } x < 0, \\ u^+ & \text{if } x > 0 \end{cases}$$

is called a *Riemann problem.* Let us write the jump $u^+ - u^-$ as a linear combination of eigenvectors:

$$u^+ - u^- = \sum_{i=1}^{n} c_i r_i.$$

Moreover, let us define the intermediate states $u^- = \omega_0, \omega_1, \ldots, \omega_n = u^+$ by setting

$$\omega_k \doteq u^- + \sum_{i=1}^{k} c_i r_i \quad k = 0, \ldots, n.$$

The solution of (3.7), still given by (3.8), can now be written more explicitly as

$$u(t,x) = \begin{cases} u^- & \text{if } x < \lambda_1 t, \\ \omega_i & \text{if } \lambda_i t < x < \lambda_{i+1} t, \ 1 \le i < n, \\ u^+ & \text{if } x > \lambda_n t. \end{cases}$$

In other words (Fig. 3.1), the solution is piecewise constant in the $t$–$x$ plane, with discontinuities along the $n$ lines $x = \lambda_i t, i = 1, \ldots, n$. Each of these jumps $\omega_i - \omega_{i-1} = c_i r_i$ is an eigenvector of the matrix $A$.

## 3.2 A single semilinear equation

This section is concerned with the Cauchy problem for the scalar, semilinear equation:

$$u_t + a(t,x) u_x = h(t,x,u), \tag{3.9}$$

$$u(0,x) = \bar{u}(x). \tag{3.10}$$

Observe that (3.9) prescribes the directional derivative of $u : \mathbb{R}^2 \mapsto \mathbb{R}$ at each point $(t,x)$ in the direction of the vector $\mathbf{v}(t,x) = (1, a(t,x))$. The integral curves of the

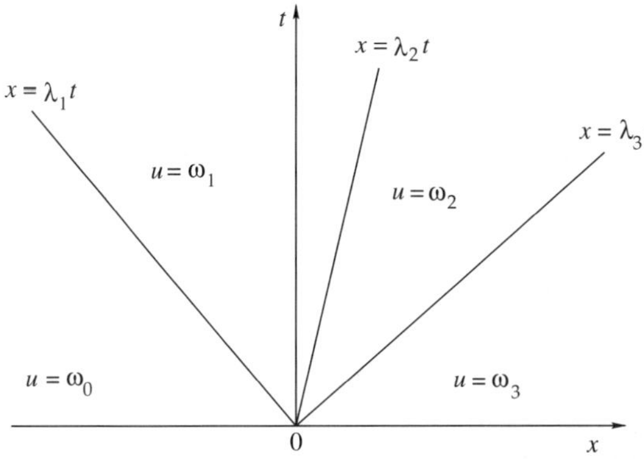


**Figure 3.1**

vector field **v** in the $t$–$x$ plane are called *characteristics*. They can be written in the form $x = x(t)$, where

$$\dot{x} \doteq \frac{dx}{dt} = a(t, x). \tag{3.11}$$

For any given point $(\tau, \xi)$, we denote by $t \mapsto x(t; \tau, \xi)$ the characteristic line through $(\tau, \xi)$, i.e. the solution of the Cauchy problem

$$\dot{x}(t) = a(t, x(t)), \qquad x(\tau) = \xi. \tag{3.12}$$

If $x = x(t)$ is any solution of (3.11), by (3.9) at every time $t$ one has

$$\frac{d}{dt} u(t, x(t)) = u_t + \dot{x}(t) u_x = u_t + a(t, x) u_x = h(t, x(t), u).$$

Therefore, the value of the solution $u$ of (3.9)–(3.10) at the point $(\tau, \xi)$ coincides with the value at time $\tau$ of the solution to the Cauchy problem for the ODE

$$\frac{du}{dt} = h(t, x(t; \tau, \xi), u), \qquad u(0) = \bar{u}(x(0; \tau, \xi)). \tag{3.13}$$

For the scalar, semilinear initial value problem (3.9)–(3.10), the value of the solution at any point $(\tau, \xi)$ can thus be determined by solving the two scalar Cauchy problems (3.12)–(3.13). If the functions $a, h$ are Lipschitz continuous, standard results in the theory of ODEs imply that the solutions of (3.12)–(3.13) are unique and depend on the point $(\tau, \xi)$ in a differentiable way. The method of characteristics, described above, thus reduces the initial value problem for the PDE (3.9)–(3.10) to a family of Cauchy problems for ODEs.

Observe that the value of $\bar{u}$ at a point $x_0$ determines the values of the solution $u$ along the entire characteristic line $t \mapsto x(t; 0, x_0)$. The 'information' contained in the initial data is thus transported along the characteristic lines. In the semilinear case, these characteristics are entirely determined by the equations in (3.12), and hence do not depend on the initial condition $\bar{u}$.

**Example 3.2.** Consider the Cauchy problem

$$u_t + 2u_x = 3u, \qquad u(0, x) = \sin x. \tag{3.14}$$

The characteristic lines are the solutions to $\dot{x} = 2$, namely $x(t; \tau, \xi) = \xi + 2(t - \tau)$. Along the characteristic through the point $(\tau, \xi)$, the function $u$ satisfies

$$\frac{du}{dt} = 3u, \qquad u(0) = \sin(x(0; \tau, \xi)).$$

The solution of (3.14) is thus given by $u(\tau, \xi) = e^{3\tau} \sin(\xi - 2\tau)$.

## 3.3 The scalar quasilinear equation

We now study the more general quasilinear Cauchy problem

$$u_t + a(t, x, u)u_x = h(t, x, u), \qquad u(0, x) = \bar{u}(x), \tag{3.15}$$

assuming that $a$, $h$ and $\bar{u}$ are continuously differentiable. In the space $\mathbb{R}^3$ with coordinates $t, x, u$, consider the integral curves of the vector field $\mathbf{v} = (1, a, h)$. These curves are obtained by solving the system of ODEs

$$\frac{dx}{dt} = a(t, x, u), \qquad \frac{du}{dt} = h(t, x, u). \tag{3.16}$$

For every $y \in \mathbb{R}$, we denote by $t \mapsto (x(t, y), u(t, y))$ the solution of (3.16) with initial conditions (at $t = 0$)

$$x(0) = y, \qquad u(0) = \bar{u}(y). \tag{3.17}$$

As $y$ varies, the graphs of all these solutions generate a two-dimensional surface $\mathcal{S}$ in $\mathbb{R}^3$, parametrized by $(t, y)$. We claim that, at least locally, the surface $\mathcal{S}$ is the graph of a function $u = u(t, x)$, which provides a classical solution to (3.15). Indeed, by classical theorems on ODEs, the map $(t, y) \mapsto (t, x(t, y), u(t, y))$ defining the surface $\mathcal{S}$ in parametric form is continuously differentiable. Fix any point $x_0 \in \mathbb{R}$. At $(t, y) = (0, x_0)$ we have

$$\frac{\partial t}{\partial t} = 1, \qquad \frac{\partial t}{\partial y} = 0, \qquad \frac{\partial x}{\partial t} = a(0, x_0, \bar{u}(x_0)), \qquad \frac{\partial x}{\partial y} = 1.$$

By the implicit function theorem, the map $(t, y) \mapsto (t, x(t, y))$ is thus locally invertible in a neighbourhood $\Omega$ of $(0, x_0)$. Therefore, $\mathcal{S}$ is locally the graph of a $\mathcal{C}^1$ function $u = u(t, x)$. Because of (3.17), the relation $u(0, x) = \bar{u}(x)$ clearly holds. Now let $(t, x)$ be any point in $\Omega$, say with $x = x(t, y)$ for some $y$. Then

$$u_t + a(t, x, u)u_x = \frac{d}{dt}u(t, x(t, y)) = h(t, x(t, y), u),$$

proving that $u$ is a solution of (3.15).

**Remark 3.1.** It is interesting to compare some basic properties of quasilinear and semilinear equations:

1. In the semilinear case, the characteristic curves in the $t$–$x$ plane are independent of the solution $u$. In particular, if the initial condition $u(0, \cdot) = \bar{u}$ is given on some interval $[\xi^-, \ \xi^+]$, then the values $u(t, x)$ can be uniquely determined on the interval $[x(t; 0, \xi^-), \ x(t; 0, \xi^+)]$. This interval is the same for every initial condition $\bar{u}$.

On the other hand, in the quasilinear case one can still define as *characteristics* the projections on the $t$–$x$ plane of the integral curves of (3.16). In this case, however, the characteristic lines depend on the particular solution under consideration. In particular, the domain where a solution is determined by the initial data actually varies with $\bar{u}$.

2. If $t \mapsto x(t)$ denotes a characteristic curve, in both the semilinear and quasilinear case we have

$$\frac{d}{dt} u(t, x(t)) = h(t, x, u).$$

Therefore, if $\bar{u} \in \mathbf{L}^\infty$ and the function $h$ satisfies a bound of the form

$$|h(t, x, u)| \leq C(1 + |u|),$$

then for all $t \geq 0$ the solution $u(t, \cdot)$ will remain bounded. On the other hand, if the perturbation $h$ is superlinear (say, $h = e^u$ or $h = u^p$ with $p > 1$), then the solution may blow up in finite time, for semilinear as well as quasilinear equations.

3. In the semilinear case, as long as the solution is bounded, its gradient also remains bounded. Indeed, differentiating (3.9) w.r.t. $x$ one obtains the equation

$$u_{xt} + a(t, x) u_{xx} = (h_u - a_x) u_x + h_x,$$

where $h_u, h_x$ are functions of $t, x, u$. If $x = x(t)$ is any characteristic line satisfying (3.11), we thus have

$$\frac{d}{dt} u_x(t, x(t)) = (h_u - a_x) u_x + h_x.$$

Observe that the above equation is linear w.r.t. $u_x$. Therefore, as long as $u$ remains bounded, the same holds for its gradient $u_x$.

The situation is quite different for quasilinear equations, where a 'gradient catastrophe' may occur within finite time. Differentiating (3.15) we now find

$$(u_x)_t + a(t, x, u)(u_x)_x = -a_u u_x^2 + (h_u - a_x) u_x + h_x.$$

If $x = x(t)$ is any characteristic line satisfying (3.16), we thus have

$$\frac{d}{dt} u_x(t, x(t)) = -a_u u_x^2 + (h_u - a_x) u_x + h_x.$$

This equation is quadratic w.r.t. $u_x$. If $a_u \neq 0$, it may well happen that $|u_x| \to \infty$ at a finite time along a characteristic line, even if the function $u$ itself remains uniformly bounded.

**Example 3.3.** For the initial value problem

$$u_t + uu_x = 0, \qquad u(0,x) = \frac{1}{1+x^2}, \tag{3.18}$$

the system (3.16) takes the simple form

$$\frac{dx}{dt} = u, \qquad \frac{du}{dt} = 0.$$

In this case, the surface $S$ is parametrized by

$$(t,y) \mapsto \left(t, y + \frac{t}{1+y^2}, \frac{1}{1+y^2}\right).$$

For $t < 8/\sqrt{27}$, the function $y \mapsto x(t,y) \doteq y + t/(1+y^2)$ has a smooth inverse, say $y = y(t,x)$. The solution of (3.18) can thus be written as

$$u(t,x) = \frac{1}{1+y^2(t,x)} \qquad t < \frac{8}{\sqrt{27}}.$$

For $t > 8/\sqrt{27}$ no classical solution defined on the entire real line exists. Observe that (3.18) implies

$$u_{xt} + uu_{xx} = -u_x^2, \qquad u_x(0,x) = \frac{-2x}{(1+x^2)^2}.$$

Hence, along any characteristic line $x(t) = x_0 + t/(1+x_0^2)$, we have

$$\frac{d}{dt}u_x(t,x(t)) = -u_x^2(t,x(t)).$$

In particular, choosing the inflection point $x_0 = \sqrt{1/3}$ where $u_x(0,\cdot)$ attains its minimum, one finds $u_x\big(t, \sqrt{1/3}+3t/4\big) = \big(t - 8/\sqrt{27}\big)^{-1}$, showing that $\sup_{x\in\mathbb{R}} \big|u_x(t,x)\big| \to \infty$ as $t \to 8/\sqrt{27}$.

## 3.4 Semilinear systems: broad solutions

We consider here the semilinear hyperbolic system

$$u_t + A(t,x)u_x = h(t,x,u), \tag{3.19}$$

with the following basic hypothesis:

(H1) The map $A : \mathbb{R} \times \mathbb{R} \mapsto \mathbb{R}^{n\times n}$ is locally Lipschitz continuous, while $h : \mathbb{R} \times \mathbb{R} \times \mathbb{R}^n \mapsto \mathbb{R}^n$ is locally bounded, measurable w.r.t. $t, x$ and locally Lipschitz continuous w.r.t. $u$.

Call $\lambda_1(t,x) < \cdots < \lambda_n(t,x)$ the eigenvalues of the matrix $A(t,x)$, and define the left and right eigenvectors $l_i(t,x)$, $r_i(t,x)$ as in (3.2)–(3.3). The assumptions on $A$ imply that $\lambda_i$, $l_i$, $r_i$ are locally Lipschitz-continuous functions of $t, x$. In the $t$–$x$ plane, the integral curves of the vector field $\mathbf{v}_i = (1, \lambda_i)$, $i = 1, \ldots, n$, are called the *$i$-characteristics*. If the eigenvalues $\lambda_i$ are uniformly bounded, then for each $i$ and each point $(\tau, \xi)$ the Cauchy problem

$$\dot{x} = \lambda_i(t,x), \qquad x(\tau) = \xi \tag{3.20}$$

has a unique solution, defined for all $t$, describing the $i$-th characteristic through $(\tau, \xi)$. Such a solution will be denoted by $t \mapsto x_i(t; \tau, \xi)$.

Now assume that an initial condition

$$u(0,x) = \bar{u}(x) \quad x \in [a,b] \tag{3.21}$$

is prescribed on an interval $[a,b]$. A closed region $\mathcal{D} \subseteq [0,\infty[\, \times \mathbb{R}$ is called a *domain of determinacy* for the initial value problem (3.19), (3.21) provided that the following holds (Fig. 3.2). For every $(\tau,\xi) \in \mathcal{D}$ and every $i \in \{1,\ldots,n\}$, the characteristic curve $\{(t, x_i(t;\tau,\xi));\ 0 \le t \le \tau\}$ is entirely contained inside $\mathcal{D}$. Moreover, $x(0;\tau,\xi) \in [a,b]$.

For example, if $|\lambda_i(t,x)| \le C$ for all $t, x, i$, one such domain of determinacy is the triangle

$$\mathcal{D}_C \doteq \{(t,x);\ t \ge 0,\ a + Ct \le x \le b - Ct\}. \tag{3.22}$$

Clearly, the largest domain of determinacy is

$$\mathcal{D}_{max} \doteq \{(\tau,\xi);\ \tau \ge 0,\ x_i(0;\tau,\xi) \in [a,b] \text{ for all } i = 1,\ldots,n\}.$$

Otherwise stated, a point $(\tau,\xi)$ belongs to $\mathcal{D}_{max}$ iff $\tau \ge 0$ and all the $n$ characteristic lines through $(\tau,\xi)$ fall inside the interval $[a,b]$ at $t = 0$. Observe that we have the representation

$$\mathcal{D}_{max} = \{(t,x);\ t \ge 0,\ x_n(t;0,a) \le x \le x_1(t;0,b)\}. \tag{3.23}$$

A subsequent analysis will show that $\mathcal{D}_{max}$ is the largest set where the solution $u$ of (3.19) is uniquely determined, if the initial data (3.21) are assigned on $[a,b]$.

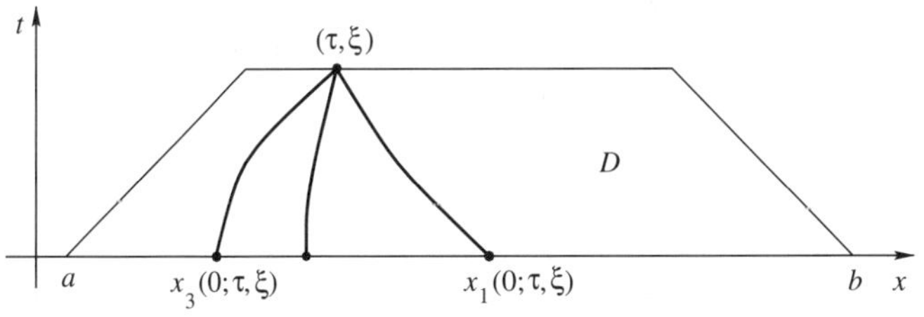


**Figure 3.2**

For $i = 1, \ldots, n$, we define the components

$$u_i(t,x) \doteq l_i(t,x) \cdot u(t,x).$$

By (3.21), we thus have

$$u_i(0,x) = \bar{u}_i(x) \doteq l_i(0,x) \cdot \bar{u}(x). \tag{3.24}$$

Multiplying (3.19) on the left by $l_1, \ldots, l_n$, one obtains the system of $n$ scalar equations

$$(u_i)_t + \lambda_i(u_i)_x = g_i(t,x,u) \quad i = 1, \ldots, n, \tag{3.25}$$

where

$$g_i = l_i \cdot h + [l_{i,t} + \lambda_i l_{i,x}] \cdot u, \quad u = \sum_{i=1}^{n} u_i r_i. \tag{3.26}$$

Here and in the sequel, we use the shorter notation $l_{i,t} \doteq (l_i)_t$, $l_{i,x} \doteq (l_i)_x$ to indicate partial derivatives of $i$-th components. The assumption (H1) implies that the left eigenvectors $l_i$ are Lipschitz-continuous functions of $t, x$. Hence, by Rademacher's theorem, their partial derivatives are defined a.e. The maps $g_i$ are thus measurable w.r.t. $t, x$ and locally Lipschitz continuous w.r.t. $u$. For almost every characteristic line $t \mapsto x_i(t;\tau,\xi)$, (3.20) and (3.25) together imply

$$\frac{d}{dt} u_i(t, x_i(t;\tau,\xi)) = g_i(t, x_i(t;\tau,\xi), u(t, x_i(t;\tau,\xi))). \tag{3.27}$$

Under suitable regularity conditions, for each $(\tau,\xi) \in \mathcal{D}$ one can integrate (3.27) from 0 to $\tau$ and obtain

$$u_i(\tau,\xi) = \bar{u}_i\,(x_i(0;\tau,\xi)) + \int_0^{\tau} g_i(t, x_i(t;\tau,\xi), u(t, x_i(t;\tau,\xi)))\,dt. \tag{3.28}$$

Motivated by (3.28), we say that a locally integrable function $u = \sum u_i r_i$ from a domain of determinacy $\mathcal{D}$ into $\mathbb{R}^n$ is a *broad solution* of the Cauchy problem (3.19), (3.21) provided that (3.28) holds at almost every point $(\tau,\xi) \in \mathcal{D}$, for every $i = 1, \ldots, n$. Equivalently, we require that each component $u_i$ be absolutely continuous and satisfy (3.27) along almost every characteristic line $x_i(\cdot\,;\tau,\xi)$.

The existence and uniqueness of such a solution, with initial data in $\mathbf{L}^1$ or in $\mathbf{L}^\infty$, will now be proved by a fixed point argument.

**Theorem 3.1 (Global existence in $\mathbf{L}^\infty$).** *Let the functions $A, h$ satisfy the basic hypothesis (H1) and let $\mathcal{D}$ be a domain of determinacy for the problem (3.19), (3.21). Assume that there exist constants $L, C$ such that*

$$|g_i(t,x,u) - g_i(t,x,u')| \le L \cdot \max_j |u_j - u'_j|, \quad |\lambda_i(t,x)| \le C, \quad |g_i(t,x,0)| \le C \tag{3.29}$$

*for every $i \in \{1, \ldots, n\}$, $u, u' \in \mathbb{R}^n$ and almost every $(t,x) \in \mathcal{D}$. Then, for every bounded measurable mapping $\bar{u} : [a,b] \mapsto \mathbb{R}^n$, the Cauchy problem (3.19), (3.21) has a unique broad solution $u : \mathcal{D} \mapsto \mathbb{R}^n$.*

*Proof.* Let $X_\infty$ be the space of all measurable functions $u = (u_1, \ldots, u_n) : \mathcal{D} \mapsto \mathbb{R}^n$ such that

$$\|u\|_* = \|(u_1, \ldots, u_n)\|_* \doteq \underset{\substack{(t,x)\in\mathcal{D}\\ i=1,\ldots,n}}{\text{ess-sup}}\ e^{-2Lt}|u_i(t,x)| < \infty. \tag{3.30}$$

Introduce the transformation $\mathcal{T} = (\mathcal{T}_1, \ldots, \mathcal{T}_n)$ given by

$$\mathcal{T}_i(u)(\tau, \xi) = \bar{u}_i(x_i(0; \tau, \xi)) + \int_0^\tau g_i(t, x_i(t; \tau, \xi), u(t, x_i(t; \tau, \xi)))\, dt. \tag{3.31}$$

By the assumptions (3.29), it is clear that $\mathcal{T}$ maps $X_\infty$ into itself. Indeed, if $u \in X_\infty$, the measurability hypotheses on $g_i$ imply that the map $\tau \mapsto \mathcal{T}_i(u)(\tau, x(\tau, 0, x_0))$ is well defined and absolutely continuous for almost every $x_0$. Moreover, if $|\bar{u}(x)| \le C'$ for every $x$, then the bounds on $g_i$ imply

$$|\mathcal{T}_i(u)(\tau, \xi)| \le C' + \int_0^\tau C + Le^{2Lt}\|u\|_*\, dt,$$

and hence $\mathcal{T}(u) \in X_\infty$.

If $\|u - v\|_* = \delta$ for some $\delta \ge 0$, the definition (3.30) implies that, for $t, x$ a.e.,

$$|u_j(t,x) - v_j(t,x)| \le \delta e^{2Lt} \quad \text{for every } j = 1, \ldots, n.$$

Together with (3.29), this yields the estimate

$$\begin{aligned}
&|\mathcal{T}_i(u)(\tau, \xi) - \mathcal{T}_i(v)(\tau, \xi)| \\
&\quad\le \int_0^\tau |g_i(t, x_i(t; \tau, \xi), u(t, x_i(t; \tau, \xi))) - g_i(t, x_i(t; \tau, \xi), v(t, x_i(t; \tau, \xi)))|\, dt \\
&\quad\le \int_0^\tau L \cdot \max_j |u_j(t, x_i(t; \tau, \xi)) - v_j(t, x_i(t; \tau, \xi))|\, dt \\
&\quad\le \int_0^\tau L\delta e^{2Lt}\, dt < \frac{\delta}{2} e^{2L\tau}.
\end{aligned} \tag{3.32}$$

Since (3.32) holds at almost every point $(\tau, \xi) \in \mathcal{D}$, recalling the definition (3.30) we have

$$\|\mathcal{T}(u) - \mathcal{T}(v)\|_* \le \frac{\delta}{2} = \frac{1}{2}\|u - v\|_*. \tag{3.33}$$

We can now use Theorem 2.7 and deduce the existence of a unique fixed point for the transformation $\mathcal{T}$, i.e. a unique broad solution to (3.19), (3.21). □

**Example 3.4.** When the matrix $A$ is constant and $h$ does not depend on $u$, the solution of the linear system

$$u_t + Au_x = h(t, x), \qquad u(0, x) = \bar{u}(x)$$

is provided by $u(t, x) = \sum u_i(t, x) r_i$, where

$$u_i(t, x) = l_i \cdot \bar{u}(x - \lambda_i t) + \int_0^t l_i \cdot h(s, x - \lambda_i(t - s))\, ds.$$

**Theorem 3.2 (Global existence in $\mathbf{L}^1$).** *Let $\mathcal{D}$ be a domain of determinacy for the problem (3.19), (3.21). Let the basic hypothesis (H1) hold, together with the global bounds (3.29), and assume*

$$|\lambda_i(t, x) - \lambda_i(t, x')| \le L'|x - x'|, \tag{3.34}$$

*for some constant $L'$ and every $x, x' \in \mathbb{R}^n$, $(t, x), (t, x') \in \mathcal{D}$, $i \in \{1, \dots, n\}$. Then, for every initial condition $\bar{u} \in \mathbf{L}^1([a, b]; \mathbb{R}^n)$, the Cauchy problem (3.19), (3.21) has a unique broad solution $u : \mathcal{D} \mapsto \mathbb{R}^n$.*

*Proof.* It is not restrictive to assume that the domain $\mathcal{D}$ is compact; more precisely, there exists some $T > 0$ such that

$$\mathcal{D} \subseteq \{(t, x) \in \mathcal{D}_{\max};\ t \le T\}. \tag{3.35}$$

The broad solution of (3.19), (3.21) will again be obtained as the unique fixed point of the transformation $\mathcal{T}$ defined in (3.31). In this case, however, we let $\mathcal{T}$ act on the space $X_1$ of all locally integrable functions $u = (u_1, \dots, u_n) : \mathcal{D} \mapsto \mathbb{R}^n$ such that

$$\|u\|_\dagger \doteq \sup_t \left\{ e^{-Kt} \sum_{i=1}^n \int_{(t,x)\in\mathcal{D}} |u_i(t, x)|\, dx \right\} < \infty, \tag{3.36}$$

with $K = 2nLe^{L'T}$. To check that $\mathcal{T}$ maps $X_1$ into itself, we first observe that the assumption (3.34) on the Lipschitz continuity of $\lambda_i$ by Lemma 2.10 implies

$$|x_i(t; \tau, \xi) - x_i(t; \tau, \xi')| \le e^{L'|t-\tau|} |\xi - \xi'|. \tag{3.37}$$

Moreover, (3.29) yields

$$|g_i(t, x, u)| \le C + L \cdot \sum_{j=1}^n |u_j(t, x)|. \tag{3.38}$$

Let $u \in X_1$. For every fixed $\tau \in [0, T]$, (3.37) and (3.38) yield

$$\begin{aligned}
&\int_{(\tau,\xi)\in\mathcal{D}} |\mathcal{T}_i(u)(\tau,\xi)|\,d\xi \\
&\quad \le \int_{(\tau,\xi)\in\mathcal{D}} |\bar{u}_i(x_i(0;\tau,\xi))|\,d\xi + \int_{(\tau,\xi)\in\mathcal{D}} \int_0^\tau |g_i(t,\ x_i(t,\tau,\xi), u(t, x_i(t;\tau,\xi)))|\,dt\,d\xi \\
&\quad \le \int_a^b e^{L'\tau}|\bar{u}_i(x)|\,dx + \int_0^\tau \int_{(t,x)\in\mathcal{D}} e^{L'|t-\tau|}\Big(C + L\sum_{i=1}^n |u_i(t,x)|\Big)\,dx\,dt \\
&\quad \le e^{L'\tau}\int_a^b |\bar{u}_i(x)|\,dx + e^{L'\tau}\int_0^\tau (C + Le^{Kt}\|u\|_\dagger)\,dt,
\end{aligned}$$

proving that $\mathcal{T}(u) \in X_1$. Next, let $u, v \in X_1$, with $\|u - v\|_\dagger = \delta$. Then, for any fixed $\tau$, using (3.29) one obtains

$$\begin{aligned}
&\int_{(\tau,\xi)\in\mathcal{D}} |\mathcal{T}_i(u)(\tau,\xi) - \mathcal{T}_i(v)(\tau,\xi)|\,d\xi \\
&\quad \le \int_{(\tau,\xi)\in\mathcal{D}} \int_0^\tau |g_i\big(t, x_i, u) - g_i(t, x_i, v)|\,dt\,d\xi \\
&\quad \le \int_{(\tau,\xi)\in\mathcal{D}} \int_0^\tau L\sum_{j=1}^n |u_j(t, x_i(t;\tau,\xi)) - v_j(t, x_i(t;\tau,\xi))|\,dt\,d\xi \\
&\quad \le \int_0^\tau \int_{(t,x)\in\mathcal{D}} e^{L'|t-\tau|}L\sum_{j=1}^n |u_j(t,x) - v_j(t,x)|\,dx\,dt \\
&\quad \le \int_0^\tau Le^{L'T}\delta e^{Kt}\,dt \\
&\quad \le \frac{\delta}{2n}e^{K\tau},
\end{aligned}$$

because of the choice of the constant $K$. In turn, this implies

$$\sup_\tau e^{-K\tau}\sum_{i=1}^n \int_{(\tau,\xi)\in\mathcal{D}} |(\mathcal{T}_i(u) - \mathcal{T}_i(v))(\tau,\xi)|\,d\xi \le \sup_\tau e^{-K\tau}\cdot n\frac{\delta}{2n}e^{K\tau} = \frac{\delta}{2},$$

proving that $\mathcal{T}$ is a strict contraction in the space $X_1$. Therefore, $\mathcal{T}$ admits a unique fixed point, which is by definition the broad solution of (3.19), (3.21). □

If the functions $A, h$ are locally Lipschitz continuous but do not satisfy the uniform bounds (3.29) on the entire space, then the existence of solutions to (3.19) can still be proved, but only locally in time.

**Theorem 3.3 (Local existence and uniqueness).** *For the system (3.19), let the assumptions (H1) hold. Then, for any bounded measurable mapping $\bar u$ from a bounded interval $[a, b]$ into $\mathbb{R}^n$, there exist constants $C, \delta > 0$ such that the Cauchy problem (3.19), (3.21) has a unique bounded broad solution defined on the set*

$$\mathcal{D}_{C,\delta} \doteq \{(t, x);\ 0 \le t \le \delta,\ a + Ct \le x \le b - Ct\}. \tag{3.39}$$

*Proof.* Assume $|\bar u(x)| \le C_0$ for every $x \in [a, b]$. Let $\varphi : \mathbb{R}^2 \mapsto [0, 1]$, $\psi : \mathbb{R}^2 \times \mathbb{R}^n \mapsto [0, 1]$ be smooth functions with compact support such that

$$\varphi(t, x) = \psi(t, x, u) = 1 \qquad \text{for all} \quad t \in [0, 1],\ \ x \in [a, b],\ \ |u| \le C_0 + 1. \tag{3.40}$$

For $i = 1, \ldots, u$, define the functions

$$g_i^\dagger(t, x, u) = \psi(t, x, u) g_i(t, x, u), \qquad \lambda_i^\dagger(t, x) = \varphi(t, x)\lambda_i(t, x)$$

and call $t \mapsto x_i^\dagger(t, \tau, \xi)$ the solution of

$$\dot x = \lambda_i^\dagger(t, x), \qquad x(\tau) = \xi.$$

Since all mappings $\lambda_i^\dagger, g_i^\dagger$ now have compact support, there exist constants $L, C$ such that

$$\begin{aligned} |g_i^\dagger(t, x, u) - g_i^\dagger(t, x, u')| &\le L \cdot \max_j |u_j - u'_j|, \\ |\lambda_i^\dagger(t, x)| \le C, \qquad & |g_i^\dagger(t, x, u)| \le C. \end{aligned} \tag{3.41}$$

Call $X_\infty$ the space of all bounded measurable functions $u = (u_1, \ldots, u_n)$ from the triangular domain $\mathcal{D}_C$ defined at (3.22) into $\mathbb{R}^n$. Define the transformation $\mathcal{T}^\dagger = (\mathcal{T}_1^\dagger, \ldots, \mathcal{T}_n^\dagger) : X_\infty \mapsto X_\infty$ by setting

$$\mathcal{T}_i^\dagger(u)(\tau, \xi) = \bar u_i(x_i^\dagger(0, \tau, \xi)) + \int_0^\tau g_i^\dagger(t, x_i^\dagger(t, \tau, \xi), u(t, x_i^\dagger(t, \tau, \xi)))\, dt.$$

Because of (3.41), the same arguments used in the proof of Theorem 3.1 yield the existence of a unique bounded measurable function $u : \mathcal{D}_C \mapsto \mathbb{R}^n$ such that $\mathcal{T}^\dagger(u) = u$. Because of the uniform Lipschitz continuity of each component $u_i$ along the corresponding $i$-characteristics, there exists $\delta \in ]0, 1]$ small enough so that $|u(t, x)| < C_0 + 1$ for all $(t, x) \in \mathcal{D}_{C,\delta}$. By (3.40), on $\mathcal{D}_{C,\delta}$ we thus have $\lambda_i^\dagger \equiv \lambda_i$, $g_i^\dagger \equiv g_i$ for all $i$; hence $u$ is a broad solution of the original problem (3.19), (3.21).

To prove uniqueness, assume that $u,\ u' : \mathcal{D} \mapsto \mathbb{R}^n$ are bounded broad solutions of the Cauchy problem (3.19), (3.21), where $\mathcal{D}$ is any compact domain of determinacy. Since $u, u'$ are bounded, there exists a constant $L$ such that the first inequality in (3.29) holds on the set

$$\Big\{(t, x, w); (t, x) \in \mathcal{D}, |w| \le \max_{(t,x)\in\mathcal{D}} \{|u(t, x)|, |u'(t, x)|\}\Big\}.$$

Define the scalar function

$$z(t) = \max_{i=1,\dots,n} \operatorname*{ess-sup}_{(t,x)\in\mathcal{D}} |u_i(t, x) - u_i'(t, x)|.$$

By (3.28) and the first relation in (3.29), $z(\cdot)$ is Lipschitz continuous and satisfies

$$\dot z(t) \le Lz(t), \qquad z(0) = 0.$$

Therefore, Lemma 2.10 implies $z(t) = 0$ for every $t \ge 0$. □

## 3.5 Continuity and continuous dependence of broad solutions

In this section we show that, if the data of the Cauchy problem (3.19), (3.21) are continuous, then the same is true for the broad solution. The dependence of the solution $u$ on the data $A, h, \bar u$ will also be investigated.

**Theorem 3.4 (Continuity of solutions).** *In addition to the assumptions of Theorem 3.1, let $h, \bar u$ be continuous. Then the broad solution of (3.19), (3.21) is a continuous function.*

*Proof.* By the proof of Theorem 3.1, the broad solution is the unique fixed point of the contractive mapping $\mathcal{T}$ on $X_\infty$. Therefore, starting with the function $w(t, x) \doteq \bar u(x)$, by the contraction mapping theorem we know that the sequence of iterates $\mathcal{T}^m(w)$ converges to the solution of (3.19), (3.21), uniformly on bounded sets. Since the uniform limit of continuous functions is continuous, it thus suffices to show that, if a function $u \in X_\infty$ is continuous, then $\mathcal{T}(u)$ is continuous as well. Fix any $\varepsilon > 0$ and let $v \in X_\infty$ be a $\mathcal{C}^1$ function such that $\|v - u\|_* \le \varepsilon$. We claim that $\mathcal{T}(v)$ is continuous. Indeed, recalling (3.26), we can write

$$\begin{aligned} \mathcal{T}_i(v)(\tau, \xi) &= \bar u_i(x_i(0; \tau, \xi)) + \int_0^\tau \Big[l_i \cdot h + \frac{dl_i}{dt} \cdot v\Big] dt \\ &= \bar u_i(x_i(0; \tau, \xi)) + \int_0^\tau \Big[l_i \cdot h - l_i \cdot \frac{dv}{dt}\Big] dt \\ &\quad + l_i \cdot v(\tau, \xi) - l_i \cdot v(0, x_i(0; \tau, \xi)), \end{aligned} \tag{3.42}$$

where all integrals are evaluated along the curve $t \mapsto x_i(t; \tau, \xi)$. Since $l_i, h$ are continuous, $v$ is $\mathcal{C}^1$ and since the points $x_i(t, \tau, \xi)$ depend continuously on $t, \tau, \xi$, the continuous dependence of the right hand side of (3.42) on $\tau, \xi$ is clear. By (3.33), $\|\mathcal{T}_i(v) - \mathcal{T}_i(u)\|_* \le \varepsilon/2$. Since $\varepsilon > 0$ was arbitrary, the above shows that $\mathcal{T}(u)$ can be approximated by continuous functions uniformly on compact sets. Therefore, $\mathcal{T}(u)$ itself is continuous. □

In the next theorem, we consider a sequence of initial value problems:

$$u_t^\nu + A^\nu(t,x)u_x^\nu = h^\nu(t,x,u^\nu), \qquad u^\nu(0,x) = \bar{u}^\nu(x). \tag{3.43$_\nu$}$$

Assuming that $A^\nu \to A^0$, $h^\nu \to h^0$ and $\bar{u}^\nu \to \bar{u}^0$ in a suitable topology, we would like to conclude the convergence of the corresponding solutions $u^\nu \to u^0$. But in aiming for this goal, a preliminary difficulty arises. Indeed, even if all initial conditions $\bar{u}^\nu$ are defined on the same interval $[a,b]$, the domains of determinacy for the problems $(3.43)_\nu$ may vary with $\nu$. Any convergence result concerning the sequence of solutions $u^\nu$ can thus be valid only on some region $\mathcal{D}$ which is a common domain of dependency for all problems $(3.43)_\nu$, with $\nu$ sufficiently large. The existence of such domains is provided by the following lemma. By $\lambda_i^\nu, l_i^\nu$ we denote here the $i$-th eigenvalue and left eigenvector of the matrix $A^\nu$, respectively.

**Lemma 3.1.** *Let all matrix-valued functions $A^\nu$ be bounded and locally Lipschitz continuous. For any fixed $\varepsilon, T > 0$, consider the compact set*

$$\mathcal{D} \doteq \{(t,x);\ t \in [0,T],\ \alpha(t) \le x \le \beta(t)\}, \tag{3.44}$$

*where $\alpha, \beta$ are solutions of the Cauchy problems*

$$\dot{\alpha} = \lambda_n^0(t,\alpha) + \varepsilon, \quad \alpha(0) = a, \qquad \dot{\beta} = \lambda_1^0(t,\beta) - \varepsilon, \quad \beta(0) = b. \tag{3.45}$$

*If $A^\nu$ converges to $A^0$ uniformly on $\mathcal{D}$, then $\mathcal{D}$ is a domain of determinacy for all problems $(3.43)_\nu$, with $\nu$ sufficiently large.*

*Proof.* Choose $\bar{\nu}$ large enough so that

$$|\lambda_i^\nu(t,x) - \lambda_i^0(t,x)| < \varepsilon \qquad \text{for all} \quad (t,x) \in \mathcal{D},\ \ i \in \{1,n\},\ \ \nu \ge \bar{\nu}. \tag{3.46}$$

If now $\nu \ge \bar{\nu}$, $i \in \{1,\dots,n\}$, the $i$-th characteristic through a point $(\tau,\xi) \in \mathcal{D}$ solves the Cauchy problem

$$\dot{x} = \lambda_i^\nu(t,x), \qquad x(\tau) = \xi. \tag{3.47}$$

From (3.46), (3.47) it follows that

$$\lambda_1^0(t,x) - \varepsilon < \lambda_1^\nu(t,x) \le \lambda_i^\nu(t,x) = \dot{x}(t) \le \lambda_n^\nu(t,x) < \lambda_n^0(t,x) + \varepsilon. \tag{3.48}$$

Since $\alpha(\tau) \le x(\tau) \le \beta(\tau)$, (3.48) implies

$$\alpha(t) \le x(t) \le \beta(t) \quad \text{for all } t \in [0,\tau].$$

Therefore, $(t, x(t)) \in \mathcal{D}$ for all $t \in [0,\tau]$, proving the lemma. □

**Theorem 3.5 (Continuous dependence).** *Let $\mathcal{D} \subset [0,\infty[\, \times \mathbb{R}$ be a compact domain. Consider a sequence $(A^\nu, h^\nu, \bar{u}^\nu)_{\nu \ge 0}$, with $A^\nu : \mathcal{D} \mapsto \mathbb{R}^{n\times n}$, $h^\nu : \mathcal{D} \times \mathbb{R}^n \mapsto \mathbb{R}^n$, $\bar{u}^\nu : [a,b] \mapsto \mathbb{R}^n$. Assume that $A^\nu, h^\nu$ satisfy the basic hypothesis (H1) and that the corresponding functions $g_i^\nu$ and eigenvalues $\lambda_i^\nu$ satisfy the uniform bounds (3.29), (3.34)* *with constants $C, L, L'$ independent of $\nu$. Let $\mathcal{D}$ be a common domain of determinacy for the problems $(3.43)_\nu$.*

*(i) If $h^\nu, \bar{u}^\nu$ are continuous, $A^\nu \to A^0$ uniformly on $\mathcal{D}$, $\bar{u}^\nu \to \bar{u}^0$ uniformly on $[a, b]$, while $h^\nu \to h^0$ uniformly on $\mathcal{D} \times K$ for every compact set $K \subset \mathbb{R}^n$, then the sequence of solutions $u^\nu$ of (3.43) converges to $u^0$ uniformly on $\mathcal{D}$.*

*(ii) Assume that $A^\nu \to A^0$ uniformly on $\mathcal{D}$, and $\bar{u}^\nu \to \bar{u}^0$ in $\mathbf{L}^1([a, b]; \mathbb{R}^n)$. Moreover, assume that, for every compact set $K \subset \mathbb{R}^n$, the functions $h^\nu$ are uniformly bounded on $\mathcal{D} \times K$ and satisfy*

$$\lim_{\nu\to\infty} h^\nu(t, x, u) = h^0(t, x, u)$$

*for all $u \in K$ and $(t, x) \in \mathcal{D} \setminus \mathcal{N}$, with $\mathcal{N}$ a set of zero measure. Then*

$$\lim_{\nu\to\infty} \sup_t \int_{(t,x)\in\mathcal{D}} |u^\nu(t, x) - u^0(t, x)|\, dx = 0. \tag{3.49}$$

*Proof.* To show (i), on the space $X$ of continuous functions $u : \mathcal{D} \mapsto \mathbb{R}^n$ with the norm (3.30), consider the sequence of transformations $\mathcal{T}^\nu$ defined as in (3.31), with $A, h, \bar{u}$ replaced with $A^\nu, h^\nu, \bar{u}^\nu$ respectively. By definition, for each $\nu \geq 0$ the solution $u^\nu$ of $(3.43)_\nu$ is the fixed point of the strict contraction $\mathcal{T}^\nu$. By the last statement of Theorem 2.7, it suffices to show that $\mathcal{T}^\nu(u) \to \mathcal{T}^0(u)$, for $u$ in a dense subset of $X$. When $u \in \mathcal{C}^1$, one can integrate by parts and obtain

$$\begin{aligned}\mathcal{T}_i^\nu(u)(\tau, \xi) &= \bar{u}_i^\nu(x_i^\nu(0; \tau, \xi)) + \int_0^\tau l_i^\nu \cdot h^\nu(t, x_i^\nu(t; \tau, \xi), u(t, x_i^\nu(t; \tau, \xi))\, dt \\ &\quad - \int_0^\tau \Big[l_i^\nu(t, x_i^\nu(t, \tau, \xi)) \cdot \frac{d}{dt} u(t, x_i^\nu(t; \tau, \xi))\Big]\, dt + l_i^\nu \cdot u(\tau, \xi) \\ &\quad - l_i^\nu \cdot u(0, x_i^\nu(0; \tau, \xi)).\end{aligned} \tag{3.50}$$

Our assumptions now imply the uniform convergence $x_i^\nu(t; \tau, \xi) \to x_i^0(t; \tau, \xi)$, $l_i^\nu \to l_i^0$. Therefore, for all $i = 1, \ldots, n$, the right hand side of (3.50) converges to $\mathcal{T}_i^0(u)$ uniformly on $\mathcal{D}_C$. Since $\mathcal{C}^1$ is dense in $\mathcal{C}^0$, this proves the first part of the theorem.

The proof of (ii) is similar. Consider the space $Y$ of all measurable maps $u = (u_1, \ldots, u_n)$ from $\mathcal{D}$ into $\mathbb{R}^n$, whose weighted norm (3.36) is finite. Again by Theorem 2.7, it suffices to show that $\|\mathcal{T}^\nu(u) - \mathcal{T}^0(u)\|_\dagger \to 0$ for $u$ in a suitable subset of $Y$. If $u \in \mathcal{C}^1(\mathcal{D}; \mathbb{R}^n)$, then $\mathcal{T}_i^\nu(u)$ is computed by (3.50). In order to prove that the right hand side of (3.50) converges to $\mathcal{T}_i^0(u)$, we first show that, for every fixed $\tau$,

$$\lim_{\nu\to\infty} \int_{(\tau,\xi)\in\mathcal{D}} |\bar{u}_i^\nu(x_i^\nu(0; \tau, \xi)) - \bar{u}_i^0(x_i^0(0, \tau, \xi))|\, d\xi = 0. \tag{3.51}$$

For any fixed $\varepsilon > 0$, choose a continuous function $\phi$ such that

$$\int_a^b |\phi(x) - \bar{u}_i^0(x)|\, dx \leq \varepsilon.$$

Then

$$\int_{(\tau,\xi)\in\mathcal{D}} |\bar{u}_i^\nu(x_i^\nu(0;\tau,\xi)) - \bar{u}_i^\nu(x_i^0(0;\tau,\xi))|\, d\xi$$

$$\leq \int_{(\tau,\xi)\in\mathcal{D}} \big\{|\bar{u}_i^\nu(x_i^\nu(0;\tau,\xi)) - \bar{u}_i^0(x_i^\nu(0;\tau,\xi))| + |\phi(x_i^\nu(0;\tau,\xi)) - \phi(x_i^0(0;\tau,\xi))|$$

$$+ |\bar{u}_i^0(x_i^\nu(0;\tau,\xi)) - \phi(x_i^\nu(0;\tau,\xi))| + |\phi(x_i^0(0;\tau,\xi)) - \bar{u}_i^\nu(x_i^0(0;\tau,\xi))|\big\}\, d\xi$$

$$\leq e^{L'\tau} \int_a^b |\bar{u}_i^\nu(x) - \bar{u}_i^0(x)|\, dx + \int_{(\tau,\xi)\in\mathcal{D}} |\phi(x_i^\nu(0;\tau,\xi)) - \phi(x_i^0(0;\tau,\xi))|\, d\xi$$

$$+ 2e^{L'\tau} \int_a^b |\phi(x) - \bar{u}_i^0(x)|\, dx. \tag{3.52}$$

As $\nu \to \infty$, the first two integrals on the right hand side of (3.52) approach zero, while the third is bounded by $\varepsilon$. Since $\varepsilon$ was arbitrary, this proves (3.51).

Next, for any fixed $\tau$ we show that

$$\lim_{\nu\to\infty} \int_{(\tau,\xi)\in\mathcal{D}} \int_0^\tau |l_i^\nu \cdot h^\nu(t, x_i^\nu(t;\tau,\xi), u(t, x_i^\nu(t;\tau,\xi)))$$

$$- l_i^0 \cdot h^0(t, x_i^0(t;\tau,\xi), u(t, x_i^0(t;\tau,\xi)))|\, dt\, d\xi = 0. \tag{3.53}$$

For any fixed $\varepsilon > 0$, choose a continuous function $\psi : \mathcal{D} \mapsto \mathbb{R}^n$ such that

$$\iint_{\mathcal{D}} |\psi(t,x) - l_i^0 \cdot h^0(t,x,u(t,x))|\, dt\, dx < \varepsilon.$$

Using the notation $h^\nu(t, x_i^\nu, u) \doteq h^\nu(t, x_i^\nu(t,\tau,\xi), u(t, x_i^\nu(t,\tau,\xi)))$, for any fixed $\tau$ one has

$$\int_{(\tau,\xi)\in\mathcal{D}} \int_0^\tau |l_i^\nu \cdot h^\nu(t, x_i^\nu, u) - l_i^0 \cdot h^0(t, x_i^0, u)|\, dt\, d\xi$$

$$\leq \int_{(\tau,\xi)\in\mathcal{D}} \int_0^\tau \big\{|l_i^\nu \cdot h^\nu(t, x_i^\nu, u) - l_i^0 \cdot h^0(t, x_i^\nu, u)| + |\psi(t, x_i^\nu) - \psi(t, x_i^0)|$$

$$+ |l_i^0 \cdot h^0(t, x_i^\nu, u) - \psi(t, x_i^\nu)| + |\psi(t, x_i^0) - l_i^0 \cdot h^0(t, x_i^0, u)|\big\}\, dt\, d\xi$$

$$\leq \int_{(\tau,\xi)\in\mathcal{D}} \int_0^\tau \{e^{L'(\tau-t)}|l_i^\nu \cdot h^\nu(t,x,u) - l_i^0 \cdot h^0(t,x,u)| + |\psi(t,x_i^\nu) - \psi(t,x_i^0)|\}\,dt\,dx$$

$$+ 2e^{L'\tau} \int_{(\tau,\xi)\in\mathcal{D}} \int_0^\tau |l_i^0 \cdot h^0(t,x,u) - \psi(t,x)|\,dt\,dx. \tag{3.54}$$

As $\nu \to \infty$, the first integral on the right hand side of (3.54) approaches zero, while the second integral is bounded by $\varepsilon$. Since $\varepsilon$ was arbitrary, this proves (3.53).

The convergence of the last three terms on the right hand side of (3.50) to the corresponding terms in the expression for $\mathcal{T}_i^0(u)$ is clear. We now observe that, for every $u \in Y$ and $\nu \geq 0$, the function $\mathcal{T}^\nu(u)$ can be uniformly approximated by a $\mathcal{C}^1$ function in the norm (3.36). By (iii) in Theorem 2.7, the fixed points of the transformations $\mathcal{T}^\nu$ thus converge to the fixed point of $\mathcal{T}^0$ in the norm of $Y$. More precisely

$$\lim_{\nu\to\infty} \sup_t \int_{(t,x)\in\mathcal{D}} |u_i^\nu(t,x) - u_i^0(t,x)|\,dx = 0 \quad i = 1,\ldots,n.$$

This establishes (3.49), completing the proof of the theorem. □

## 3.6 Classical solutions

The main goal of this section is to prove that, if the functions $A$, $h$ and the initial condition $\bar{u}$ are continuously differentiable, then the broad solution of the Cauchy problem is actually a classical solution. Two technical results will be needed.

**Lemma 3.2.** *Let $u = u(t,\tau,\xi)$, $l = l(t,\tau,\xi)$ be continuously differentiable functions. Then the integral*

$$I(\tau,\xi) = \int_0^\tau \frac{\partial l(t,\tau,\xi)}{\partial t} u(t,\tau,\xi)\,dt \tag{3.55}$$

*is continuously differentiable.*

*Proof.* For $\varepsilon \neq 0$ we have

$$\frac{1}{\varepsilon}[I(\tau,\xi+\varepsilon) - I(\tau,\xi)]$$

$$= \frac{1}{\varepsilon}\int_0^\tau \left\{\frac{\partial l(t,\tau,\xi+\varepsilon)}{\partial t} u(t,\tau,\xi+\varepsilon) - \frac{\partial l(t,\tau,\xi)}{\partial t} u(t,\tau,\xi)\right\} dt$$

$$= \int_0^\tau \frac{\partial l(t,\tau,\xi+\varepsilon)}{\partial t} \cdot \left\{\frac{u(t,\tau,\xi+\varepsilon) - u(t,\tau,\xi)}{\varepsilon}\right\} dt$$

$$+ \frac{1}{\varepsilon}\int_0^\tau \frac{\partial}{\partial t}\left\{l(t,\tau,\xi+\varepsilon) - l(t,\tau,\xi)\right\} \cdot u(t,\tau,\xi)\,dt.$$

Integrating by parts the last integral and passing to the limit as $\varepsilon \to 0$, one obtains

$$\frac{\partial I}{\partial \xi}(\tau, \xi) = \int_0^\tau \big\{ l_t(t, \tau, \xi) u_\xi(t, \tau, \xi) - l_\xi(t, \tau, \xi) u_t(t, \tau, \xi) \big\}\, dt$$
$$+ l_\xi(\tau, \tau, \xi) u(\tau, \tau, \xi) - l_\xi(0, \tau, \xi) u(0, \tau, \xi). \tag{3.56}$$

The computation of $\partial I / \partial \tau$ is entirely similar. □

**Lemma 3.3.** *Let $(Y_\nu)_{\nu \geq 0}$, $(Z_\nu)_{\nu \geq 0}$ be sequences of non-negative functions such that*

$$Y_0(\tau) \leq \alpha, \qquad Y_{\nu+1}(\tau) \leq \alpha + \beta \int_0^\tau Y_\nu(t)\, dt, \tag{3.57}$$

$$Z_0(\tau) \leq \alpha, \qquad Z_{\nu+1}(\tau) \leq \alpha 2^{-\nu} + \beta \int_0^\tau Z_\nu(t)\, dt, \tag{3.58}$$

*for some constants $\alpha, \beta > 0$. Then, for all $\nu, \tau \geq 0$ one has*

$$Y_\nu(\tau) \leq \alpha e^{\beta \tau}, \qquad Z_\nu(\tau) \leq 2^{-\nu} K e^{\gamma \tau}, \tag{3.59}$$

*with $K = 2\alpha + 1$, $\gamma = 2\beta$. In particular, the series $\sum_{\nu=0}^\infty Z_\nu(\tau)$ is uniformly convergent on every bounded interval* $[0, T]$.

*Proof.* When $\nu = 0$ the bounds (3.59) are trivially satisfied. Assuming that $Y_\nu$, $Z_\nu$ satisfy (3.59), by (3.57)–(3.58) it follows that the same is true for $Y_{\nu+1}$, $Z_{\nu+1}$. By induction, this proves the lemma. □

**Theorem 3.6 (Existence of classical solutions).** *In addition to the hypothesis of Theorem 3.1, let $A$, $h$ and $\bar{u}$ in (3.19), (3.21) be continuously differentiable. Then the broad solution $u : \mathcal{D} \mapsto \mathbb{R}^n$ of (3.19), (3.21) is continuously differentiable and actually provides a classical solution. Its partial derivatives $u_t$, $u_x$ are broad solutions respectively of the semilinear systems*

$$(u_x)_t + A(t, x)(u_x)_x = h_x + h_u u_x - A_x u_x, \tag{3.60}$$
$$(u_t)_t + A(t, x)(u_t)_x = h_t + h_u u_t - A_t u_x. \tag{3.61}$$

*Proof.* It is not restrictive to assume that $[a, b]$ is bounded and that $\mathcal{D}$ is compact. By Theorem 3.1, the broad solution $u$ of (3.19), (3.21) is the uniform limit of the sequence of continuous functions $\overset{(\nu)}{u}$, defined inductively as

$$\overset{(0)}{u}(t, x) = \bar{u}(x), \qquad \overset{(\nu+1)}{u} = \mathcal{T}(\overset{(\nu)}{u}). \tag{3.62}$$

In the first part of the proof, under the auxiliary assumption that $A, h \in \mathcal{C}^2$, we will show that all functions $\overset{(\nu)}{u}$ are $\mathcal{C}^1$, and that their first-order derivatives converge to the derivatives of $u$, uniformly on $\mathcal{D}$. In the second part of the proof, by an approximation argument, we prove that the theorem still holds when $A, h \in \mathcal{C}^1$.

1. Assume that the functions $A, h$ are $\mathcal{C}^2$. Recalling (3.26), (3.31), the sequence $\overset{(\nu)}{u}$ can be written in the form

$$\overset{(\nu+1)}{u}(\tau,\xi) = \sum_{i=1}^{n} \overset{(\nu+1)}{u_i}(\tau,\xi) r_i(\tau,\xi), \tag{3.63}$$

$$\begin{aligned} \overset{(\nu+1)}{u_i}(\tau,\xi) = \bar{u}_i(x_i(0;\tau,\xi)) &+ \int_0^\tau l_i(t, x_i(t;\tau,\xi)) \cdot h(t, x_i(t;\tau,\xi), \overset{(\nu)}{u}(t, x_i(t;\tau,\xi)))\,dt \\ &+ \int_0^\tau \frac{d}{dt}[l_i(t, x_i(t;\tau,\xi))] \cdot \overset{(\nu)}{u}(t, x_i(t;\tau,\xi))\,dt. \end{aligned} \tag{3.64}$$

Since $\bar{u} \in \mathcal{C}^1$ while $l_i, r_i, \lambda_i, h \in \mathcal{C}^2$, it is clear that $\overset{(\nu)}{u} \in \mathcal{C}^1$ implies $\overset{(\nu+1)}{u} \in \mathcal{C}^1$ for every $\nu \geq 0$.

2. Differentiating (3.64) w.r.t. $\xi$ and using integration by parts as in Lemma 3.2 to handle the last integral, we obtain

$$\begin{aligned} \frac{\partial}{\partial \xi} \overset{(\nu+1)}{u_i}(\tau,\xi) = {}& \frac{\partial \bar{u}_i}{\partial x} \frac{\partial x_i(0,\tau,\xi)}{\partial \xi} + \int_0^\tau \left[ \frac{\partial l_i}{\partial x} \frac{\partial x_i}{\partial \xi} \cdot h(t, x_i, \overset{(\nu)}{u}) \right] dt \\ &+ \int_0^\tau l_i \cdot \left[ \frac{\partial h}{\partial x}\frac{\partial x_i}{\partial \xi} + \frac{\partial h}{\partial u} \cdot \frac{\partial \overset{(\nu)}{u}}{\partial x} \frac{\partial x_i}{\partial \xi} \right] dt + \int_0^\tau \frac{dl_i}{dt} \cdot \frac{\partial \overset{(\nu)}{u}}{\partial x}\frac{\partial x_i}{\partial \xi}\,dt \\ &- \int_0^\tau \frac{\partial l_i}{\partial x}\frac{\partial x_i}{\partial \xi} \cdot \frac{d\overset{(\nu)}{u}}{dt}\,dt + \frac{\partial l_i(\tau,\xi)}{\partial \xi} \overset{(\nu)}{u}(\tau,\xi) \\ &- \frac{\partial l_i}{\partial x}\frac{\partial x_i(0;\tau,\xi)}{\partial \xi} \bar{u}(x_i(0;\tau,\xi)), \end{aligned} \tag{3.65}$$

where all integrals are evaluated along the curve $t \mapsto x_i(t;\tau,\xi)$.

3. For each integer $\nu \geq 1$ and every $\tau \geq 0$, we define the quantities

$$Y_\nu(\tau) = \sup_{\substack{(\tau,\xi)\in\mathcal{D} \\ i=1,\dots,n}} \left| \frac{\partial \overset{(\nu)}{u}_i(\tau,\xi)}{\partial \xi} \right|, \qquad Z_\nu(\tau) = \sup_{\substack{(\tau,\xi)\in\mathcal{D} \\ i=1,\dots,n}} \left| \frac{\partial \overset{(\nu+1)}{u_i}(\tau,\xi)}{\partial \xi} - \frac{\partial \overset{(\nu)}{u_i}(\tau,\xi)}{\partial \xi} \right|.$$

The proof of Theorem 3.1 implies that, for some constant $C_1$, we have

$$\Big|\overset{(\nu+1)}{u}(\tau,\xi) - \overset{(\nu)}{u}(\tau,\xi)\Big| \leq C_1 2^{-\nu} \quad \text{for all } (\tau,\xi)\in\mathcal{D},\ \ \nu\geq 0. \tag{3.66}$$

Moreover, all functions $\overset{(\nu)}{u}$ are uniformly bounded. From (3.65) and the smoothness of $l_i, h, x_i$ we deduce the existence of some constants $C_2, C_3$ such that

$$Y_{\nu+1} \leq C_2 + C_3 \int_0^\tau Y_\nu(t)\,dt.$$

4. By Lemma 3.3, the functions $Y_\nu$ are uniformly bounded. Using (3.65) again, together with (3.66) and the uniform boundedness of the derivatives $\partial\overset{(\nu)}{u_i}/\partial\xi$, we also deduce the existence of new constants $C_4, C_5$ such that

$$Z_{\nu+1}(\tau) \leq C_4 2^{-\nu} + C_5 \int_0^\tau Z_\nu(t)\,dt.$$

By Lemma 3.3, this implies the uniform convergence of the derivatives $\partial\overset{(\nu)}{u_i}/\partial x$ to the derivative $\partial u_i/\partial x$. The uniform convergence of the sequence of derivatives $\partial\overset{(\nu)}{u_i}/\partial t$ is proved by an entirely similar argument.

5. Since $u = \sum u_i r_i$, the continuous differentiability of $u_i, r_i$ now implies that $u\in\mathcal{C}^1$. To show that $u$ is a classical solution of (3.19), we use (3.25), (3.26) together with the identities

$$(l_i\cdot r_j)_x = l_{i,x}\cdot r_j + l_i\cdot r_{j,x} \equiv 0, \qquad (l_i\cdot r_j)_t = l_{i,t}\cdot r_j + l_i\cdot r_{j,t}\equiv 0, \tag{3.67}$$

and obtain

$$\begin{aligned}
u_t + Au_x &= \sum_{i=1}^n (u_{i,t}r_i + u_i r_{i,t}) + A\Big(\sum_{i=1}^n u_{i,x}r_i + \sum_{i=1}^n u_i r_{i,x}\Big)\\
&= \sum_{i=1}^n (u_{i,t} + \lambda_i u_{i,x})r_i + \sum_{j=1}^n u_j(r_{j,t} + Ar_{j,x})\\
&= \sum_{i=1}^n [l_i\cdot h + (l_{i,t} + \lambda_i l_{i,x})\cdot u]r_i + \sum_{i,j=1}^n [l_i\cdot(r_{j,t} + Ar_{j,x})u_j]r_i\\
&= h + \sum_{i=1}^n [(l_{i,t}+\lambda_i l_{i,x})\cdot u]r_i - \sum_{i,j=1}^n (l_{i,t}\cdot r_j u_j + \lambda_i l_{i,x}\cdot r_j u_j)r_i\\
&= h + \sum_{i=1}^n [(l_{i,t}+\lambda_i l_{i,x})\cdot u]r_i - \sum_{i=1}^n (l_{i,t}\cdot u + \lambda_i l_{i,x}\cdot u)r_i = h.
\end{aligned} \tag{3.68}$$

Hence $u$ solves (3.19) in the classical sense.

6. Now let $u$ be a classical solution of (3.19), still assuming $A, h \in \mathcal{C}^2$. Since $u$ is clearly also a broad solution, for any fixed time $\bar{\tau}$ we can write

$$\begin{aligned} u_i(\tau,\xi) &= u_i(\bar{\tau}, x_i(\bar{\tau};\tau,\xi)) \\ &\quad + \int_{\bar{\tau}}^{\tau} l_i(t, x_i(t;\tau,\xi)) \cdot h(t, x_i(t;\tau,\xi), u(t, x_i(t;\tau,\xi)))\,dt \\ &\quad + \int_{\bar{\tau}}^{\tau} \left(\frac{d}{dt} l_i(t, x_i(t;\tau,\xi))\right) \cdot u(t, x_i(t;\tau,\xi))\,dt. \end{aligned} \tag{3.69}$$

Differentiating (3.69) w.r.t. $\xi$ and using integration by parts as in Lemma 3.2 to handle the last integral, we obtain

$$\begin{aligned} \frac{\partial}{\partial\xi} u_i(\tau,\xi) &= \frac{\partial}{\partial\xi} u_i(\bar{\tau}, x_i(\bar{\tau};\tau,\xi)) + \int_{\bar{\tau}}^{\tau} \left[\frac{\partial l_i}{\partial x}\frac{\partial x_i}{\partial\xi} \cdot h(t, x_i, u)\right] dt \\ &\quad + \int_{\bar{\tau}}^{\tau} l_i \cdot \left[\frac{\partial h}{\partial x}\frac{\partial x_i}{\partial\xi} + \frac{\partial h}{\partial u} \cdot \frac{\partial u}{\partial x}\frac{\partial x_i}{\partial\xi}\right] dt + \int_{\bar{\tau}}^{\tau} \frac{dl_i}{dt} \cdot \frac{\partial u}{\partial x}\frac{\partial x_i}{\partial\xi}\,dt \\ &\quad - \int_{\bar{\tau}}^{\tau} \frac{\partial l_i}{\partial x}\frac{\partial x_i}{\partial\xi} \cdot \frac{du}{dt}\,dt + \frac{\partial l_i}{\partial x}(\tau,\xi) \cdot u(\tau,\xi) \\ &\quad - \frac{\partial l_i}{\partial x}(\bar{\tau}, x_i(\bar{\tau};\tau,\xi)) \frac{\partial x_i(\bar{\tau};\tau,\xi)}{\partial\xi} \cdot u(\bar{\tau}; x_i(\bar{\tau};\tau,\xi)) \end{aligned} \tag{3.70}$$

where all integrals are evaluated along the curve $t \mapsto x_i(t;\tau,\xi)$.

For a fixed $(\bar{\tau},\bar{\xi})$, consider the characteristic line $t \mapsto (\tau(t),\xi(t)) \doteq (t, x_i(t;\bar{\tau},\bar{\xi}))$. When $t = \bar{\tau}$, we have $\partial x_i(\bar{\tau};\bar{\tau},\xi)/\partial\xi = 1$. With this in mind, we now compute the derivative of (3.70) along this characteristic line:

$$\begin{aligned} \frac{d}{dt}\left\{\frac{\partial}{\partial\xi} u_i(t, x_i(t;\bar{\tau},\bar{\xi}))\right\}_{t=\bar{\tau}} &= -\lambda_{i,x}(l_i \cdot u)_x + l_{i,x} \cdot h + l_i \cdot h_x + l_i \cdot h_u u_x \\ &\quad + [l_{i,t} + \lambda_i l_{i,x}] u_x - l_{i,x} \cdot [u_t + \lambda_i u_x] \\ &\quad + \frac{d}{dt}(l_{i,x} \cdot u) + \lambda_{i,x} l_{i,x} \cdot u. \end{aligned} \tag{3.71}$$

7. To prove that $u_x$ provides a broad solution of (3.60), we first observe that

$$u_x^i \doteq l_i \cdot u_x = (l_i \cdot u)_x - l_{i,x} \cdot u \tag{3.72}$$

is continuously differentiable along each $i$-th characteristic line, because of (3.71) and the regularity of the left eigenvectors $l_i$. For every fixed $\bar{\tau}, \bar{\xi}$, we have to show that

$$\frac{d}{dt}\{(l_i \cdot u_x)(t, x_i(t;\bar{\tau},\bar{\xi}))\}_{t=\bar{\tau}} = l_i \cdot [h_x + h_u \cdot u_x - A_x u_x] + [l_{i,t} + \lambda_i l_{i,x}] u_x. \tag{3.73}$$

Using the identities

$$(l_i A)_x u_x = l_{i,x} A u_x + l_i A_x u_x = (\lambda_i l_i)_x u_x = \lambda_{i,x} l_i u_x + \lambda_i l_{i,x} u_x,$$

$$l_i A_x u_x = \lambda_{i,x} l_i u_x + \lambda_i l_{i,x} u_x - l_{i,x}[h - u_t],$$

because of (3.71), the proof of (3.73) is obtained by checking the identity

$$\begin{aligned}
&\frac{d}{dt}\{(l_i \cdot u)_x - l_{i,x} u\} \\
&\quad = -\lambda_{i,x} l_i u_x + l_{i,x} h + l_i h_x + l_i h_u u_x + [l_{i,t} + \lambda_i l_{i,x}] u_x - l_{i,x}[u_t + \lambda_i u_x] \\
&\quad = l_i h_x + l_i h_u u_x - l_i A_x u_x + [l_{i,t} + \lambda_i l_{i,x}] u_x.
\end{aligned}$$

The proof that $u_t$ provides a broad solution to (3.61) is entirely analogous.

8. To cover the case where $A, h \in \mathcal{C}^1$, let $A_\nu, h_\nu$ be sequences of $\mathcal{C}^2$ functions approaching $A, h$ in $\mathcal{C}^1$. Let $(u^\nu)_{\nu \geq 1}$ be the corresponding sequence of solutions to the Cauchy problems (3.43), with $\bar{u}^\nu = \bar{u}$ for all $\nu$. By the previous analysis, each $u^\nu$ is continuously differentiable. Moreover, the derivatives $u_x^\nu$, $u_t^\nu$ are broad solutions of the systems

$$\begin{aligned}
(u_x)_t + A^\nu(t, x)(u_x)_x &= h_x^\nu + h_u^\nu u_x - A_x^\nu u_x, \\
(u_t)_t + A^\nu(t, x)(u_t)_x &= h_t^\nu + h_u^\nu u_t - A_t^\nu u_x.
\end{aligned} \tag{3.74}$$

Applying Theorem 3.5, part (i), to the systems (3.43) and (3.74), we conclude that $u^\nu \to u$, $u_x^\nu \to u_x$ and $u_t^\nu \to u_t$ uniformly on $\mathcal{D}$. Hence $u$ is continuously differentiable w.r.t. $t$, $x$, and its partial derivatives are broad solutions of the hyperbolic systems (3.60), (3.61). □

## 3.7 Lipschitz solutions

Consider again the semilinear initial value problem (3.19), (3.21). In this section we prove that, if the function $h$ and the initial condition $\bar{u}$ are Lipschitz continuous, then the same is true of the solution $u$. In particular, by Rademacher's theorem the partial derivatives $u_x, u_t$ exist a.e. Using an approximation argument, it will be shown that these derivatives provide a broad solution to the systems (3.60), (3.61), respectively.

**Theorem 3.7.** *In addition to the hypothesis of Theorem 3.1, let $A, h, \bar{u}$ be locally Lipschitz continuous w.r.t. all variables. Then the broad solution $u$ of (3.19), (3.21) is locally Lipschitz continuous. If, in addition, the function $h$ is continuously differentiable, then the partial derivatives $u_x, u_t$ are broad solutions of the systems (3.60), (3.61), respectively.*

*Proof.* It clearly suffices to prove that, for every $\varepsilon, T > 0$, the solution $u$ is Lipschitz continuous on the domain of dependence $\mathcal{D}$ defined as in (3.44), (3.45).

1. By Theorem 3.4, the unique broad solution $u : \mathcal{D} \mapsto \mathbb{R}^n$ is continuous. Hence, we can assume the bound

$$|u(t, x)| < C_0 \quad \text{for all } (t, x) \in \mathcal{D}. \tag{3.75}$$

The Lipschitz continuity of the functions $A, h, \bar{u}$ implies that their partial derivatives exist a.e, and remain bounded on compact sets. Moreover, the same is true of the eigenvalues and eigenvectors of the matrix $A$. In particular, there exist constants $C_1, C_2, C_3$ such that

$$|\bar{u}_x(x)| \leq C_1, \qquad |h(0, x, \bar{u}(x)) - A(0, x)\bar{u}_x(x)| \leq C_1, \tag{3.76}$$

for almost every $x \in [a, b]$,

$$|l_i \cdot h_x| \leq C_2, \qquad |l_i \cdot h_t| \leq C_2, \tag{3.77}$$

$$|l_i h_u - l_i A_x| + |l_{i,t} + \lambda_i l_{i,x}| \leq C_3, \qquad |l_i h_u - l_i A_t| + \left|l_{i,t} + \lambda_i l_{i,x}\right| \leq C_3, \tag{3.78}$$

for $i = 1, \ldots, n$ and almost all $t, x, u$ with $(t, x) \in \mathcal{D}$ and $|u| \leq C_0$.

2. We now construct sequences $A^\nu, h^\nu, \bar{u}^\nu$ of $\mathcal{C}^1$ functions with the following properties:

(I) Calling $\lambda_i^\nu, l_i^\nu$ the corresponding eigenvalues and left eigenvectors of the matrix $A^\nu$, normalized as in (3.3), for all $\nu \geq 1$ and $x \in [a, b]$, one has

$$|\bar{u}_x^\nu(x)| \leq C_1', \qquad |h^\nu(0, x, \bar{u}^\nu(x)) - A^\nu(0, x)\bar{u}_x^\nu(x)| \leq C_1'. \tag{3.79}$$

Moreover, for all $(t, x) \in \mathcal{D}$, $|u| \leq C_0$, one has

$$|l_i^\nu \cdot h_x^\nu| \leq C_2', \qquad |l_i^\nu \cdot h_t^\nu| \leq C_2', \tag{3.80}$$

$$|l_i^\nu h_u^\nu - l_i^\nu A_x^\nu| + |l_{i,t}^\nu + \lambda_i^\nu l_{i,x}^\nu| \leq C_3', \qquad |l_i^\nu h_u^\nu - l_i^\nu A_t^\nu| + |l_{i,t}^\nu + \lambda_i^\nu l_{i,x}^\nu| \leq C_3', \tag{3.81}$$

for some constants $C_i' > C_i$, $i = 1, 2, 3$.

(II) As $\nu \to \infty$, we have uniform convergence:

$$\sup_{(t,x)\in\mathcal{D}} |A^\nu(t, x) - A(t, x)| \to 0, \qquad \sup_{\substack{(t,x)\in\mathcal{D} \\ |u|\leq C_0}} |h^\nu(t, x, u) - h(t, x, u)| \to 0,$$

$$\sup_{x\in[a,b]} |\bar{u}^\nu(x) - \bar{u}(x)| \to 0. \tag{3.82}$$

(III) In the case where $h$ is continuously differentiable, we also require that $A_x^\nu \to A_x$, $A_t^\nu \to A_t^\nu$ pointwise a.e. on $\mathcal{D}$, while $\bar{u}_x^\nu \to \bar{u}_x$ a.e. on $[a, b]$, and simply take $h^\nu = h$ for all $\nu$.

Observe that the above conditions are all satisfied if we let $A^\nu, h^\nu, \bar{u}^\nu$ be suitable mollifications of the Lipschitz functions $A, h, \bar{u}$.

3. By Theorem 3.6, for every $\nu$ the initial value problem

$$u_t + A^\nu(t, x)u_x = h^\nu(t, x, u), \qquad u(0, x) = \bar{u}^\nu(x) \tag{3.83}$$

has a unique classical solution defined on some maximal domain of dependency $\mathcal{D}^\nu$. By Lemma 3.1, we have $\mathcal{D} \subset \mathcal{D}^\nu$ for all $\nu$ sufficiently large.

Since the derivatives $u_x^\nu \doteq v^\nu$ are broad solutions of the systems

$$v_t + A^\nu v_x = h_x^\nu + h_u^\nu v - A_x^\nu v, \qquad v(0, x) = \bar{u}_x^\nu(x), \tag{3.84}$$

their components satisfy

$$\frac{d}{dt}|l_i^\nu \cdot u_x^\nu(t, x_i(t, 0, \xi))| \le |l_i^\nu \cdot h_x^\nu| + |l_i^\nu h_u^\nu - l_i^\nu A_x^\nu||u_x^\nu| + |l_{i,t}^\nu + \lambda_i^\nu l_{i,x}^\nu||u_x^\nu|$$
$$\le C_2' + C_3'|u_x^\nu|. \tag{3.85}$$

We introduce the scalar function

$$Y_\nu(t) \doteq \sup_{i,x} |l_i^\nu \cdot u_x^\nu(t, x)|.$$

Recalling that

$$|u_x| = \left| \sum_{i=1}^n (l_i \cdot u_x) r_i \right| \le n \max_i |l_i \cdot u_x|, \tag{3.86}$$

by (3.85) the function $Y_\nu$ satisfies the differential inequality

$$\dot{Y}_\nu(t) \le C_2' + C_3' n Y_\nu(t), \qquad Y_\nu(0) \le C_1'.$$

Therefore, a standard comparison argument yields

$$Y_\nu(t) \le e^{nC_3't} C_1' + (e^{nC_3't} - 1) \cdot \frac{C_2'}{nC_3'} \doteq \widehat{Y}_\nu(t),$$
$$|u_x^\nu(t, x)| \le nY_\nu(t) \le n\widehat{Y}_\nu(t). \tag{3.87}$$

In order to estimate $|u_t^\nu|$, we observe that $u_t^\nu$ provides a broad solution to the problem

$$w_t + A^\nu w_x = h_t^\nu + h_u^\nu w - A_t^\nu w, \qquad w(0, x) = h^\nu(0, x, \bar{u}^\nu(x)) - A^\nu(0, x)\bar{u}_x^\nu(x). \tag{3.88}$$

An entirely similar argument now yields the estimate

$$|u_t^\nu(t, x)| \le n\widehat{Y}_\nu(t). \tag{3.89}$$

4. By Theorem 3.5, part (i), the sequence of solutions $u^\nu$ converges to $u$ uniformly on $\mathcal{D}$. By (3.87), (3.89), all derivatives $u_x^\nu$, $u_t^\nu$ are uniformly bounded. Therefore, the function $u$ is Lipschitz continuous on $\mathcal{D}$. This proves the first part of the theorem.

5. In the case where $h$ is continuously differentiable, we can apply part (ii) of Theorem 3.5 and conclude that the solutions $v^\nu$, $w^\nu$ of the systems (3.84), (3.88) converge in $\mathbf{L}^1$ respectively to the broad solutions $v$, $w$ of the systems (3.60), (3.61). Recalling that $v^\nu = u_x^\nu$, $w^\nu = u_t^\nu$ and $u^\nu \to u$ uniformly, we conclude that the functions $v$, $w$ yield the distributional derivatives $u_x$, $u_t$. Since $u$ is Lipschitz continuous, by Rademacher's theorem we have $u_x = v$, $u_t = w$ at almost every point $(t, x) \in \mathcal{D}$. □

## 3.8 Quasilinear systems

This section is concerned with the quasilinear system

$$u_t + A(t, x, u)u_x = h(t, x, u), \tag{3.90}$$

under the basic hypothesis

(H2) Each matrix $A(t, x, u)$ has $n$ real distinct eigenvalues. The functions $A : [0, \infty[ \times \mathbb{R} \times \mathbb{R}^n \mapsto \mathbb{R}^{n\times n}$ and $h : [0, \infty[ \times \mathbb{R} \times \mathbb{R}^n \mapsto \mathbb{R}^n$ are locally Lipschitz continuous w.r.t. all variables.

A continuously differentiable function which satisfies (3.90) at every point of its domain is called a *classical solution*. Clearly, $u$ is then a solution of the semilinear system

$$u_t + \hat{A}(t, x)u_x = h(t, x, u), \tag{3.91}$$

with

$$\hat{A}(t, x) \doteq A(t, x, u(t, x)). \tag{3.92}$$

A locally Lipschitz-continuous function $u$, defined on a closed domain $\mathcal{D} \subset [0, \infty[ \times \mathbb{R}$, is called a *broad solution* of the quasilinear problem (3.90) with initial data

$$u(t, x) = \bar{u}(x) \quad x \in [a, b] \tag{3.93}$$

provided that the following conditions hold:

(i) The matrix-valued function $\hat{A}$ in (3.92) is locally Lipschitz continuous.
(ii) $\mathcal{D}$ is a domain of determinacy for the initial value problem (3.91), (3.93).
(iii) $u$ is a broad solution of the semilinear problem (3.91), (3.93).

Observe that the Lipschitz continuity of the matrix $\hat{A}$ is essential in order to define uniquely the characteristic lines, and hence the transformation $\mathcal{T}$ in (3.31) whose fixed point yields the broad solution. Since $\hat{A}$ in (3.92) depends on $u$, the study of broad solutions to quasilinear systems is thus restricted within the class of Lipschitz-continuous functions. This is in contrast to the semilinear case, where the broad solutions were defined within the larger set of locally integrable functions.

Another major difference between the semilinear and the quasilinear case is that, in the latter, the characteristic lines, and hence the domains of determinacy, also depend on the solution $u$ itself. In particular, different initial conditions $\bar{u}$ may well yield solutions defined on distinct domains of determinacy.

We also remark that, in the semilinear case, solutions can be prolonged forward in time as long as they remain bounded. This is no longer true for quasilinear problems. Indeed, the partial derivatives $u_x$, $u_t$ of a smooth solution $u$ of (3.90) satisfy the equations

$$\begin{aligned}
(u_x)_t + A(t, x, u)(u_x)_x &= -[A_x(t, x, u) + A_u(t, x, u)u_x]u_x + h_x(t, x, u) \\
&\quad + h_u(t, x, u)u_x, \\
(u_t)_t + A(t, x, u)(u_t)_x &= -[A_t(t, x, u) + A_u(t, x, u)u_t]u_x + h_t(t, x, u) \\
&\quad + h_u(t, x, u)u_t.
\end{aligned}$$

In contrast to (3.60), (3.61), here the right hand sides of the equations are quadratic w.r.t. the derivatives $u_x$, $u_t$. Therefore, it may happen that a solution $u$ remains uniformly bounded while its gradient approaches infinity in finite time. An example of this 'gradient catastrophe', for a scalar equation, is provided by initial value problem (3.18).

The main theorem in this section provides the existence and uniqueness of a local solution to the quasilinear initial value problem. A preliminary lemma will first be proved.

**Lemma 3.4.** *Let $(Z_\nu)_{\nu\geq 0}$ be a sequence of continuous, non-negative functions on the interval $[0, T]$. Assume that, for some constants $\alpha, \beta, \gamma \geq 0$ and for all $\nu \geq 1$, one has*

$$Z_\nu(t) \leq \int_0^s \{\alpha Z_\nu(s) + \beta Z_{\nu-1}(s) + 2^{-\nu}\gamma\}\, ds. \tag{3.94}$$

*Then there exists a constant $C$ such that $Z_\nu(t) \leq C \cdot 2^{-\nu}$. In particular, the series $\sum_{\nu\geq 1} Z_\nu(t)$ is uniformly convergent.*

*Proof.* Choose constants $K, L$ such that

$$Z_0(t) \leq K \quad \text{for all } t \in [0, T],\ K > 3\gamma, \qquad L > \alpha, \quad \frac{e^{\alpha T}\beta}{L-\alpha} \leq \frac{1}{3}. \tag{3.95}$$

With the above choices, we claim that

$$Z_\nu(t) \leq K e^{Lt} \cdot 2^{-\nu}. \tag*{$(3.96)_\nu$}$$

Indeed, $(3.96)_0$ is obvious. Assuming that $(3.96)_{\nu-1}$ holds, by (3.94) we have

$$\begin{aligned}
Z_\nu(t) &\leq \int_0^t e^{\alpha(t-s)}[\beta K e^{Ls} \cdot 2^{1-\nu} + 2^{-\nu}\gamma]\, ds \\
&= \int_0^t e^{\alpha t}\beta K e^{(L-\alpha)s} 2^{1-\nu}\, ds + \frac{e^{\alpha t}-1}{\alpha} 2^{-\nu}\gamma \\
&< \frac{e^{\alpha T}\beta K}{L-\alpha} 2^{1-\nu} e^{(L-\alpha)t} + e^{Lt} 2^{-\nu}\gamma \\
&< 2^{-\nu}\frac{2K}{3} e^{Lt} + 2^{-\nu}\frac{K}{3} e^{Kt}.
\end{aligned}$$

By induction, the estimate (3.96) holds for all $\nu$, proving the lemma. □

**Theorem 3.8 (Local existence).** *In connection with the problem (3.90), (3.93), let the basic hypothesis (H2) hold and let $\bar{u}$ be Lipschitz continuous. Then there exist constants $C, \delta > 0$ and a Lipschitz-continuous function $u$ which is the unique broad solution of (3.90), (3.93) on the domain*

$$\mathcal{D} = \mathcal{D}_{C,\delta} \doteq \{(t, x);\ t \in [0, \delta],\ a + Ct \leq x \leq b - Ct\}. \tag{3.97}$$

*If the functions $A$, $h$ and $\bar{u}$ are continuously differentiable, then $u$ is actually a classical solution.*

*Proof.* The solution will be obtained as the uniform limit of the sequence $\overset{(\nu)}{u} : \mathcal{D} \mapsto \mathbb{R}^n$, where $\overset{(0)}{u}(t,x) \doteq \bar{u}(x)$ and $\overset{(\nu)}{u}$ is inductively defined as the solution of the semilinear problem

$$\overset{(\nu)}{u}_t + A(t,x,\overset{(\nu-1)}{u}(t,x))\,\overset{(\nu)}{u}_x = h(t,x,\overset{(\nu)}{u}), \qquad \overset{(\nu)}{u}(0,x) = \bar{u}(x). \tag{3.98$_\nu$}$$

By suitably choosing the constants $\delta$, $C$ in (3.97), we will prove the following:

(i) $\mathcal{D}$ is a common domain of determinacy for all problems $(3.98)_\nu$.

(ii) The Lipschitz constant of the functions $\overset{(\nu)}{u}$ remains uniformly bounded on $\mathcal{D}$.

(iii) The sequence $\overset{(\nu)}{u}$ converges to some function $u$, uniformly on $\mathcal{D}$.

1. We consider first the case where $A, h, \bar{u}$ are $\mathcal{C}^1$. We define the constants

$$C_0 \doteq \max_{x\in[a,b]} |\bar{u}(x)|, \tag{3.99}$$

$$C = \max\left\{|\lambda_i(t,x,u)|;\ t\in[0,1], x\in[a,b], |u|\le C_0+1, i=1,\ldots,n\right\}, \tag{3.100}$$

where $\lambda_i(t,x,u)$ denotes the $i$-th eigenvalue of the matrix $A(t,x,u)$. With this choice of $C$, the set $\mathcal{D}$ in (3.97) will be a domain of determinacy for $(3.98)_\nu$ provided that

$$\delta \le 1, \qquad |\overset{(\nu-1)}{u}(t,x)| \le C_0 + 1 \quad \text{for all } (t,x)\in\mathcal{D}. \tag{3.101$_\nu$}$$

2. Let $C_1$ be a constant such that

$$|\bar{u}_x(x)| < C_1, \quad |h(0,x,\bar{u}(x)) - A(t,x,\bar{u}(x))\bar{u}_x(x)| < C_1 \qquad \text{for all } x\in[a,b]. \tag{3.102}$$

Then choose constants $C_2, C_3, C_4$ such that

$$|l_i\cdot h_x| \le C_2, \qquad |l_i\cdot h_t| \le C_2, \tag{3.103}$$

$$|l_i h_u| + |l_i A_x| + |l_{i,t} + \lambda_i l_{i,x}| \le C_3, \qquad |l_i h_u| + |l_i A_t| + |l_{i,t} + \lambda_i l_{i,x}| \le C_3, \tag{3.104}$$

$$|l_i A_u| + |l_{i,u}| + |\lambda_i l_{i,u}| \le C_4, \tag{3.105}$$

for all $t\in[0,1]$, $x\in[a,b]$, $|u|\le C_0+1$. Of course, $l_i = l_i(t,x,u)$ denotes here the $i$-th left eigenvector of the matrix $A$.

Now set $C_5 = \max\{C_2, C_3, C_4\}$ and call $Y^\dagger$ the solution of the scalar Cauchy problem

$$\dot{Y} = C_5(1+nY)^2, \qquad Y(0) = C_1. \tag{3.106}$$

Explicitly,

$$Y^\dagger(t) = \frac{n^{-1}}{(1+nC_1)^{-1} - nC_5 t} - \frac{1}{n} \quad \text{for all } t\in\left[0, \frac{1}{nC_5(1+nC_1)}\right[. \tag{3.107}$$

We now choose $\delta > 0$ small enough so that

$$\int_0^\delta nY^\dagger(t)\,dt \le 1. \tag{3.108}$$

We claim that, with this choice of $\delta$ in (3.97), the properties (i)–(iii) hold.

3. By induction, for every $\nu \ge 0$ we now prove that $(3.101)_\nu$ holds, together with the bounds

$$|\overset{(\nu)}{u}_x(t,x)| \le nY^\dagger(t), \quad |\overset{(\nu)}{u}_t(t,x)| \le nY^\dagger(t) \qquad \text{for all } (t,x) \in \mathcal{D}. \tag{3.109$_\nu$}$$

Indeed, for $\nu = 0$, $(3.109)_0$ is obvious. Next, assume that $(3.109)_{\nu-1}$ holds. Then for $(t,x) \in \mathcal{D}$, since $t \in [0,\delta]$, from (3.99), (3.108) it follows that

$$\begin{aligned}|\overset{(\nu-1)}{u}(t,x)| &\le |\overset{(\nu-1)}{u}(0,x)| + \int_0^t |\overset{(\nu-1)}{u_t}(s,x)|\,ds \\ &\le |\bar u(0,x)| + \int_0^t nY^\dagger(s)\,ds \le C_0 + 1.\end{aligned} \tag{3.110}$$

This proves $(3.101)_\nu$. By Theorem 3.6, the initial value problem $(3.98)_\nu$ therefore has a classical solution $\overset{(\nu)}{u}$ on $\mathcal{D}$. Calling $v \doteq \overset{(\nu)}{u}_x$, $w \doteq \overset{(\nu)}{u}_t$, these partial derivatives are respectively broad solutions of the systems

$$\begin{aligned} v_t + A(t,x,\overset{(\nu-1)}{u}(t,x))v_x &= h_x + h_u v - (A_x + A_u \overset{(\nu-1)}{u_x})v, \\ w_t + A(t,x,\overset{(\nu-1)}{u}(t,x))w_x &= h_t + h_u w - (A_t + A_u \overset{(\nu-1)}{u_t})v. \end{aligned} \tag{3.111}$$

By the definition of broad solution, if $x_i(\cdot\,;\tau,\xi)$ denotes any characteristic line of the $i$-th family, the equations in (3.111) imply

$$\begin{aligned} \frac{d}{dt}\big\{(l_i\cdot v)(t,x_i(t;\tau,\xi))\big\} &= l_i\cdot[h_x + h_u v - (A_x + A_u \overset{(\nu-1)}{u_x})v] \\ &\quad + [l_{i,t} + l_{i,u}\overset{(\nu-1)}{u_t} + \lambda_i(l_{i,x} + l_{i,u}\overset{(\nu-1)}{u_x})]v, \end{aligned} \tag{3.112}$$

$$\begin{aligned} \frac{d}{dt}\big\{(l_i\cdot w)(t,x_i(t;\tau,\xi))\big\} &= l_i\cdot[h_t + h_u w - (A_t + A_u \overset{(\nu-1)}{u_t})w] \\ &\quad + [l_{i,t} + l_{i,u}\overset{(\nu-1)}{u_t} + \lambda_i(l_{i,x} + l_{i,u}\overset{(\nu-1)}{u_x})]v. \end{aligned} \tag{3.113}$$

Consider the scalar function

$$Y(t) \doteq \text{ess-sup}\big\{|l_i\cdot \overset{(\nu)}{u}_x(t,x)|,\ |l_i\cdot \overset{(\nu)}{u}_t(t,x)|;\ x \in [a+Ct,\ b-Ct],\ i = 1,\dots,n\big\}. \tag{3.114}$$

Observe that $|\overset{(\nu)}{u}_x|, |\overset{(\nu)}{u}_t| \leq nY(t)$ and

$$|\overset{(\nu)}{u}(t,x)| \leq |\bar{u}(x)| + \int_0^t |\overset{(\nu)}{u}_t(s,x)|\, ds \leq C_0 + \int_0^t nY(s)\, ds.$$

Calling $Y^\dagger$ the solution of (3.106), from (3.112)–(3.113) and the bounds (3.101)–(3.105) we deduce that

$$\dot{Y}(t) < C_2 + C_3 nY(t) + C_4 n^2 Y^\dagger(t) Y(t) \leq \dot{Y}^\dagger(t), \qquad Y(0) < C_1, \tag{3.115}$$

is valid as long as

$$Y(t) \leq Y^\dagger(t), \qquad \int_0^t nY(s)\, ds \leq 1.$$

Recalling (3.108), a straightforward comparison argument now yields

$$Y(t) < Y^\dagger(t) \quad \text{for all } t \in [0, \delta]. \tag{3.116}$$

By induction, this establishes (i) and (ii).

4. In order to prove the uniform convergence of the sequence $\overset{(\nu)}{u}$, for every $\nu$ define

$$w^\nu = \overset{(\nu)}{u} - \overset{(\nu-1)}{u}, \qquad A^{\nu-1} = A(t,x,\overset{(\nu-1)}{u}), \qquad h^\nu = h(t,x,\overset{(\nu)}{u}). \tag{3.117}$$

From (3.98) it follows that

$$w_t^\nu + A^{\nu-1} w_x^\nu = h^\nu - h^{\nu-1} - (A^{\nu-1} - A^{\nu-2})\, \overset{(\nu-1)}{u_x}. \tag{3.118}$$

Since $A = A(t,x,u)$ and $h = h(t,x,u)$ are Lipschitz continuous, there exists a constant $C_6$ such that

$$|A(t,x,u) - A(t,x,u')| \leq C_6 |u - u'|, \qquad |h(t,x,u) - h(t,x,u')| \leq C_6 |u - u'|, \tag{3.119}$$

as long as $(t,x) \in \mathcal{D}$, $|u|, |u'| \leq C_0 + 1$. Since $w^\nu$ is a broad solution of (3.118), it satisfies

$$\begin{aligned} \frac{d}{dt}\big\{(l_i \cdot w^\nu)(t, x_i(t;\tau,\xi))\big\} &= l_i[(h^\nu - h^{\nu-1}) - (A^{\nu-1} - A^{\nu-2})\, \overset{(\nu-1)}{u_x}] \\ &\quad + [l_{i,t} + \lambda_i l_{i,x}] w^\nu, \end{aligned} \tag{3.120}$$

where $l_i, \lambda_i$ are eigenvectors and eigenvalues of the matrix $A^{\nu-1}$. The a priori bounds (3.109) on the derivatives of $\overset{(\nu-1)}{u}$ imply the existence of a constant $C_7$ such that

$$|l_i| \leq C_7, \qquad |l_{i,t} + \lambda_i l_{i,x}| \leq C_7, \qquad |\overset{(\nu)}{u_x}| \leq C_7, \tag{3.121}$$

for all $(t,x) \in \mathcal{D}$ and all $\nu \geq 0$.

We introduce the function

$$Z_\nu(t) \doteq \max\big\{|l_i \cdot w^\nu(t,x)|;\ a + Ct \leq x \leq b - Ct,\ i = 1, \ldots, n\big\}. \tag{3.122}$$

From (3.120) and the bounds (3.119), (3.121) it now follows that

$$\dot{Z}_\nu \leq C_7 \cdot C_6 n Z_\nu(t) + C_7[C_6 n Z_{\nu-1}(t)]C_7 + C_7 \cdot n Z_\nu(t), \qquad Z_\nu(0) = 0.$$

We can now apply Lemma 3.4 with $\alpha = nC_7(C_6 + 1)$, $\beta = nC_6C_7^2$, $\gamma = 0$, and deduce the uniform convergence of the series $\sum Z_\nu(t)$, for $t \in [0, \delta]$. This in turn implies the uniform convergence of the sequence $\overset{(\nu)}{u}$ on $\mathcal{D}$.

5. By part (i) of Theorem 3.5, the function $u = \lim \overset{(\nu)}{u}$ is a Lipschitz-continuous, broad solution of (3.90), (3.93) on $\mathcal{D}$. In order to prove that $u$ is continuously differentiable, and thus provides a classical solution, we have to show the uniform convergence of the partial derivatives $\overset{(\nu)}{u}_x, \overset{(\nu)}{u}_t$. By Lemma 3.4, there exists a constant $C_8$ such that

$$\max_{(t,x)\in\mathcal{D}} |\overset{(\nu)}{u}(t,x) - \overset{(\nu-1)}{u}(t,x)| \leq C_8 2^{-\nu} \quad \text{for all } \nu \geq 1. \tag{3.123}$$

For some constant $C_9$ we thus have

$$\max_{(t,x)\in\mathcal{D}} \big\{|A^{\nu-1} - A^{\nu-2}|,\ |A_t^{\nu-1} - A_t^{\nu-2}|,\ |A_x^{\nu-1} - A_x^{\nu-2}|,\ |A_u^{\nu-1} - A_u^{\nu-2}|,$$
$$|h_t^\nu - h_t^{\nu-1}|,\ |h_x^\nu - h_x^{\nu-1}|,\ |h_u^\nu - h_u^{\nu-1}|\big\} \leq C_9 2^{-k} \quad \text{for all } \nu. \tag{3.124}$$

By Theorem 3.6, the derivatives $v^\nu \doteq \overset{(\nu)}{u}_x$ are broad solutions of the system

$$v_t^\nu + A^{\nu-1} v_x^\nu = h_x^\nu + h_u^\nu v^\nu - A_x^{\nu-1} v^\nu - (A_u^{\nu-1} v^{\nu-1}) v^\nu. \tag{3.125}$$

Therefore, the differences $w^\nu \doteq v^\nu - v^{\nu-1}$ provide a broad solution to the system

$$\begin{aligned} w_t^\nu + A^{\nu-1} w_x^\nu = {} & (A^{\nu-2} - A^{\nu-1}) v^{(\nu-1)} + (h_x^\nu - h_x^{\nu-1}) + h_u^\nu w^\nu \\ & + (h_u^\nu - h_u^{\nu-1}) v^{\nu-1} - A_x^{\nu-1} w^\nu - (A_x^{\nu-1} - A_x^{\nu-2}) v^{\nu-1} \\ & - (A_u^{\nu-1} w^{\nu-1}) v^\nu - ((A_u^{\nu-1} - A_u^{\nu-2}) v^{\nu-2}) v^\nu - (A_u^{\nu-2} v^{\nu-2}) w^\nu, \end{aligned} \tag{3.126}$$

with $w^\nu(0, x) = 0$.

Defining the scalar functions $Z_\nu$ as in (3.122), from (3.126) and (3.125) we derive for the sequence $(Z_\nu)_{\nu\geq 1}$ a set of inequalities of the form (3.94). By Lemma 3.4, the series $\sum Z_\nu$ is thus uniformly convergent. In turn, this establishes the uniform convergence of the sequence $\overset{(\nu)}{u}_x$. The convergence of the sequence $\overset{(\nu)}{u}_t$ is proved by an entirely similar argument.

6. In the Lipschitz-continuous case, the solution can be constructed following exactly the same steps 1–4. By Rademacher's theorem, the functions $A, h, \bar{u}_x$ are differentiable a.e. Choosing the constants $C_1, \ldots, C_4$ so that the estimates (3.99)–(3.102) hold a.e., the bounds in (3.106) remain valid in the a.e. sense. The existence of the sequence of Lipschitz-continuous broad solutions $\overset{(\nu)}{u}$ is now provided by Theorem 3.7. Their uniform convergence on $\mathcal{D}$ to a solution $u$ of (3.90), (3.93) is proved as in step 4.

7. Concerning the uniqueness of the solution, let $u, u'$ be two broad solutions of (3.90), (3.93), defined on a common domain of determinacy $\mathcal{D}$. Their difference $w \doteq u - u'$ is then a solution of the semilinear problem

$$\begin{aligned} w_t + A(t, x, u(t, x))w_x &= [A(t, x, u' + w) - A(t, x, u')]u'_x \\ &\quad + h(t, x, u' + w) - h(t, x, u'), \\ w(0, x) &= 0. \end{aligned} \tag{3.127}$$

Clearly, $w \equiv 0$ is a solution of (3.127). Observing that the right hand side of (3.127) is Lipschitz continuous w.r.t. $w$, by Theorem 3.3 we conclude that $w \equiv 0$ is the unique broad solution of (3.127). This establishes the uniqueness of the local solution, completing the proof of the theorem. □

## Problems

(1) Call $t \mapsto x(t; \tau, \xi)$ the solution of the Cauchy problem

$$\dot{x} = \lambda(t, x), \quad x(\tau) = \xi, \quad t \in [0, T].$$

Assume that the function $\lambda = \lambda(t, x)$ is Lipschitz continuous, and let $\mathcal{N}$ be a set of measure zero. Prove that for almost every $(\tau, \xi)$ in the strip $[0, T] \times \mathbb{R}$,

$$\text{meas}\,(\{t \in [0, T]; x(t; \tau, \xi) \in \mathcal{N}\}) = 0.$$

This result was tacitly used in the proofs of Theorems 3.1 and 3.2.

(2) Solve the Cauchy problem

$$u_t + xu_x = 2xu, \qquad u(0, x) = 1.$$

(3) Show that, for every $\delta > 0$, the Cauchy problem

$$u_t + x^2 u_x = 0, \qquad u(0, x) = 0$$

has infinitely many classical solutions defined on the domain $[0, \delta] \times \mathbb{R}$.

(4) Let $u = u(t, x)$ be a bounded, smooth solution of the quasilinear equation $u_t + a(u)u_x = h(u)$, defined on the strip $[0, T] \times \mathbb{R}$. Assume that the functions $a, h$ are smooth, with $h' \leq 0$. Prove that the map $t \mapsto$ Tot. Var. $\{u(t, \cdot)\}$ is non-increasing. If $h \equiv 0$, show that the total variation of $u(t, \cdot)$ remains constant in time.

(5) Let the function $a : \mathbb{R} \mapsto \mathbb{R}$ be continuously differentiable, with $da(s)/ds > 0$ for every $s \in \mathbb{R}$. Assume that $\bar{u} \in \mathcal{C}^1$. Prove that the scalar Cauchy problem $u_t + a(u)u_x = 0$, $u(0, x) = \bar{u}(x)$ has a classical solution defined for all $t \in [0, \infty[$ iff $\bar{u}$ is non-decreasing.

(6) Let $u = u(t, x)$ be a $\mathcal{C}^1$ solution of the strictly hyperbolic quasilinear system

$$u_t + A(u)u_x = 0.$$

As usual, for $i = 1, \ldots, n$ call $\lambda_i(u)$, $r_i(u)$ and $l_i(u)$ respectively the eigenvalues and the right and left eigenvectors of the matrix $A(u)$ normalized as in (3.3). For every $v \in \mathbb{R}^n$, this implies

$$A(u)v = \sum_i \lambda_i(u)(l_i(u) \cdot v)r_i(u).$$

Prove that the gradient components $u_x^i \doteq l_i(u) \cdot u_x$ are broad solutions of the system

$$(u_x^i)_t + (\lambda_i u_x^i)_x = \sum_{j \neq k} \lambda_j (l_i \cdot [r_j, r_k]) u_x^j u_x^k \quad i = 1, \ldots, n.$$

Here $[r_j, r_k] \doteq (Dr_k) \cdot r_j - (Dr_j) \cdot r_k$ denotes the Lie bracket of the vector fields $r_j, r_k$.

(7) Consider the initial value problem for the wave equation

$$w_{tt} - c^2 w_{xx} = 0, \quad w(0, x) = \phi(x), \quad w_t(0, x) = \psi(x). \tag{3.128}$$

Introducing the variables $U = w_x$, $V = w_t$, one obtains the system

$$\begin{cases} U_t - V_x = 0, \\ V_t - c^2 U_x = 0, \end{cases} \qquad \begin{cases} U(0, x) = \phi'(x), \\ V(0, x) = \psi(x). \end{cases}$$

Write the explicit solution of this system, using (3.8). Then, observing that

$$w(t, x) = w(0, x) + \int_0^t w_t(s, x)\, ds,$$

prove that the solution of (3.128) is given by the formula of d'Alambert:

$$w(t, x) = \frac{1}{2}[\phi(x - ct) + \phi(x + ct)] + \frac{1}{2c} \int_{x-ct}^{x+ct} \psi(y)\, dy.$$

# 4
# Discontinuous solutions

In this chapter we begin the study of discontinuous solutions, satisfying the system of conservation laws in the distributional sense. In this case, no a priori regularity is required: for example, the set of points of discontinuity may well be everywhere dense.

The qualitative properties of general weak solutions to the Cauchy problem with small $BV$ initial data will be studied in Chapter 10. In the present chapter, to provide some basic understanding of the problem, we begin by looking at a particular class of solutions whose structure is easy to describe: that is, the functions $u = u(t, x)$ which are piecewise Lipschitz continuous in the $t$–$x$ plane, with jumps along a finite number of curves of the form $x = \gamma_j(t)$, $j = 1, \ldots, N$. For such solutions, we derive the basic Rankine–Hugoniot equations, relating the values of $u$ across a jump to the speed $\dot{\gamma}_j$ of the discontinuity. We will show that a piecewise Lipschitz function provides a solution to the system of conservation laws if it satisfies the corresponding quasilinear system pointwise a.e., and if the Rankine–Hugoniot equations hold a.e. along each jump curve $\gamma_j$.

By simple examples, one can see that the concept of weak solution is not stringent enough to determine a unique solution to a general Cauchy problem, as soon as discontinuities are present. For this reason, one has to impose further conditions in order to select a (hopefully) unique solution. The second part of this chapter is concerned with various admissibility conditions which appear in the literature, formulated in terms of vanishing viscosity, convex entropies and characteristic speeds. These conditions will play a key role in the proofs of uniqueness of weak solutions, given in Chapter 6 for the scalar case and in Chapter 9 for $n \times n$ systems of equations.

## 4.1 Weak solutions

Let $f : \mathbb{R}^n \mapsto \mathbb{R}^n$ be a smooth vector field.

**Definition 4.1.** A measurable function $u = u(t, x)$ from an open set $\Omega \subseteq \mathbb{R} \times \mathbb{R}$ into $\mathbb{R}^n$ is a *distributional solution* to the system of conservation laws

$$u_t + f(u)_x = 0 \tag{4.1}$$

if, for every $C^1$ function $\phi : \Omega \mapsto \mathbb{R}^n$ with compact support, one has

$$\iint_\Omega \{u\, \phi_t + f(u)\, \phi_x\}\, dx\, dt = 0. \tag{4.2}$$

Observe that no continuity assumption is made on $u$. We only require $u$ and $f(u)$ to be locally integrable in $\Omega$. An easy consequence of this definition is the closure of the set of solutions w.r.t. convergence in $\mathbf{L}^1_{\text{loc}}$, stated below.

**Lemma 4.1.** *Let $\{u_\nu\}_{\nu\geq 1}$ be a sequence of distributional solutions of (4.1).*

*(i) If*

$$u_\nu \to u, \quad f(u_\nu) \to f(u) \text{ in } \mathbf{L}^1_{\text{loc}}, \tag{4.3}$$

*then the limit function $u$ is itself a solution of (4.1).*

*(ii) The same conclusion holds if $u_\nu \to u$ in $\mathbf{L}^1_{\text{loc}}$ and if all functions $u_\nu$ take values within a fixed compact set.*

*Proof.* Indeed, if (4.3) holds, then for every $\phi \in \mathcal{C}^1_c$

$$\iint_\Omega \{u\,\phi_t + f(u)\,\phi_x\}\,dx\,dt = \lim_{\nu\to\infty} \iint_\Omega \{u_\nu\,\phi_t + f(u_\nu)\,\phi_x\}\,dx\,dt = 0,$$

showing that $u$ itself is a solution.

Next, if $u_\nu \to u$ in $\mathbf{L}^1_{\text{loc}}$, by possibly taking a subsequence we have $u_\nu(t,x) \to u(t,x)$ a.e. Hence also $f(u_\nu(t,x)) \to f(u(t,x))$ a.e. The boundedness assumption on $u$ now implies $f(u_\nu) \to f(u)$ in $\mathbf{L}^1_{\text{loc}}$. Part (ii) of the lemma is thus a consequence of part (i). □

**Definition 4.2.** Given an initial condition

$$u(0,x) = \bar{u}(x), \tag{4.4}$$

with $\bar{u} \in \mathbf{L}^1_{\text{loc}}(\mathbb{R};\mathbb{R}^n)$, we say that a function $u : [0,T]\times\mathbb{R} \mapsto \mathbb{R}^n$ is a distributional solution to the Cauchy problem (4.1), (4.4) if

$$\int_0^T\int_{-\infty}^{\infty} \{u\,\phi_t + f(u)\,\phi_x\}\,dx\,dt + \int_{-\infty}^{\infty} \bar{u}(x)\,\phi(0,x)\,dx = 0 \tag{4.5}$$

for every $\mathcal{C}^1$ function $\phi$ with compact support contained in the set $]-\infty, T[\,\times\mathbb{R}$.

Throughout the following, we shall use a somewhat stronger concept of solution, requiring the continuity of $u$ as a function of time, with values into $\mathbf{L}^1_{\text{loc}}(\mathbb{R})$.

**Definition 4.3.** A function $u : [0,T]\times\mathbb{R} \mapsto \mathbb{R}^n$ is a *weak solution* of the Cauchy problem (4.1), (4.4) if $u$ is continuous as a function from $[0,T]$ into $\mathbf{L}^1_{\text{loc}}$, the initial condition (4.4) holds and the restriction of $u$ to the open strip $]0,T[\,\times\mathbb{R}$ is a distributional solution of (4.1).

**Remark 4.1.** Every weak solution is also a solution in the distributional sense. Indeed, let $\phi \in \mathcal{C}^1_c$ be given. Let $\alpha : \mathbb{R} \mapsto [0,2]$ be a smooth function with support contained inside $]0,1[$ and such that

$$\int \alpha(t)\,dt = 1.$$

For every $\varepsilon > 0$, we define the functions

$$\alpha^\varepsilon(t) \doteq \frac{1}{\varepsilon}\alpha(t/\varepsilon), \qquad \beta^\varepsilon(t) \doteq \int_0^t \alpha^\varepsilon(s)\,ds,$$

and set $\phi^\varepsilon \doteq \phi \cdot \beta^\varepsilon$. Observing that $\beta^\varepsilon_t = \alpha^\varepsilon$ and that the support of $\phi^\varepsilon$ is contained in $]0, \infty[\,\times\mathbb{R}$, we can use the continuity of the map $t \mapsto u(t, \cdot)$ to compute

$$\begin{aligned}
0 &= \lim_{\varepsilon\to 0}\int_0^\infty\int_{-\infty}^\infty \{u\phi^\varepsilon_t + f(u)\phi^\varepsilon_x\}\,dx\,dt \\
&= \lim_{\varepsilon\to 0}\int_0^\infty\int_{-\infty}^\infty \{u\phi_t + f(u)\phi_x\}\beta^\varepsilon\,dx\,dt + \lim_{\varepsilon\to 0}\int_0^\infty\int_{-\infty}^\infty u\phi\alpha^\varepsilon\,dx\,dt \\
&= \int_0^T\int_{-\infty}^\infty \{u\cdot\phi_t + f(u)\cdot\phi_x\}\,dx\,dt + \int_{-\infty}^\infty \bar{u}(x)\cdot\phi(0,x)\,dx = 0.
\end{aligned}$$

Therefore, every weak solution also satisfies (4.5).

The converse is clearly false. Indeed, let $u = u(t, x)$ be any non-constant distributional solution of (4.1), (4.4). We can change its values at all rational times and define the function

$$v(t,x) \doteq \begin{cases} 0 & \text{if } t > 0,\ t \in \mathbb{Q}, \\ u(t,x) & \text{otherwise.} \end{cases}$$

Since $v$ coincides with $u$ a.e. in the $t$–$x$ plane, this new function is a distributional solution as well. But of course $t \mapsto v(t, \cdot)$ is not continuous as a map with values in $\mathbf{L}^1_{\mathrm{loc}}$.

**Remark 4.2.** Consider a domain of the form

$$\Omega \doteq \{(t,x);\, t \in [\tau_0,\ \tau],\, \gamma_1(t) \le x \le \gamma_2(t)\}.$$

If $u$ is a smooth solution of the system of conservation laws (4.1), an application of the divergence theorem to the vector $\Psi \doteq (u,\ f(u))$ on the domain $\Omega$ yields

$$\begin{aligned}
0 &= \iint_\Omega \{u_t + f(u)_x\}\,dx\,dt \\
&= \int_{\gamma_1(\tau)}^{\gamma_2(\tau)} u(\tau,x)\,dx - \int_{\gamma_1(\tau_0)}^{\gamma_2(\tau_0)} u(\tau_0,x)\,dx + \int_{\tau_0}^{\tau}\{\dot\gamma_1(t)u(t,\gamma_1(t)) - f(u(t,\gamma_1(t)))\}\,dt \\
&\quad - \int_{\tau_0}^{\tau}\{\dot\gamma_2(t)u(t,\gamma_2(t)) - f(u(t,\gamma_2(t)))\}\,dt. \qquad (4.6)
\end{aligned}$$

Clearly, (4.6) need not hold for arbitrary weak solutions. Indeed, such solutions are defined only up to equivalence in $\mathbf{L}^1_{\text{loc}}$. One can thus arbitrarily assign the point values of $u$ along the two curves $x = \gamma_1(t)$, $x = \gamma_2(t)$ and arrange things so that (4.6) fails.

However, assume that the map $t \mapsto u(t, \cdot)$ is continuous with values in $\mathbf{L}^1_{\text{loc}}$, that $u$ is uniformly bounded and that its point values are chosen so that $u(t, x) = u(t, x+) \doteq \lim_{y \to x+} u(t, y)$. In this case, we claim that the identity (4.6) holds. Indeed, let $\beta : \mathbb{R} \mapsto [0, 1]$ be a smooth, non-decreasing function such that $\beta(s) = 0$ for $s \leq 0$ and $\beta(s) = 1$ for $s \geq 1$. For $\varepsilon > 0$, define the rescaled map $\beta^\varepsilon(s) \doteq \beta(s/\varepsilon)$ and consider the test function

$$\phi^\varepsilon(t, x) = \left[\beta^\varepsilon(x - \gamma_1(t)) - \beta^\varepsilon(x - \gamma_2(t))\right] \cdot \left[\beta^\varepsilon(t - \tau_0) - \beta^\varepsilon(t - \tau)\right],$$

providing an approximation to the characteristic function of the set $\Omega$. Letting $\varepsilon \to 0$ in the identity

$$\int_0^\infty \int_{-\infty}^\infty \left\{u\phi^\varepsilon_t + f(u)\phi^\varepsilon_x\right\} dx\, dt = 0$$

and using the above regularity assumptions on $u$, we now obtain (4.6).

## 4.2 Rankine–Hugoniot conditions

In this section we consider a function $u$ which is piecewise Lipschitz continuous, with jumps on a finite number of curves in the $t$–$x$ plane. Our goal is to derive conditions which ensure that $u$ is a solution of (4.1). We start with the simple case of a piecewise constant function, say

$$U(t, x) = \begin{cases} u^+ & \text{if } x > \lambda t, \\ u^- & \text{if } x < \lambda t, \end{cases} \tag{4.7}$$

for some $u^-, u^+ \in \mathbb{R}^n$, $\lambda \in \mathbb{R}$.

**Lemma 4.2.** *The function $U$ in (4.7) is a solution of (4.1) iff*

$$\lambda\,(u^+ - u^-) = f(u^+) - f(u^-). \tag{4.8}$$

Indeed, for every continuously differentiable $\phi$ with compact support, an application of the divergence theorem to the vector field $(U \cdot \phi, f(U) \cdot \phi)$ on the two domains

$$\Omega^+ \doteq \{x > \lambda t\}, \qquad \Omega^- \doteq \{x < \lambda t\}$$

yields

$$0 = \iint \{U\,\phi_t + f(U)\,\phi_x\}\,dx\,dt = \int \left\{\lambda\,[u^+ - u^-] - [f(u^+) - f(u^-)]\right\} \cdot \phi(t, \lambda t)\,dt. \tag{4.9}$$

Since (4.9) holds for every $\phi \in \mathcal{C}^1_c$, this implies (4.8).

The vector equations (4.8) are the famous *Rankine–Hugoniot conditions*. They form a set of $n$ scalar equations relating the right and left states $u^+, u^- \in \mathbb{R}^n$ and the speed $\lambda$ of the discontinuity.

**Remark 4.3.** Denote by $A(u) = Df(u)$ the $n \times n$ Jacobian matrix of $f$ at $u$. Assuming that the segment joining the two states $u, v$ is entirely contained in the domain of $f$, we define the averaged matrix

$$A(u, v) \doteq \int_0^1 A(\theta u + (1-\theta)v)\, d\theta \tag{4.10}$$

and call $\lambda_i(u, v)$, $i = 1, \ldots, n$, its eigenvalues. We can then write (4.8) in the equivalent form

$$\begin{aligned} \lambda\,(u^+ - u^-) = f(u^+) - f(u^-) &= \int_0^1 Df(\theta u^+ + (1-\theta)u^-) \cdot (u^+ - u^-)\, d\theta \\ &= A(u^+, u^-) \cdot (u^+ - u^-). \end{aligned} \tag{4.11}$$

In other words, the Rankine–Hugoniot equations hold iff the jump $u^+ - u^-$ is an eigenvector of the averaged matrix $A(u^+, u^-)$ and the speed $\lambda$ coincides with the corresponding eigenvalue.

**Remark 4.4.** In the case where $f$ is Lipschitz continuous, denoting by $Lip(f)$ the Lipschitz constant of $f$, we obtain from (4.8) an upper bound on the speed of the jump:

$$|\lambda| \leq Lip(f). \tag{4.12}$$

**Remark 4.5.** If $u^+ \neq u^-$, the function

$$U(t, x) \doteq \begin{cases} u^+ & \text{if } t > 0, \\ u^- & \text{if } t < 0 \end{cases} \tag{4.13}$$

cannot be a solution of (4.1). Indeed, in this case an application of the divergence theorem yields

$$0 = \iint \{U \cdot \phi_t + f(U) \cdot \phi_x\}\, dx\, dt = -\int [u^+ - u^-] \cdot \phi(0, x)\, dx. \tag{4.14}$$

If (4.14) were to hold for all $\phi \in \mathcal{C}_c^1$, then $u^+ = u^-$, contrary to the assumptions.

We now consider a more general solution $u = u(t, x)$ of (4.1) and show that the Rankine–Hugoniot equations (4.8) are still satisfied at every point $(\tau, \xi)$ where $u$ has an approximate jump discontinuity, in the sense of Definition 2.1.

**Theorem 4.1.** *Let $u$ be a bounded distributional solution of (4.1) and fix a point $(\tau, \xi)$. For some states $u^-, u^+$ and a speed $\lambda$, and defining $U$ as in (4.7), assume that*

$$\lim_{r \to 0} \frac{1}{r^2} \int_{-r}^{r} \int_{-r}^{r} |u(\tau + t, \xi + x) - U(t, x)|\, dx\, dt = 0. \tag{4.15}$$

*Then the Rankine–Hugoniot equations (4.8) hold.*

*On the other hand, the limit (4.15) can never hold with $U$ as in (4.13) and $u^+ \neq u^-$.*

*Proof.* For every $\eta > 0$, one can easily check that the rescaled function

$$u^\eta(t,x) \doteq u(\tau + \eta t, \xi + \eta x) \tag{4.16}$$

is also a solution to the system of conservation laws. We claim that, as $\eta \to 0$, one has

$$u^\eta \to U, \qquad f(u^\eta) \to f(U)$$

in $\mathbf{L}^1_{\text{loc}}(\mathbb{R}^2;\mathbb{R}^n)$. Indeed, given an arbitrarily large constant $R$, one has

$$\int_{-R}^{R}\int_{-R}^{R} |u^\eta(t,x) - U(t,x)|\,dx\,dt = \frac{1}{\eta^2}\int_{-\eta R}^{\eta R}\int_{-\eta R}^{\eta R} |u(\tau+t,\xi+x) - U(t,x)|\,dx\,dt. \tag{4.17}$$

Letting $\eta \to 0$ in (4.17), by (4.15) it follows the convergence $u^\eta \to U$ in $\mathbf{L}^1_{\text{loc}}$. Since $u$ is bounded and $f$ is uniformly Lipschitz continuous on bounded sets, the same argument yields the convergence $f(u^\eta) \to f(U)$. By Lemma 4.1, $U$ itself is a distributional solution of (4.1); hence by Lemma 4.2 the Rankine–Hugoniot equations (4.8) hold.

To prove the last statement, let the limit (4.15) hold with $U$ as in (4.13). In this case, as $\eta \to 0$, the rescaled functions $u^\eta$ in (4.16) converge in $\mathbf{L}^1_{\text{loc}}$ to the function $U$. Hence, $U$ itself is a distributional solution of (4.1). By Remark 4.5 this can hold only if $u^+ = u^-$. □

We conclude this section by proving a necessary and sufficient condition for a piecewise Lipschitz-continuous function to be a solution of (4.1). Consider the following (piecewise Lipschitz, PL) regularity assumptions:

(PL) The function $u = u(t,x)$ is measurable and bounded. Moreover, there exist a finite number of points $P_i = (t_i, x_i)$ and finitely many disjoint Lipschitz-continuous curves $\gamma_j : ]a_j, b_j[ \mapsto \mathbb{R}$ such that the following hold:

- (a) Every point $P = (t,x)$, not coinciding with some $P_i$ and lying outside the curves $\gamma_j$, has a neighbourhood $V$ where the function $u$ is Lipschitz continuous.
- (b) Every point $Q = (t, \gamma_j(t))$ on each curve $\gamma_j$ has a neighbourhood $V$ such that the restrictions of $u$ to the subsets $V^+ = V \cap \{x > \gamma_j(t)\}$, $V^- = V \cap \{x < \gamma_j(t)\}$ are both Lipschitz continuous.

In the following $A \doteq Df$ denotes the Jacobian matrix of $f$, as usual. The right and left limits of $u(t,\cdot)$ along the curve $\gamma_j$ are denoted by

$$u_j^+(t) \doteq \lim_{x\to\gamma_j(t)+} u(t,x), \qquad u_j^-(t) \doteq \lim_{x\to\gamma_j(t)-} u(t,x). \tag{4.18}$$

**Theorem 4.2.** *Let $\Omega$ be an open set in $\mathbb{R}^2$ and let $u : \Omega \mapsto \mathbb{R}^n$ be a function with the regularity properties* (PL). *Then the following are equivalent:*

*(i) $u$ is a distributional solution of (4.1).*

*(ii) The quasilinear equation*

$$u_t + A(u)u_x = 0 \tag{4.19}$$

*holds at almost every point* $(t, x)$. *Moreover, for each* $j$, *at almost every* $t \in ]a_j, b_j[$ *one has*

$$\dot{\gamma}_j(t) \cdot \left[u_j^+(t) - u_j^-(t)\right] = f(u_j^+(t)) - f(u_j^-(t)). \tag{4.20}$$

*If the above conditions hold, then the curves* $\gamma_j$ *are continuously differentiable.*

*Proof.* As a preliminary, observe that by Rademacher's theorem the functions $u$ and $f(u)$ are differentiable a.e. in the $t$–$x$ plane. Similarly, every $\gamma_j$ is a.e. differentiable as a function of time.

To prove the implication (i)⇒(ii), let $u$ be a distributional solution. If $\Omega' \subset \Omega$ is any open set on which $u$ is Lipschitz continuous, for every $\phi \in \mathcal{C}_c^1$ with support inside $\Omega'$ we have

$$\iint \{u_t + A(u)u_x\}\phi \, dx \, dt = -\iint \{u\phi_t + f(u)\phi_x\} \, dx \, dt = 0. \tag{4.21}$$

Since (4.21) holds for all $\phi \in \mathcal{C}_c^1(\Omega')$, this implies (4.19).

Next, if $\gamma_j$ is differentiable at some time $\tau$, then $u$ has an approximate jump discontinuity at the point $(\tau, \gamma_j(\tau))$. That is, (4.15) holds with $U$ defined at (4.7), where $u^- = u_j^-(\tau), u^+ = u_j^+(\tau), \lambda = \dot{\gamma}_j(\tau)$. By Theorem 4.1, the Rankine–Hugoniot relations (4.20) hold.

To prove that (ii)⇒(i), let $\phi \in \mathcal{C}_c^1(\Omega)$. Let $\alpha : [0, \infty[ \mapsto [0, 1]$ be a smooth scalar function such that

$$\alpha(r) = \begin{cases} 1 & \text{if } r \leq 1/2, \\ 0 & \text{if } r \geq 1, \end{cases}$$

and define the rescaled functions

$$\alpha^\varepsilon(r) \doteq \alpha(r/\varepsilon).$$

For $\varepsilon > 0$ sufficiently small, the circles $B(P_i, \varepsilon)$ centred at the points $P_i$ with radius $\varepsilon$ are mutually disjoint. For any point $P = (t, x)$, we now define

$$\widetilde{\phi}^\varepsilon(P) \doteq \phi(P) \cdot \sum_i \alpha^\varepsilon(|P - P_i|), \qquad \phi^\varepsilon(P) \doteq \phi(P) - \widetilde{\phi}^\varepsilon(P).$$

Observe that each $P_i$ lies outside the support of $\phi^\varepsilon$. Repeating the argument at (2.12), for every $\varepsilon > 0$ an application of the divergence theorem yields

$$\begin{aligned} &\iint \{u\phi_t^\varepsilon + f(u)\phi_x^\varepsilon\} \, dx \, dt \\ &\quad = -\iint \{u_t + A(u)u_x\}\phi^\varepsilon \, dx \, dt \\ &\qquad + \sum_j \int_{a_j}^{b_j} \{\dot{\gamma}_j[u_j^+ - u_j^-] - [f(u_j^+) - f(u_j^-)]\}\phi^\varepsilon(t, \gamma_j) \, dt \\ &\quad = 0. \end{aligned} \tag{4.22}$$

On the other hand, we have

$$\left| \iint \{u\tilde{\phi}_t^\varepsilon + f(u)\tilde{\phi}_x^\varepsilon\}\, dx\, dt \right| \le \left(\|u\|_{\mathbf{L}^\infty} + \|f(u)\|_{\mathbf{L}^\infty}\right) \cdot \left\|\nabla\widetilde{\phi}^\varepsilon\right\|_{\mathbf{L}^1}. \tag{4.23}$$

Observing that

$$\lim_{\varepsilon\to 0} \left\|\nabla\widetilde{\phi}^\varepsilon\right\|_{\mathbf{L}^1} = 0,$$

by (4.22)–(4.23) we can write

$$\iint \{u\phi_t + f(u)\phi_x\}\, dx\, dt = \lim_{\varepsilon\to 0+} \iint \{u(\phi_t^\varepsilon + \tilde{\phi}_t^\varepsilon) + f(u)(\phi_x^\varepsilon + \tilde{\phi}_x^\varepsilon)\}\, dx\, dt = 0,$$

proving (4.2).

Since the speed $\dot{\gamma}_j$ of a shock is a continuous function of its right and left states $u_j^+, u_j^-$, the last statement of the theorem is clear. □

**Example 4.1.** For Burgers' equation

$$u_t + \left(\frac{u^2}{2}\right)_x = 0, \tag{4.24}$$

along any shock line $x = \gamma(t)$ the Rankine–Hugoniot conditions reduce to

$$\dot{\gamma}(t) = \frac{[(u^+)^2/2] - [(u^-)^2/2]}{u^+ - u^-} = \frac{u^+ + u^-}{2}. \tag{4.25}$$

A weak solution of (4.24) with initial condition

$$u(0,x) = \begin{cases} 1 - |x| & \text{if } x \in [-1, 1], \\ 0 & \text{otherwise,} \end{cases}$$

is thus provided by

$$u(t,x) = \begin{cases} \dfrac{x+1}{t+1} & \text{if } x \in [-1, t], \\ \dfrac{1-x}{1-t} & \text{if } x \in\, ]t, 1], \\ 0 & \text{otherwise,} \end{cases} \tag{4.26$_1$}$$

when $0 \le t < 1$, while for $t \ge 1$ one has

$$u(t,x) = \begin{cases} \dfrac{x+1}{t+1} & \text{if } x \in \left[-1, -1 + \sqrt{2t+2}\right], \\ 0 & \text{otherwise.} \end{cases} \tag{4.26$_2$}$$

## 4.3 Coordinate transformations

Let $u = u(t, x)$ be a $\mathcal{C}^1$ solution of (4.1), and hence of (4.19). Consider a change of variable, described by $u = h(v)$. An elementary computation then shows that $v(t, x) \doteq h^{-1}(u(t, x))$ provides a $\mathcal{C}^1$ solution to the system

$$v_t + B(v)v_x = 0, \tag{4.27}$$

where the $n \times n$ matrix $B$ is given by

$$B(v) = [Dh(v)]^{-1} A(h(v)) Dh(v). \tag{4.28}$$

Assume that there exists a map $g$ whose Jacobian is $B$, i.e. $B(v) = Dg(v)$ for all $v$. The system (4.27) can then be written in conservation form

$$v_t + [g(v)]_x = 0. \tag{4.29}$$

Clearly, a $C^1$ function $v = v(t, x)$ is a classical solution of (4.29) iff $u = h(v)$ provides a classical solution to (4.19) and hence to (4.1). The same holds if $u, v$ are Lipschitz continuous. One should be aware, however, that this equivalence between (4.1) and (4.29) does not carry over to discontinuous solutions. In other words, the notion of weak solution is not invariant under non-linear transformations of the dependent variables.

**Example 4.2.** Let $u$ be the solution of Burgers' equation defined at $(4.26)_{1,2}$ in Example 4.1. Consider the change of variable $v = u^3$, with corresponding flux function $g(v) = \frac{3}{4}v^{4/3}$. For $t \leq 1$ the solution $u$ is continuous, and hence $v$ is a weak solution to

$$v_t + \left[\tfrac{3}{4}v^{4/3}\right]_x = 0. \tag{4.30}$$

However, $v$ is no longer a weak solution to (4.30) when $t > 1$, owing to the appearance of a shock. Indeed, if $x = \gamma(t)$ is the location of the shock, the Rankine–Hugoniot conditions applied to (4.24) yield (4.25), while from (4.30) we deduce that

$$\dot{\gamma}(t) = \frac{\frac{3}{4}(v^+)^{4/3} - \frac{3}{4}(v^-)^{4/3}}{v^+ - v^-} = \frac{3}{4} \cdot \frac{(u^+)^4 - (u^-)^4}{(u^+)^3 - (u^-)^3},$$

which is clearly not equivalent to (4.25).

We remark that, for the solution considered in Example 4.1, when $t \leq 1$, both integrals

$$I_1(t) \doteq \int_{-\infty}^{\infty} u(t, x)\, dx, \qquad I_2(t) \doteq \int_{-\infty}^{\infty} u^3(t, x)\, dx$$

remain constant in time. However, when $t > 1$, there exists no speed $\dot{\gamma}(t)$ for the discontinuity, which can keep the values of both $I_1$ and $I_2$ constant.

## 4.4 Admissibility conditions

Given an initial value problem for a system of conservation laws, Definition 4.3 for the weak solution is usually not stringent enough to single out a unique solution. In some cases, an entire family of weak solutions can be found, continuously depending on a parameter $\alpha \in [0, 1]$.

**Example 4.3.** For Burgers' equation (4.24), consider the Cauchy problem with initial data

$$u(0, x) = \begin{cases} 1 & \text{if } x \geq 0, \\ 0 & \text{if } x < 0. \end{cases}$$

For every $\alpha \in ]0, 1[$, define the piecewise constant function $u_\alpha : [0, \infty) \times \mathbb{R} \mapsto \mathbb{R}$ as

$$u_\alpha(t, x) = \begin{cases} 0 & \text{if } x < \alpha t/2, \\ \alpha & \text{if } \alpha t/2 \le x < (1+\alpha)t/2, \\ 1 & \text{if } x \ge (1+\alpha)t/2. \end{cases}$$

Then each $u_\alpha$ is a solution to the Cauchy problem, because it satisfies the equation a.e. and the Rankine–Hugoniot conditions hold along the two lines of discontinuity $\gamma_1(t) = \alpha t/2$, $\gamma_2(t) = (\alpha + 1)t/2$.

From the previous example it is clear that, in order to achieve the uniqueness and continuous dependence on the initial data, the notion of weak solution must be supplemented with further 'admissibility conditions', possibly motivated by physical considerations. Some of these conditions will be discussed in the present section.

**Admissibility condition 1 (Vanishing viscosity).** A weak solution $u$ of (4.1) is admissible if there exists a sequence of smooth solutions $u^\varepsilon$ to

$$u_t^\varepsilon + A(u^\varepsilon)u_x^\varepsilon = \varepsilon u_{xx}^\varepsilon \quad (A = Df) \tag{4.31}$$

which converge to $u$ in $\mathbf{L}^1_{\text{loc}}$ as $\varepsilon \to 0+$.

Unfortunately, it is often very difficult to provide uniform estimates on solutions to the above parabolic system, and to characterize the corresponding limits as $\varepsilon \to 0$. From the above condition, however, one can deduce other conditions which can be more easily verified in practice.

**Definition 4.4.** A continuously differentiable function $\eta : \mathbb{R}^n \mapsto \mathbb{R}$ is called an *entropy* for the system (4.1), with *entropy flux* $q : \mathbb{R}^n \mapsto \mathbb{R}$, if

$$D\eta(u) \cdot Df(u) = Dq(u) \quad u \in \mathbb{R}^n. \tag{4.32}$$

Observe that (4.32) implies that, if $u = u(t, x)$ is a $\mathcal{C}^1$ solution of (4.1), then

$$\eta(u)_t + q(u)_x = 0. \tag{4.33}$$

Indeed, (4.19) yields

$$D\eta(u)u_t + Dq(u)u_x = D\eta(u)(-Df(u)u_x) + Dq(u)u_x = 0.$$

In other words, whenever we have a smooth solution to (4.1), not only the quantities $u_1, \ldots, u_n$ are conserved, but the additional conservation law (4.33) holds as well. However, one should be aware that, when $u$ is discontinuous, in general it does not provide a weak solution to (4.33), i.e. $\eta = \eta(u)$ is not a conserved quantity. This can be seen in Example 4.2, taking $\eta(u) = u^3$ and $q(u) = (3/4)u^4$.

We now study how a convex entropy behaves in the presence of a small diffusion term. Assume $\eta, q \in \mathcal{C}^2$, with $\eta$ convex. Multiplying both sides of (4.31) on the left by $D\eta(u^\varepsilon)$ one finds

$$\eta(u^\varepsilon)_t + q(u^\varepsilon)_x = \varepsilon D\eta(u^\varepsilon)u_{xx}^\varepsilon = \varepsilon\{\eta(u^\varepsilon)_{xx} - D^2\eta(u^\varepsilon) \cdot (u_x^\varepsilon \otimes u_x^\varepsilon)\}. \tag{4.34}$$

Observe that the last term in (4.34) satisfies

$$D^2\eta(u^\varepsilon)\cdot(u_x^\varepsilon\otimes u_x^\varepsilon)=\sum_{i,j=1}^{n}\frac{\partial^2\eta(u^\varepsilon)}{\partial u_i\,\partial u_j}\cdot\frac{\partial u_i^\varepsilon}{\partial x}\frac{\partial u_j^\varepsilon}{\partial x}\geq 0.$$

Indeed, $\eta$ is convex, and hence its second derivative at any point $u^\varepsilon$ is a positive semi-definite quadratic form. Multiplying (4.34) by a non-negative smooth function $\varphi$ with compact support and integrating by parts, we thus have

$$\iint\{\eta(u^\varepsilon)\varphi_t+q(u^\varepsilon)\varphi_x\}\,dx\,dt\geq-\varepsilon\iint\eta(u^\varepsilon)\varphi_{xx}\,dx\,dt.$$

Now assume that, as $\varepsilon\to 0$, the solutions $u^\varepsilon$ remain uniformly bounded and converge to a function $u$ in $\mathbf{L}^1_{\mathrm{loc}}$. The above inequality now yields

$$\iint\{\eta(u)\varphi_t+q(u)\varphi_x\}\,dx\,dt\geq 0 \tag{4.35}$$

whenever $\varphi\geq 0$. The above can be restated by saying that $\eta(u)_t+q(u)_x\leq 0$ in a distributional sense, i.e. any convex entropy does not increase in time. The above analysis motivates the following definition:

**Admissibility condition 2 (Entropy inequality).** A weak solution $u$ of (4.1) is *entropy admissible* if

$$\eta(u)_t+q(u)_x\leq 0 \tag{4.36}$$

in the distributional sense, for every pair $(\eta,q)$, where $\eta$ is a convex entropy for (4.1) and $q$ is the corresponding entropy flux.

Of course, the above condition can be useful only if some non-trivial convex entropy for the system (4.1) is known. Observe that (4.32) can be regarded as a first-order system of $n$ equations for the two scalar variables $\eta$, $q$. For $n\geq 3$, this system is overdetermined. In general, one should thus expect to find solutions only in the case $n\leq 2$. In mathematical physics, however, one encounters several special cases of systems where a non-trivial convex entropy exists.

**Example 4.4.** Denote by $|\cdot|$ and $\langle\cdot,\cdot\rangle$ respectively the Euclidean norm and inner product on $\mathbb{R}^n$. Let $f$ be a gradient field, i.e. $f(u)=\nabla\varphi(u)$ for some smooth function $\varphi:\mathbb{R}^n\mapsto\mathbb{R}$. Then the function $\eta(u)=|u|^2/2$ is a convex entropy with flux $q(u)=\langle u,\nabla\varphi\rangle-\varphi$. Indeed, if $u=u(t,x)$ is a smooth solution of (4.1), then

$$\begin{aligned}\eta(u)_t&=\langle u,u_t\rangle=-\langle u,(\nabla\varphi)_x\rangle,\\ q(u)_x&=\langle u_x,\nabla\varphi\rangle+\langle u,(\nabla\varphi)_x\rangle-\langle\nabla\varphi,u_x\rangle.\end{aligned}$$

Hence (4.33) holds.

For a piecewise Lipschitz solution, we now show that the entropy admissibility condition (4.36) poses additional restrictions only along shock lines. As in Theorem 4.2, we use the notation (4.18) to denote the left and right limits of $u$ along a shock.

**Theorem 4.3.** *Let $u=u(t,x)$ be a solution of the system (4.1) satisfying the piecewise Lipschitz (PL) regularity assumptions. Let $\eta$ be a convex entropy with flux $q$.*

*Then the admissibility condition (4.36) holds iff, along every jump curve $\gamma_j$, for every $t$ one has*

$$\dot\gamma_j(t)[\eta(u_j^+(t)) - \eta(u_j^-(t))] \ge q(u_j^+(t)) - q(u_j^-(t)). \tag{4.37}$$

*Proof.* For every non-negative $\phi \in \mathcal{C}_c^1$ the same arguments used to prove Theorem 4.2 yield

$$\begin{aligned}\iint \{\eta(u)\phi_t + q(u)\phi_x\}\,dx\,dt &\\ = -\iint \{\eta(u)_t + q(u)_x\}\phi\,dx\,dt&\\ + \sum_j \int_{a_j}^{b_j} \{\dot\gamma_j[\eta(u_j^+) - \eta(u_j^-)] - [q(u_j^+) - q(u_j^-)]\}\phi(t,\gamma_j)\,dt.&\end{aligned}$$

By the assumptions, outside the jump curves one has

$$\eta(u)_t + q(u)_x = D\eta(u)u_t + Dq(u)q_x = D\eta(u)[u_t + A(u)u_x] = 0.$$

Therefore, the entropy condition (4.36) will hold iff

$$\begin{aligned}\iint \{\eta(u)\phi_t + q(u)\phi_x\}\,dx\,dt = \sum_j \int_{a_j}^{b_j} \{\dot\gamma_j[\eta(u_j^+) - \eta(u_j^-)] - [q(u_j^+) - q(u_j^-)]\}&\\ \cdot \phi(t,\gamma_j)\,dt \ge 0&\end{aligned}$$

for every non-negative $\phi \in \mathcal{C}_c^1$. This is the case iff (4.37) holds. □

**Remark 4.6.** In the definition of entropy functions, one can more generally consider locally Lipschitz-continuous maps $\eta, q$ that satisfy (4.32) a.e. In particular, assume that, for each $\nu \ge 1$, we are given a smooth convex entropy $\eta_\nu$ for the system (4.1), with entropy flux $q_\nu$. As $\nu \to \infty$, assume that $\eta_\nu \to \eta$, $q_\nu \to q$ uniformly on bounded sets. If $u^\varepsilon$ is a bounded sequence of solutions to the viscous system (4.31), with $u^\varepsilon \to u$ in $\mathbf{L}^1_{\mathrm{loc}}$ as $\varepsilon \to 0+$, then the previous analysis yields

$$\iint \{\eta_\nu(u)\,\varphi_t + q_\nu(u)\,\varphi_x\}\,dx\,dt \ge 0$$

for every $\nu$ and every non-negative function $\varphi \in \mathcal{C}_c^1$. Letting $\nu \to \infty$ we obtain (4.35). It is thus appropriate to require the entropy inequality (4.36) also for the pair $\eta, q$, even if these functions are only Lipschitz continuous, possibly not $\mathcal{C}^1$.

A third admissibility condition, due to Lax (1957), is particularly useful because it can be applied to any system and has a simple geometrical meaning. Consider a line of discontinuity $x = \gamma(t)$, where the solution jumps from a left state $u^-$ to a right state $u^+$. According to (4.11), this discontinuity must travel with a speed $\dot\gamma = \lambda_i(u^-, u^+)$ equal to an eigenvalue of the averaged matrix $A(u^-, u^+)$. In this setting, the Lax condition requires that the $i$-characteristics, travelling on the left and on the right of the jump with speeds

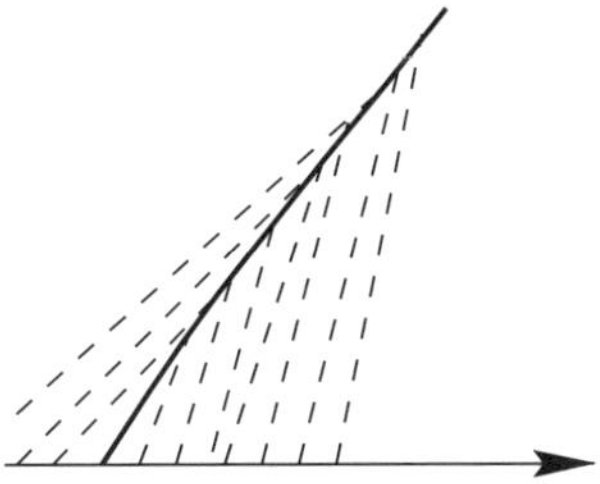


**Figure 4.1**

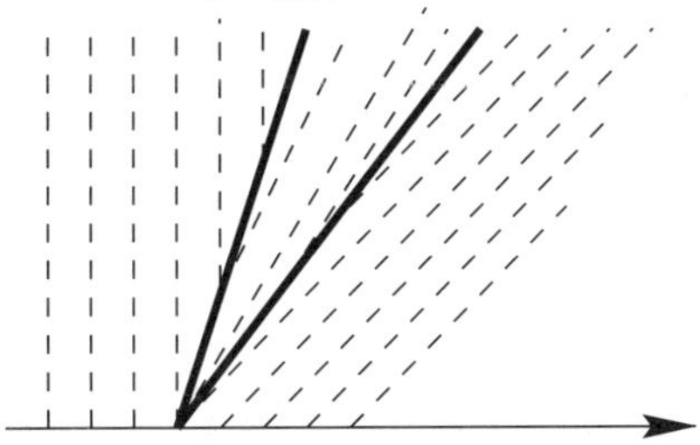


**Figure 4.2**

$\lambda_i(u^-)$, $\lambda_i(u^+)$ respectively, both run towards the line of discontinuity. This situation is illustrated in Fig. 4.1. On the other hand, one can check that for the solutions constructed in Example 4.3, neither of the two shocks satisfies the Lax condition (Fig. 4.2).

A more precise formulation of this condition is given below. We recall that a function $u$ has an approximate jump at a point $(\tau, \xi)$ if (4.15) holds, with $U$ as in (4.7). In this case, by Theorem 4.1 the right and left states $u^-$, $u^+$ and the speed $\lambda$ of the jump satisfy the Rankine–Hugoniot equations. In particular, $\lambda$ must be an eigenvalue of the averaged matrix $A(u^-, u^+)$ defined at (4.11), i.e. $\lambda = \lambda_i(u^-, u^+)$ for some $i \in \{1, \dots, n\}$.

**Admissibility condition 3 (Lax condition).** A weak solution $u = u(t, x)$ of (4.1) is admissible if, at every point $(\tau, \xi)$ of approximate jump, the left and right states $u^-$, $u^+$ and the speed $\lambda = \lambda_i(u^-, u^+)$ of the jump satisfy

$$\lambda_i(u^-) \geq \lambda \geq \lambda_i(u^+). \tag{4.38}$$

## 4.5 The scalar case

Let $f : \mathbb{R} \mapsto \mathbb{R}$, so that (4.1) represents a single conservation law. In this case, the Rankine–Hugoniot equations (4.8) can be solved for the shock speed $\lambda$ as a function of the left and right states $u^-$, $u^+$:

$$\lambda = \frac{f(u^+) - f(u^-)}{u^+ - u^-}. \tag{4.39}$$

By (4.39), the Rankine–Hugoniot speed of a shock is thus given by the slope of the secant line to the graph of $f$ through the points $u^-$, $u^+$. Calling $a(u) = f'(u)$ the wave speed at $u$, the above equation can be rewritten as

$$\lambda = \frac{1}{u^+ - u^-} \int_{u^-}^{u^+} a(v)\, dv. \tag{4.40}$$

In other words, the speed of a shock connecting $u^-$ with $u^+$ is the average of the wave speeds $a(u)$, as $u$ ranges in the interval $[u^-, u^+]$.

In the scalar case, the search for convex entropies is straightforward. Indeed, any convex function $\eta : \mathbb{R} \mapsto \mathbb{R}$ provides an entropy. The condition (4.32) now reduces to the single ODE $q'(u) = \eta'(u) f'(u)$. As entropy flux one can thus take

$$q(u) = \int_*^u \eta'(v) f'(v)\, dv.$$

The lower limit of the integral here is an arbitrary constant.

In particular, for each $k \in \mathbb{R}$, consider the functions

$$\eta(u) = |u - k|, \qquad q(u) = \operatorname{sgn}(u - k) \cdot (f(u) - f(k)). \tag{4.41}$$

It is easily checked that $\eta$, $q$ are locally Lipschitz continuous and satisfy (4.32) at every $u \neq k$. According to Remark 4.6 we still regard $\eta$ as a convex entropy with entropy flux $q$. Motivated by the previous analysis, we shall say that a locally integrable function $u : [0, \infty) \times \mathbb{R} \mapsto \mathbb{R}$ is an *entropy solution* of (4.1) if

$$\iint \{|u - k|\varphi_t + \operatorname{sgn}(u - k)(f(u) - f(k))\varphi_x\}\, dx\, dt \geq 0 \tag{4.42}$$

for every constant $k$ and every $\mathcal{C}^1$ function $\varphi \geq 0$ with compact support contained in $[0, \infty) \times \mathbb{R}$. In Chapter 6 we will prove that the conditions (4.42), introduced by Volpert (1967), suffice to single out a unique solution to the scalar conservation law (4.1), for all initial data $\bar{u} \in \mathbf{L}^\infty$. From (4.42) we presently derive a simple geometric condition, valid on each line of discontinuity of $u$.

**Theorem 4.4.** *The piecewise constant function $U$ defined at (4.7) is an entropy solution of (4.1) iff the Rankine–Hugoniot equations (4.8) hold and for every $\alpha \in [0, 1]$ one has*

$$\begin{cases} f(\alpha u^+ + (1-\alpha)u^-) \geq \alpha f(u^+) + (1-\alpha) f(u^-) & \text{if } u^- < u^+, \\ f(\alpha u^+ + (1-\alpha)u^-) \leq \alpha f(u^+) + (1-\alpha) f(u^-) & \text{if } u^- > u^+. \end{cases} \tag{4.43}$$

The conditions (4.43) have a simple geometrical interpretation. When $u^- < u^+$ the graph of $f$ should remain above the secant line (Fig. 4.3). When $u^- > u^+$, the graph of $f$ should remain below the secant line (Fig. 4.4).

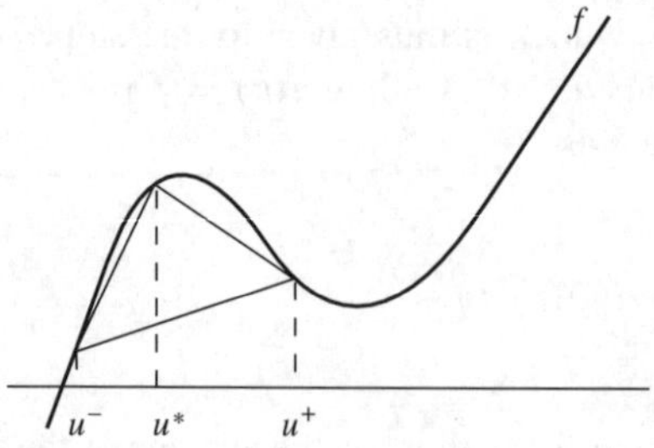


**Figure 4.3**

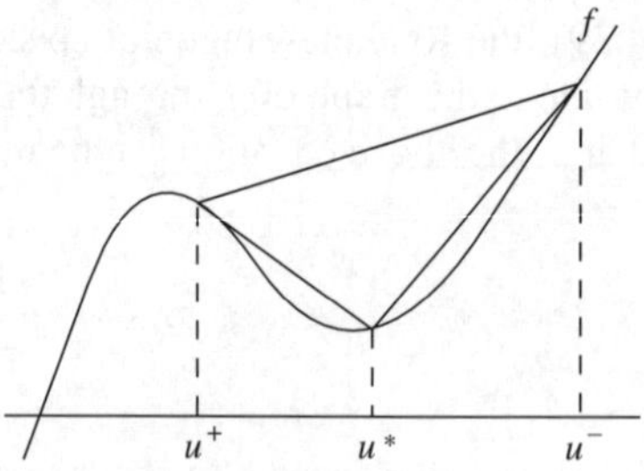


**Figure 4.4**

*Proof of Theorem 4.4.* Fix $k \in \mathbb{R}$ and consider the entropy function $\eta$ with flux $q$ defined at (4.41). By Theorem 4.3, the solution $U$ is thus entropy admissible if

$$\lambda(\eta(u^+) - \eta(u^-)) \geq q(u^+) - q(u^-). \tag{4.44}$$

By (4.41), the inequality (4.44) becomes

$$\begin{aligned} \lambda\{|u^+ - k| - |u^- - k|\} \geq {} & [(f(u^+) - f(k))\text{sgn}(u^+ - k)] \\ & - [(f(u^-) - f(k))\text{sgn}(u^- - k)]. \end{aligned} \tag{4.45}$$

When $k \leq \min\{u^-, u^+\}$ or $k \geq \max\{u^-, u^+\}$ respectively, from (4.45) it follows that

$$\lambda(u^+ - u^-) \geq f(u^+) - f(u^-), \qquad \lambda(u^+ - u^-) \leq f(u^+) - f(u^-).$$

These two inequalities are clearly equivalent to the Rankine–Hugoniot equation (4.39).

On the other hand, when $k$ lies inside the interval $[u^-, u^+]$, from (4.45) it follows that

$$\lambda(u^+ + u^- - 2k)\text{sgn}(u^+ - u^-) \geq [f(u^+) + f(u^-) - 2f(k)]\text{sgn}(u^+ - u^-). \tag{4.46}$$

Observe that, in this case, we can choose $\alpha \in [0, 1]$ so that

$$k = \alpha u^+ + (1 - \alpha)u^-.$$

Multiplying both sides of (4.46) by the positive quantity $|u^+ - u^-|$ and using (4.39) we obtain

$$\begin{aligned} (f(u^+) - f(u^-))(u^+ + u^- - 2k) &\geq (u^+ - u^-)[f(u^+) + f(u^-) - 2f(k)], \\ [f(u^+) - f(u^-)](1 - 2\alpha)(u^+ - u^-) &\geq (u^+ - u^-)[f(u^+) + f(u^-) - 2f(k)], \\ -[2\alpha f(u^+) + (2 - 2\alpha)f(u^-)](u^+ - u^-) &\geq -2(u^+ - u^-)f(\alpha u^+ + (1 - \alpha)u^-), \end{aligned}$$

which are equivalent to (4.43). □

**Remark 4.7.** The conditions (4.43) can be rewritten as

$$\frac{f(u^*) - f(u^-)}{u^* - u^-} \geq \frac{f(u^+) - f(u^*)}{u^+ - u^*} \tag{4.47}$$

for every $u^* = \alpha u^+ + (1 - \alpha)u^-$, with $0 < \alpha < 1$. This inequality can be interpreted as a stability condition. Indeed, let $u^* \in [u^-, u^+]$ be an intermediate state and consider a slightly perturbed solution, where the shock $(u^-, u^+)$ is decomposed as two separate jumps, $(u^-, u^*)$ and $(u^*, u^+)$, located say at $\gamma_1(t) < \gamma_2(t)$ respectively. Notice that, by the Rankine–Hugoniot conditions, the two sides of (4.47) yield precisely the speeds of these jumps. These coincide with the slopes of the secant lines to the graph of $f$ through the points $u^-, u^*$ and through the points $u^*, u^+$, respectively (Figs 4.3–4.4). If (4.47) holds, then $\dot{\gamma}_1 \geq \dot{\gamma}_2$, so that the backward shock travels at least as fast as the forward one (Figs 4.5–4.6). Therefore, the two shocks will not split apart as time increases, and the perturbed solution will remain close to the original solution possessing a single shock.

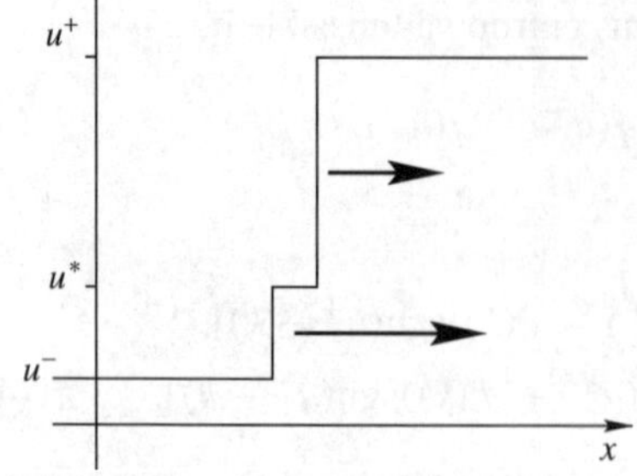


**Figure 4.5**

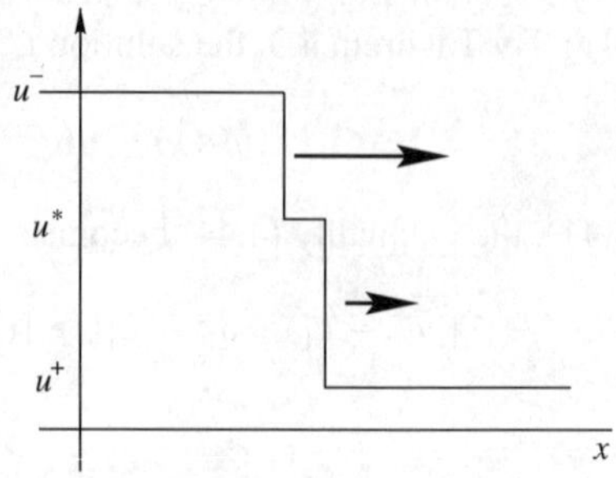


**Figure 4.6**

Taking the limits $u^* \to u^-$ and $u^* \to u^+$ in (4.47) one obtains

$$f'(u^-) \geq \frac{f(u^+) - f(u^-)}{u^+ - u^-} \geq f'(u^+).$$

Therefore, the entropy conditions (4.43) imply the Lax conditions (4.38). However, the reader should be aware that the converse implication does not hold in this case of a general (non-convex) flux function $f$.

## Problems

(1) Let (4.1) be a scalar conservation law, with $f''(u) > 0$ for every $u \in \mathbb{R}$. Consider a piecewise constant solution $U$ as in (4.7), with $\lambda$ given by (4.39). Prove that the following are equivalent: (i) $U$ is an entropy solution. (ii) The Lax condition (4.38) holds. (iii) $u^+ \leq u^-$.

(2) Consider the scalar conservation law $u_t + (u^4 - u^2)_x = 0$. Show that the function

$$u(t, x) = \bar{u}(x) = \begin{cases} -3/4 & \text{if } x > 0, \\ 3/4 & \text{if } x < 0 \end{cases}$$

provides a weak solution which satisfies the Lax condition (4.38). However, $u$ is not an entropy solution. Indeed, (4.43) fails for $\alpha = 1/2$.

(3) Construct a (non-entropic) weak solution of Burgers' equation (4.24) on the strip $[0, 2] \times \mathbb{R}$ such that

$$u(0, x) \equiv 0, \qquad u(3, x) \equiv 0, \qquad u(1, x) = \begin{cases} 1 & \text{if } x \in \left[k, k + \frac{1}{2}\right[ \\ -1 & \text{if } x \in \left[k - \frac{1}{2}, k\right[ \end{cases} \quad k \in \mathbb{Z}$$

(4) Let $u = u(t, x)$ be an entropy solution of Burgers' equation (4.24). Let $u$ be piecewise Lipschitz continuous in the $t$–$x$ plane, with jumps along the differentiable

curves $\gamma_1, \ldots, \gamma_N$, and assume that $u(t, \cdot)$ has bounded support for every $t$. Given the convex entropy $\eta(u) \doteq u^2/2$, with corresponding flux $q(u) = u^3/3$, show that

$$\frac{d}{dt} \int_{-\infty}^{\infty} \eta(u(t,x))\, dx = \sum_{i=1}^{N} \tfrac{1}{12}[u(t, \gamma_i+) - u(t, \gamma_i-)]^3 < 0.$$

(5) Let $u$ be a piecewise smooth entropy solution of (4.24), having a single shock along the curve $x = \xi(t)$. Call $u^-(t)$ and $u^+(t)$ the left and right states along the shock. Prove the identity

$$\int_{\xi(t)-tu^-(t)}^{\xi(t)-tu^+(t)} u(0,x)\, dx = \frac{t}{2}\big[(u^-)^2 - (u^+)^2\big].$$

Hint: apply the divergence theorem to a suitable triangle with vertex at $(t, \xi(t))$.

(6) Consider the sequence of functions

$$u_n(t,x) = \bar u_n(x) \doteq \begin{cases} 1 & \text{if } x < 0 \text{ or if } x \in \left[\frac{k}{n}, \frac{k+1}{n}\right[, \\ & k = 0, 2, 4, \ldots, 2n-2, \\ -1 & \text{if } x \geq 2 \text{ or if } x \in \left[\frac{k}{n}, \frac{k+1}{n}\right[, \\ & k = 1, 3, 5, \ldots, 2n-1. \end{cases}$$

Show that every $u_n$ is a weak (non-entropic) solution of Burgers' equation (4.24). Moreover, show that there exists the weak limit $u_n \rightharpoonup u$, so that

$$\lim_{n\to\infty} \int u_n \phi\, dx\, dt = \int u \phi\, dx\, dt \quad \text{for all } \phi \in \mathcal{C}_c^0.$$

Check whether $u$ is a weak solution. Compare this example with Lemma 4.1.

(7) Let $u = u(t,x)$ be a weak solution of the system (4.1) having an approximate jump at the point $(\tau, \xi)$. More precisely, assume that (4.15) holds with $U$ as in (4.7). Let $\eta$ be a convex entropy with entropy flux $q$. If $u$ is entropy admissible, show that

$$\lambda\,[\eta(u^+) - \eta(u^-)] \geq q(u^+) - q(u^-).$$

Hint: mimic the proof of Theorem 4.1.

# 5

# The Riemann problem

Let $\Omega \subseteq \mathbb{R}^n$ be an open set and let $f : \Omega \mapsto \mathbb{R}^n$ be a smooth vector field. The *Riemann problem* for the system of conservation laws

$$u_t + f(u)_x = 0 \tag{5.1}$$

consists in finding a weak solution of (5.1) with piecewise constant initial data of the form

$$u(0,x) = \begin{cases} u^- & \text{if } x < 0, \\ u^+ & \text{if } x > 0, \end{cases} \tag{5.2}$$

with $u^-, u^+ \in \Omega$.

As is customary, we denote by $A(u) = Df(u)$ the $n \times n$ Jacobian matrix of partial derivatives of $f$ at the point $u$ so that any smooth solution of (5.1) satisfies

$$u_t + A(u)u_x = 0. \tag{5.3}$$

We now introduce some basic definitions, which will be used throughout the sequel.

**Definition 5.1.** We say that the system (5.1) is *strictly hyperbolic* if, for every $u \in \Omega$, the matrix $A(u)$ has $n$ real distinct eigenvalues $\lambda_1(u) < \cdots < \lambda_n(u)$.

For strictly hyperbolic systems, one can find bases of right and left eigenvectors $\{r_1(u), \ldots, r_n(u)\}$, $\{l_1(u), \ldots, l_n(u)\}$, depending smoothly on $u$, normalized so that

$$|r_i| \equiv 1, \qquad l_i \cdot r_j = \begin{cases} 1 & \text{if } i = j, \\ 0 & \text{if } i \neq j, \end{cases} \tag{5.4}$$

for every $u \in \Omega$. In the following, the directional derivative of a function $\phi = \phi(u)$, in the direction of the vector $r_i(u)$, is written

$$r_i \bullet \phi(u) \doteq D\phi(u) \cdot r_i(u) = \lim_{\varepsilon \to 0} \frac{\phi(u + \varepsilon r_i(u)) - \phi(u)}{\varepsilon}.$$

**Definition 5.2.** For $i \in \{1, \ldots, n\}$, we say that the $i$-th characteristic field is *genuinely non-linear* if

$$r_i \bullet \lambda_i(u) \neq 0 \quad \text{for all } u \in \Omega.$$

If instead

$$r_i \bullet \lambda_i(u) = 0 \quad \text{for all } u \in \Omega,$$

we say that the $i$-th characteristic field is *linearly degenerate*.

Observe that, by continuity of the first derivatives, in the genuinely non-linear case the value of $\lambda_i$ is strictly monotonic (increasing or decreasing) along each integral curve of the vector field $r_i$. By possibly changing the sign of $r_i$ we can thus assume that

$$r_i \bullet \lambda_i(u) > 0 \quad \text{for all } u \in \Omega. \tag{5.5}$$

On the other hand, in the linearly degenerate case the eigenvalue $\lambda_i = \lambda_i(u)$ is constant along every integral curve of $r_i$.

In this chapter we shall construct weak solutions of the Riemann problem (5.1)–(5.2), for $u^-$, $u^+$ sufficiently close, under the following basic hypothesis:

(♣) The system (5.1) is strictly hyperbolic with smooth coefficients, defined on an open set $\Omega \subseteq \mathbb{R}^n$. For each $i \in \{1, \ldots, n\}$, the $i$-th characteristic field is either genuinely non-linear or linearly degenerate.

The solutions will be self-similar, having the form $u(t, x) = \psi(x/t)$, with $\psi : \mathbb{R} \mapsto \mathbb{R}^n$ possibly discontinuous. In the next two sections we discuss two special cases where the solution of the Riemann problem has a very simple form.

## 5.1 Centred rarefaction waves

In the following, by $\sigma \mapsto R_i(\sigma)(u_0)$ we mean the parametrized integral curve of the eigenvector $r_i$ through the point $u_0$. More precisely, $R_i(\sigma)(u_0)$ is the value at time $t = \sigma$ of the solution to the Cauchy problem

$$\frac{du}{dt} = r_i(u(t)), \qquad u(0) = u_0. \tag{5.6}$$

The curve $R_i$ will be called the *$i$-rarefaction curve* through $u_0$. Having defined the curve $R_i$ in terms of the solution of an ODE with smooth coefficients, we have the obvious identities

$$\frac{d}{d\sigma} R_i(\sigma)(u_0) = r_i(R_i(\sigma)(u_0)), \tag{5.7}$$

$$R_i(\sigma')(R_i(\sigma)(u_0)) = R_i(\sigma + \sigma')(u_0), \tag{5.8}$$

for all $u_0, \sigma, \sigma'$.

Now let $u^-$, $u^+$ be given. Assume that there exists a genuinely non-linear characteristic field of eigenvectors $r_i$ satisfying (5.5), and some $\bar{\sigma} \geq 0$ such that

$$u^+ = R_i(\bar{\sigma})(u^-). \tag{5.9}$$

In this case, a solution of the Riemann problem can be constructed as follows. Observe that, by genuine non-linearity, the function

$$\sigma \mapsto \lambda_i(R_i(\sigma)(u^-)) \tag{5.10}$$

is strictly increasing and maps the interval $[0, \bar{\sigma}]$ onto $\big[\lambda_i(u^-),\ \lambda_i(u^+)\big]$. For $t > 0$, we can thus define the piecewise smooth function

$$u(t, x) = \begin{cases} u^- & \text{if } x/t < \lambda_i(u^-), \\ u^+ & \text{if } x/t = \lambda_i(u^+), \\ R_i(\sigma)(u^-) & \text{if } x/t \in [\lambda_i(u^-),\ \lambda_i(u^+)],\quad x/t = \lambda_i(R_i(\sigma)(u^-)). \end{cases} \tag{5.11}$$

$x/t > \lambda_i(u^+)$ ?

We claim that $u$ is a weak solution of (5.1)–(5.2). Indeed, $u(t, \cdot) \to u(0, \cdot)$ in $\mathbf{L}^1_{\text{loc}}$ as $t \to 0+$. Moreover, the equation in (5.3) is trivially satisfied in the regions where $x/t < \lambda_i(u^-)$ or $x/t > \lambda_i(u^+)$. Next, consider the intermediate sector where $x/t \in [\lambda_i(u^-),\ \lambda_i(u^+)]$. By definition, $u$ is constant along each ray through the origin. Hence

$$u_t(t, x) + \frac{x}{t}\, u_x(t, x) = 0. \tag{5.12}$$

By (5.11), in this sector we have $x/t = \lambda_i(u(t, x))$. We now observe that $u_x = (r_i \bullet \lambda_i(u))^{-1} t^{-1} r_i(u)$ is parallel to $r_i(u)$ and hence it is an eigenvector of $A(u)$ with eigenvalue $\lambda_i(u)$. Therefore (5.3) holds also in the region where $\lambda_i(u^-) < x/t < \lambda_i(u^+)$.

A solution of the Riemann problem having the form (5.11) will be called a *centred rarefaction wave*.

**Remark 5.1.** For $t > 0$, the function $u(t, \cdot)$ defined at (5.11) is continuous and piecewise smooth, because of the assumptions $r_i \bullet \lambda_i > 0$ and $\bar{\sigma} \geq 0$. As a consequence, for any entropy $\eta$ with flux $q$, we have

$$\eta(u)_t + q(u)_x = D\eta(u) u_t + Dq(u) u_x = D\eta(u)[u_t + Df(u) u_x] = 0.$$

Hence $u$ is entropy admissible.

On the other hand, if (5.9) holds but $\bar{\sigma} < 0$, then the above construction breaks down. Indeed, one then has $\lambda_i(u^-) > \lambda_i(u^+)$ and the definition (5.11) would yield a triple-valued function on the region where $\lambda_i(u^+) < x/t < \lambda_i(u^-)$.

## 5.2 Shocks

Given the system (5.1) and a left state $u^- \in \Omega$, we wish to describe the family of all states $u^+$ for which the Riemann problem (5.1)–(5.2) has a piecewise constant solution of the form

$$u(t, x) = \begin{cases} u^- & \text{if } x < \lambda t, \\ u^+ & \text{if } x > \lambda t, \end{cases} \tag{5.13}$$

for some $\lambda$. By Lemma 4.2, this set can be determined by solving the Rankine–Hugoniot equations

$$f(u^+) - f(u^-) = \lambda(u^+ - u^-) \tag{5.14}$$

in terms of $u^+, \lambda$. Observe that the vector equation (5.14) is equivalent to a system of $n$ scalar equations in $n + 1$ variables (the $n$ components of $u^+$ together with the scalar quantity $\lambda$). For $u^+$ close to $u^-$, the set of solutions is described by the following theorem, due to Lax (1957).

**Theorem 5.1.** *Let the system (5.1) be strictly hyperbolic. Then, for every* $u_0 \in \Omega$, *there exist* $\sigma_0 > 0$ *and n smooth curves* $S_i : [-\sigma_0, \sigma_0] \mapsto \Omega$, *together with scalar functions* $\lambda_i : [-\sigma_0, \sigma_0] \mapsto \mathbb{R}$ $(i = 1, \ldots, n)$, *such that*

$$f(S_i(\sigma)) - f(u_0) = \lambda_i(\sigma)(S_i(\sigma) - u_0) \quad \sigma \in [-\sigma_0, \sigma_0]. \tag{5.15}$$

*Moreover, the parametrization can be chosen so that* $|dS_i/d\sigma| \equiv 1$ *and, at* $\sigma = 0$, *one has*

$$S_i(0) = u_0, \qquad \lambda_i(0) = \lambda_i(u_0), \tag{5.16}$$

$$\left.\frac{dS_i(\sigma)}{d\sigma}\right|_{\sigma=0} = r_i(u_0), \tag{5.17}$$

$$\left.\frac{d\lambda_i(\sigma)}{d\sigma}\right|_{\sigma=0} = \tfrac{1}{2} r_i \bullet \lambda_i(u_0), \tag{5.18}$$

$$\left.\frac{d^2 S_i(\sigma)}{d\sigma^2}\right|_{\sigma=0} = r_i \bullet r_i(u_0). \tag{5.19}$$

*Proof.* 1. As a preliminary, we introduce some notation. For any two points $u, u' \in \Omega$ such that the segment joining $u$ with $u'$ lies entirely inside $\Omega$, we define the averaged matrix

$$A(u, u') \doteq \int_0^1 A(\theta u + (1-\theta)u')\, d\theta. \tag{5.20}$$

If $u'$ is sufficiently close to $u$, the matrix $A(u, u')$ is close to $A(u)$, and hence it also has $n$ real distinct eigenvalues. Call $\lambda_i(u, u'), r_i(u, u'), l_i(u, u')$ the corresponding eigenvalues and right and left eigenvectors, normalized so that

$$|r_i(u, u')| \equiv 1, \qquad l_i(u, u') \cdot r_j(u, u') = \begin{cases} 1 & \text{if } i = j, \\ 0 & \text{if } i \neq j. \end{cases} \tag{5.21}$$

Clearly $A(u, u') = A(u', u)$, $A(u, u) = A(u)$. Since

$$f(u) - f(u_0) = \int_0^1 Df(\theta u + (1-\theta)u_0)(u - u_0)\, d\theta = A(u, u_0)(u - u_0), \tag{5.22}$$

the Rankine–Hugoniot equations

$$f(u) - f(u_0) = A(u, u_0)(u - u_0) = \lambda(u - u_0) \tag{5.23}$$

hold whenever $u - u_0$ is a right eigenvector of the matrix $A(u, u_0)$ and $\lambda = \lambda_i(u, u_0)$ is the corresponding eigenvalue, for some $i$.

2. For a fixed $i \in \{1, \ldots, n\}$, the $i$-th eigenvectors of $A(u, u_0)$ are precisely the non-trivial solutions of

$$\Phi_j(u) \doteq l_j(u, u_0)(u - u_0) = 0 \quad \text{for all } j \neq i. \tag{5.24$_i$}$$

Observe that the system (5.24)$_i$ consists of $n-1$ scalar equations in the $n$ unknowns $u = (u_1, \ldots, u_n)$. A trivial solution is $u = u_0$. In order to apply the implicit function theorem, it suffices to show that the $(n-1) \times n$ matrix $D\Phi$ computed at $u = u_0$ has rank $n-1$. This is certainly true, because the rows of $D\Phi(u_0)$ are precisely the gradients of the $n-1$ functions $\Phi_j$, $j \neq i$, computed by

$$\nabla\Phi_j(u_0) = l_j(u_0). \tag{5.25}$$

By the implicit function theorem, the system (5.24)$_i$ therefore has a one-dimensional smooth curve of solutions, which we can parametrize by arc-length, say $\sigma \mapsto S_i(\sigma)$, with $S_i(0) = u_0$. This choice of parametrization of course yields $|dS_i/d\sigma| \equiv 1$. Taking $\lambda_i(\sigma) \doteq \lambda_i(S_i(\sigma), u_0)$, we have (5.15) as well as (5.16).

3. A vector $\mathbf{v}$ is tangent to the curve $S_i$ at $u_0$ iff $\mathbf{v}$ is perpendicular to the $n-1$ gradients $\nabla\Phi_j(u_0)$, $j \neq i$. By (5.25), this is the case for the vector $r_i(u_0)$. Therefore, denoting the differentiation w.r.t. $\sigma$ by an upper dot, we have $\dot{S}_i(0) = c\, r_i(u_0)$ for some constant $c$. Observing that $|\dot{S}_i(0)| = |r_i(u_0)| = 1$, by possibly changing the orientation of the curve $S_i$ we achieve (5.17).

4. Set $A_i(\sigma) \doteq A\big(S_i(\sigma)\big)$. Differentiating (5.15) twice w.r.t. $\sigma$ and denoting these derivatives by upper dots, we find

$$\begin{aligned} A_i \dot{S}_i &= \dot{\lambda}_i (S_i - u_0) + \lambda_i \dot{S}_i, \\ \dot{A}_i \dot{S}_i + A_i \ddot{S}_i &= \ddot{\lambda}_i (S_i - u_0) + 2\dot{\lambda}_i \dot{S}_i + \lambda_i \ddot{S}_i. \end{aligned} \tag{5.26}$$

At $\sigma = 0$, $\dot{S}_i(0) = r_i(u_0)$ hence

$$A_i \ddot{S}_i + \dot{A}_i r_i = 2\dot{\lambda}_i r_i + \lambda_i \ddot{S}_i. \tag{5.27}$$

Differentiating the identity

$$A_i(\sigma) r_i(S_i(\sigma)) = \lambda_i(S_i(\sigma)) r_i(S_i(\sigma))$$

w.r.t. $\sigma$, at $\sigma = 0$ we find

$$\dot{A}_i r_i = -A_i (r_i \bullet r_i) + (r_i \bullet \lambda_i) r_i + \lambda_i (r_i \bullet r_i). \tag{5.28}$$

Observe that, in general, $\lambda_i(\sigma) \neq \lambda_i(S_i(\sigma))$. Indeed, by (5.15), $\lambda_i(\sigma)$ is the $i$-th eigenvalue of the averaged matrix $A(S_i(\sigma), u_0)$ defined by (5.20) with $u \doteq S_i(\sigma)$ and $u' \doteq u_0$. On the other hand, $\lambda_i(S_i(\sigma))$ is the $i$-th eigenvalue of $A(S_i(\sigma))$. Of course, when $\sigma = 0$ these two eigenvalues coincide. We can now substitute in (5.27) the expression for $\dot{A}_i r_i$ found in (5.28) and obtain

$$A_i \ddot{S}_i + (r_i \bullet \lambda_i) r_i + (\lambda_i - A_i)(r_i \bullet r_i) = 2\dot{\lambda}_i r_i + \lambda_i \ddot{S}_i. \tag{5.29}$$

Multiplying (5.29) on the left by $l_i(u_0)$, we obtain $r_i \bullet \lambda_i = 2\dot{\lambda}_i$, proving (5.18).

5. Using (5.18), it now follows from (5.29) that, at $\sigma = 0$,

$$(A_i - \lambda_i)(\ddot{S}_i - r_i \bullet r_i) = 0,$$

showing that the vector $\ddot{S}_i - r_i \bullet r_i$ either vanishes, or else is an eigenvector of the matrix $A_i(0) = A(u_0)$ corresponding to the eigenvalue $\lambda_i(u_0)$. For some $\beta \in \mathbb{R}$ we thus have

$$\ddot{S}_i - r_i \bullet r_i = \beta r_i. \tag{5.30}$$

However, recalling that $\dot{S}_i(0) = r_i(u_0)$, $|\dot{S}_i| \equiv |r_i| \equiv 1$, the computation of the inner products

$$\langle r_i, \ddot{S}_i \rangle - \langle r_i, r_i \bullet r_i \rangle = \langle \dot{S}_i, \ddot{S}_i \rangle - \frac{1}{2} r_i \bullet \langle r_i, r_i \rangle = \frac{1}{2}\frac{d}{d\sigma}|\dot{S}_i|^2 - \frac{1}{2} r_i \bullet (|r_i|^2) = 0$$

shows that the left hand side of (5.30) is perpendicular to $r_i$. Hence $\beta = 0$. This proves (5.19). □

**Remark 5.2.** If the $i$-th characteristic field is genuinely non-linear, the orientation of the eigenvector $r_i(u_0)$ is uniquely determined by the inequality $r_i \bullet \lambda_i(u_0) > 0$. In turn, this uniquely determines the parametrization of $S_i$. In the linearly degenerate case, however, the choice of the orientation is arbitrary.

**Remark 5.3.** If the $i$-th characteristic field is linearly degenerate, then the $i$-th shock and rarefaction curves coincide, i.e.

$$S_i(\sigma) = R_i(\sigma)(u_0) \quad \text{for all } \sigma. \tag{5.31}$$

Indeed, in this case one has $\lambda_i(R_i(\sigma)(u_0)) = \lambda_i(u_0)$ for every $\sigma$. Therefore

$$\begin{aligned} f(R_i(\sigma)(u_0)) - f(u_0) &= \int_0^\sigma \left[\frac{d}{ds} f(R_i(s)(u_0))\right] ds \\ &= \int_0^\sigma A(R_i(s)(u_0)) \cdot r_i(R_i(s)(u_0))\, ds \\ &= \int_0^\sigma \lambda_i(R_i(s)(u_0)) \frac{d(R_i(s)(u_0))}{ds}\, ds \\ &= \lambda_i(u_0)\big[R_i(\sigma)(u_0) - u_0\big]. \end{aligned} \tag{5.32}$$

By (5.32), the point $R_i(\sigma)(u_0)$ satisfies the Rankine–Hugoniot conditions; hence it lies on the $i$-th shock curve through $u_0$, as claimed. One should be aware, however, that (5.31) does not hold in the general case, when the map $\sigma \mapsto \lambda_i(R_i(\sigma)(u_0))$ is not constant.

By construction, taking $u_- = u_0, u^+ = S_i(\sigma), \lambda = \lambda_i(\sigma)$, for every $\sigma$ the Rankine–Hugoniot equations (5.14) are satisfied. Therefore, the function

$$u(t,x) = \begin{cases} u_0 & \text{if } x < t\lambda_i(\sigma), \\ S_i(\sigma) & \text{if } x > t\lambda_i(\sigma) \end{cases} \tag{5.33}$$

is a weak solution of the system (5.1). If the $i$-th characteristic field is genuinely non-linear, a solution of the form (5.33) is called a *compressive shock* when $\sigma < 0$ and a *rarefaction shock* when $\sigma > 0$. If the $i$-th characteristic field is linearly degenerate, the solution (5.33) is called a *contact discontinuity*.

At this stage, we must examine whether the piecewise constant solution (5.33) satisfies the various admissibility conditions introduced in Section 4.4. The following result shows that compressive shocks and contact discontinuities satisfy both the Lax and the entropy admissibility conditions, while rarefaction shocks violate them. For a proof that small compressive shocks satisfy the vanishing viscosity admissibility condition, see Foy (1964).

**Theorem 5.2.** *Let the system (5.1) satisfy the basic assumptions (♣). Then all contact discontinuities and all compressive shocks of suitably small strength satisfy the Lax admissibility conditions (4.38), while rarefaction shocks violate them.*

*In addition, let $\eta$ be a smooth entropy for the system (5.1) with flux $q$, and assume that the Hessian $D^2\eta(u_0)$ is strictly positive definite. If the $i$-th characteristic field is linearly degenerate, then for every $\sigma$ the solution (5.33) satisfies the additional conservation law $\eta(u)_t + q(u)_x = 0$ in a distributional sense. On the other hand, if the $i$-th characteristic field is genuinely non-linear, then there exists $\sigma_0 > 0$ such that the solution (5.33) satisfies the entropy inequality (4.36) for all $\sigma \in [-\sigma_0, 0]$, but not for $\sigma \in ]0, \sigma_0]$.*

*Proof.* 1. Consider first the Lax condition. Let $u = u(t,x)$ be a piecewise constant solution of the form (5.33). If the $i$-th field is linearly degenerate, then by (5.31)–(5.32) we have

$$\lambda_i(u_0) = \lambda_i(\sigma) = \lambda_i(R_i(\sigma)(u_0)) = \lambda_i(S_i(\sigma)).$$

Hence, for any value of $\sigma$ (positive or negative), the equality signs hold in (4.38).

On the other hand, if the $i$-th field is genuinely non-linear, by (5.17)–(5.18) we obtain

$$\left.\frac{d\lambda_i(\sigma)}{d\sigma}\right|_{\sigma=0} = \tfrac{1}{2} r_i \bullet \lambda_i(u_0), \qquad \left.\frac{d\lambda_i(S_i(\sigma))}{d\sigma}\right|_{\sigma=0} = r_i \bullet \lambda_i(u_0).$$

Recalling that $r_i \bullet \lambda_i(u_0) > 0$, from the above relations we deduce that

$$\begin{aligned} \lambda_i(u_0) < \lambda_i(\sigma) < \lambda_i(R_i(\sigma)(u_0)) \quad & \text{if } \sigma \in ]0, \sigma_0], \\ \lambda_i(u_0) > \lambda_i(\sigma) > \lambda_i(R_i(\sigma)(u_0)) \quad & \text{if } \sigma \in [-\sigma_0, 0[, \end{aligned}$$

for some $\sigma_0 > 0$ suitably small. This establishes the first part of the theorem.

2. Next, let $\eta : \Omega \mapsto \mathbb{R}$ be a smooth convex entropy with flux $q$. According to (4.37), we have to show that

$$\chi(\sigma) \doteq \lambda_i(\sigma)[\eta(S_i(\sigma)) - \eta(u_0)] - [q(S_i(\sigma)) - q(u_0)] \geq 0 \tag{5.34}$$

when $\sigma \in [-\sigma_0, 0]$.

3. Consider the linearly degenerate case first, where, by (5.32), we have $\lambda_i(\sigma) = \lambda_i(u_0)$ for all $\sigma$. Recalling that $D\eta \cdot Df = Dq$, we obtain

$$\begin{aligned}
&\lambda_i(\sigma)[\eta(S_i(\sigma)) - \eta(u_0)] - [q(S_i(\sigma)) - q(u_0)] \\
&\quad = \int_0^\sigma [\lambda_i(u_0) D\eta(S_i(s)) - Dq(S_i(s))] \frac{dS_i(s)}{ds}\, ds \\
&\quad = \int_0^\sigma [\lambda_i(u_0) D\eta(S_i(s)) - D\eta(S_i(s)) A(S_i(s))] r_i(S_i(s))\, ds \\
&\quad = 0.
\end{aligned}$$

4. Now consider the genuinely non-linear case. Differentiating the left hand side of (5.34) w.r.t. $\sigma$ we obtain

$$\dot\chi(\sigma) = \dot\lambda_i(\sigma)[\eta(S_i(\sigma)) - \eta(u_0)] + \lambda_i(\sigma) D\eta(S_i(\sigma))\dot S_i(\sigma) - Dq(S_i(\sigma))\dot S_i(\sigma)).$$

Using the relation $Dq = D\eta \cdot Df$ and observing that, by (5.26), $\dot\lambda_i(S_i - u_0) = (Df - \lambda_i)\dot S_i$, from the above equality we deduce that

$$\dot\chi(\sigma) = \dot\lambda_i(\sigma)\omega(\sigma),$$

with

$$\omega(\sigma) \doteq \eta(S_i(\sigma)) - \eta(u_0) - D\eta(S_i(\sigma))(S_i(\sigma) - u_0).$$

Hence, $\dot\chi(0) = 0$. A further differentiation yields

$$\ddot\chi = \ddot\lambda_i \omega + \dot\lambda_i \dot\omega,$$

with

$$\dot\omega = D\eta \dot S_i - D^2\eta[\dot S_i \otimes (S_i - u_0)] - D\eta \dot S_i = -D^2\eta[\dot S_i \otimes (S_i - u_0)].$$

Hence, $\ddot\chi(0) = 0$. Finally,

$$\dddot\chi(\sigma) = \dddot\lambda_i\, \omega + 2\ddot\lambda_i \dot\omega + \dot\lambda_i \ddot\omega,$$

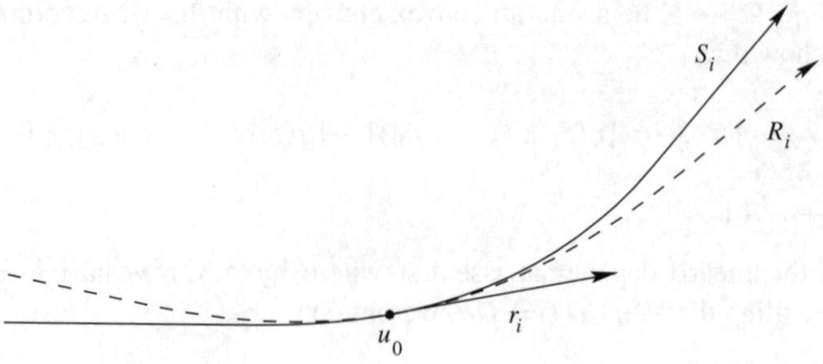


**Figure 5.1**

and hence $\ddot{\chi}(0) = \dot{\lambda}_i(0)\ddot{\omega}(0)$, with

$$\ddot{\omega}(0) = -D^2\eta(u_0)[\dot{S}_i(0) \otimes \dot{S}_i(0)] = -D^2\eta(u_0)[r_i(u_0) \otimes r_i(u_0)].$$

The Taylor approximation for $\chi$ thus takes the form

$$\chi(\sigma) = -\frac{1}{6}\dot{\lambda}_i(0) \cdot D^2\eta(u_0)[r_i(u_0) \otimes r_i(u_0)] \cdot \sigma^3 + \mathcal{O}(\sigma^4).$$

By assumption, the matrix $D^2\eta(u_0)$ is strictly positive definite. Moreover, by (5.18) it follows that $\dot{\lambda}_i(0) = \frac{1}{2} r_i \bullet \lambda_i(u_0) > 0$. Therefore, for some $\sigma_0 > 0$ sufficiently small, we have $\chi(\sigma) \geq 0$ for $\sigma \in [-\sigma_0, 0]$ while $\chi(\sigma) < 0$ for $\sigma \in ]0, \sigma_0]$. □

From now on, we shall use the longer notation $S_i(\sigma)(u_0)$ to indicate points on the $i$-shock curve originating at $u_0$. As a result of the previous analysis, through each point $u_0 \in \Omega$, for every $i \in \{1, \ldots, n\}$ we constructed the two smooth curves (Fig. 5.1)

$$\sigma \mapsto R_i(\sigma)(u_0), \qquad \sigma \mapsto S_i(\sigma)(u_0). \tag{5.35}$$

We shall refer to $R_i, S_i$ as the $i$-rarefaction and the $i$-shock curve through $u_0$, respectively. When both curves are parametrized by arc-length, (5.17) and (5.19) show that these curves have a tangency of second order at the point $u_0$, i.e.

$$R_i(\sigma)(u_0) - S_i(\sigma)(u_0) = \mathcal{O}(1) \cdot \sigma^3. \tag{5.36}$$

In the special case where the $i$-th field is genuinely non-linear, instead of (5.4) one can normalize the eigenvectors $r_i$ so that $r_i \bullet \lambda_i \equiv 1$. In this case, solving the Cauchy problem (5.6) yields a different parametrization of $i$-rarefaction curve $R_i$, say $s \mapsto R_i(s)$, characterized by the property $(d/ds)\lambda_i(R_i(s)) \equiv 1$. Since this alternative parametrization offers some technical advantages, it will be sometimes useful to adopt the following:

*Parametrization choice (♠)*

- If the $i$-th field is linearly degenerate, we choose the right eigenvectors $r_i(u)$ having unit length, and parametrize the (coinciding) shock and rarefaction curves in (5.35) by arc-length.

- If the $i$-th field is genuinely non-linear, we choose the eigenvectors $r_i(u)$ so that $r_i \bullet \lambda_i \equiv 1$. Moreover, we choose the parametrization of the $i$-shock and $i$-rarefaction curves in (5.35) so that

$$\begin{aligned} &\frac{d}{d\sigma}\lambda_i(S_i(\sigma)(u_0)) \equiv 1, \qquad \frac{d}{d\sigma}\lambda_i(R_i(\sigma)(u_0)) \equiv 1, \\ &\lambda_i(S_i(\sigma)(u_0)) = \lambda_i(R_i(\sigma)(u_0)) = \lambda_i(u_0) + \sigma. \end{aligned} \tag{5.37}$$

A straightforward computation shows that, also with this new parametrization (♠), the shock and rarefaction curves have a second-order tangency at $u_0$; hence (5.36) remains valid. As a consequence, the composite function

$$\Psi_i(\sigma)(u_0) = \begin{cases} R_i(\sigma)(u_0) & \text{if } \sigma \geq 0, \\ S_i(\sigma)(u_0) & \text{if } \sigma < 0 \end{cases} \tag{5.38}$$

is smooth for $\sigma \neq 0$, and twice continuously differentiable at $\sigma = 0$. Moreover, its second derivatives are Lipschitz-continuous functions of $\sigma$ and $u_0$. By Remark 5.3, the function $\Psi_i$ is uniquely defined even in the linearly degenerate case, when the orientation of the vector field $r_i$ can be selected arbitrarily.

**Remark 5.4.** A useful property of the parametrization (♠) is the identity

$$u_0 = S_i(-\sigma) \circ S_i(\sigma)(u_0) \quad \text{for all } \sigma, u_0. \tag{5.39}$$

Observe that (5.39) may not hold if the shock curves $S_i$ are parametrized by arc-length. Indeed, if $u = S_i(\sigma)(u_0)$, then the states $u, u_0$ satisfy the Rankine–Hugoniot equations, and hence also $u_0 = S_i(\mu)(u)$ for some $\mu$. However, the set of points on the $i$-shock curve through $u_0$ may not coincide with the set of points on the $i$-shock curve through $u$. Therefore, $\mu \neq -\sigma$ in general.

## 5.3 General solution of the Riemann problem

In the previous sections, we have constructed particular solutions of the Riemann problem, valid in the special cases where the state $u^+$ lies along the $i$-shock or the $i$-rarefaction curve through $u^-$. These cases will now be used as building blocks towards the construction of a solution with general initial data $u^-, u^+$.

Recalling (5.38), for $u^- \in \Omega$ and $(\sigma_1, \ldots, \sigma_n)$ in a neighbourhood of $0 \in \mathbb{R}^n$, we define the composite map

$$\Lambda(\sigma_1, \ldots, \sigma_n)(u^-) \doteq \Psi_n(\sigma_n) \circ \cdots \circ \Psi_1(\sigma_1)(u^-). \tag{5.40}$$

Equivalently, if the points $\omega_0, \ldots, \omega_n$ are inductively defined by

$$\omega_0 = u^-, \qquad \omega_i = \begin{cases} R_i(\sigma_i)(\omega_{i-1}) & \text{if } \sigma_i \geq 0, \\ S_i(\sigma_i)(\omega_{i-1}) & \text{if } \sigma_i < 0, \end{cases} \tag{5.41}$$

then $\Lambda(\sigma_1, \ldots, \sigma_n)(u^-) = \omega_n$.

Now assume that

$$u^+ = \Lambda(\sigma_1, \ldots, \sigma_n)(u^-).$$

By the analysis in the two previous sections, each Riemann problem with initial data

$$u(0,x) = \begin{cases} \omega_{i-1} & \text{if } x < 0, \\ \omega_i & \text{if } x > 0 \end{cases} \tag{5.42$_i$}$$

with $\omega_i$ as in (5.41), has an entropy-admissible solution consisting of a simple wave of the $i$-th characteristic family. More precisely:

CASE 1: The $i$-th characteristic field is genuinely non-linear and $\sigma_i > 0$. Then the solution of $(5.42)_i$ consists of a centred rarefaction wave. Its $i$-th characteristic speeds range over the interval $[\lambda_i^-, \lambda_i^+]$ defined by

$$\lambda_i^- \doteq \lambda_i(\omega_{i-1}), \qquad \lambda_i^+ \doteq \lambda_i(\omega_i). \tag{5.43}$$

CASE 2: Either the $i$-th characteristic field is genuinely non-linear and $\sigma_i \le 0$, or else the $i$-th characteristic field is linearly degenerate (with $\sigma_i$ arbitrary). Then the solution of $(5.42)_i$ consists of an admissible shock or a contact discontinuity, travelling with speed

$$\lambda_i^- \doteq \lambda_i^+ \doteq \lambda_i(\omega_{i-1}, \omega_i). \tag{5.44}$$

We recall that $\lambda_i(u, u')$ denotes the $i$-th eigenvalue of the averaged matrix $A(u, u')$ defined at (5.20).

The solution to the original problem (5.1)–(5.2) can now be constructed by piecing together the solutions of the $n$ Riemann problems $(5.42)_i$, $i = 1, \ldots, n$, on different sectors of the $t$–$x$ plane. Indeed, for $\sigma_1, \ldots, \sigma_n$ sufficiently small, the speeds $\lambda_i^-, \lambda_i^+$ introduced at (5.43) or (5.44) remain close to the corresponding eigenvalues $\lambda_i(u^-)$ of the matrix $A(u^-)$. By strict hyperbolicity and continuity, we can thus assume that the $n$ intervals $[\lambda_i^-, \lambda_i^+]$ are disjoint, i.e.

$$\lambda_1^- \le \lambda_1^+ < \lambda_2^- \le \lambda_2^+ < \cdots < \lambda_n^- \le \lambda_n^+.$$

Therefore, a piecewise smooth function $u : [0, \infty) \times \mathbb{R} \mapsto \mathbb{R}^n$ is well defined by the following assignment:

$$u(t,x) = \begin{cases} u^- & \text{if } x/t \in ]-\infty, \lambda_1^-[, \\ u^+ & \text{if } x/t \in ]\lambda_n^+, \infty[, \\ \omega_i & \text{if } x/t \in ]\lambda_i^+, \lambda_{i+1}^-[, \quad i = 1, \ldots, n-1, \\ R_i(s)(\omega_{i-1}) & \text{if } x/t \in [\lambda_i^-, \lambda_i^+], \quad x/t = \lambda_i(R_i(s)(\omega_{i-1})). \end{cases} \tag{5.45}$$

A typical example is shown in Fig. 5.2. We consider here a $3 \times 3$ system, assuming that the first and the third characteristic fields are genuinely non-linear, while the second is linearly degenerate. The diagram shows the structure of the solution $u = u(t, x)$ in

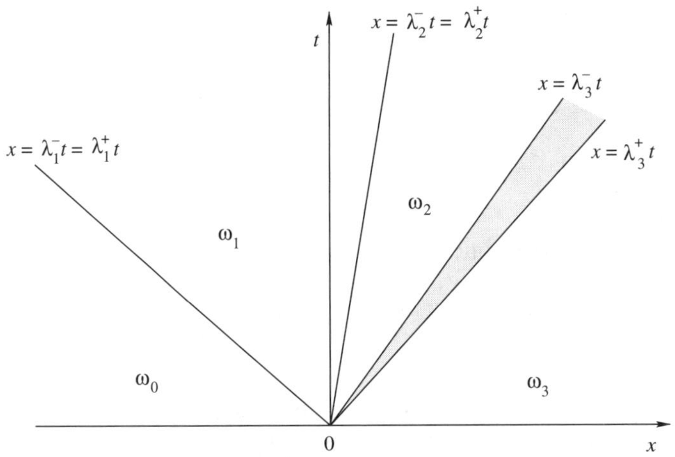


**Figure 5.2**

the case where $\sigma_1 < 0$, $\sigma_2 \neq 0$ and $\sigma_3 > 0$. In this case the solution contains a shock of the first family, travelling with speed $\lambda_1^\pm = \lambda_1(\omega_0, \omega_1)$, and a contact discontinuity of the second family, travelling with speed $\lambda_2^\pm = \lambda_2(\omega_1) = \lambda_2(\omega_2)$. Moreover, inside the sector where $x/t \in [\lambda_3^-, \lambda_3^+] = [\lambda_3(\omega_2), \lambda_3(\omega_3)]$ (shaded area), the solution contains a centred rarefaction wave of the third family.

We can now state the main result of this section.

**Theorem 5.3.** *Let the assumptions (♣) hold. Then, for every compact set $K \subset \Omega$, there exists $\delta > 0$ such that the Riemann problem (5.1)–(5.2) has a unique weak solution of the form (5.45), whenever $u^- \in K$, $|u^+ - u^-| \leq \delta$.*

*Proof.* Let $u^- \in \Omega$ be fixed. The map $\Lambda$ defined at (5.40) is twice continuously differentiable with Lipschitz-continuous second derivatives and satisfies

$$\left.\frac{\partial \Lambda}{\partial \sigma_i}\right|_{\sigma_1 = \cdots = \sigma_n = 0} = r_i(u_0). \tag{5.46}$$

Since the $n$ eigenvectors $r_1(u_0), \ldots, r_n(u_0)$ are linearly independent, by the implicit function theorem the map

$$(\sigma_1, \ldots, \sigma_n) \mapsto \Lambda(\sigma_1, \ldots, \sigma_n)(u^-)$$

is a $\mathcal{C}^2$ homeomorphism of a neighbourhood of $0 \in \mathbb{R}^n$ onto a neighbourhood of $u^-$. By Theorem 2.2 there exists $\delta > 0$ such that, for all $u^- \in K$, if $|u^+ - u^-| \leq \delta$ then $u^+ = \Lambda(\sigma_1, \ldots, \sigma_n)(u^-)$ for some $\sigma_1, \ldots, \sigma_n$. In this case, by Theorem 4.2 and the analysis in the previous sections, the function $u$ in (5.45) provides a weak solution to the Riemann problem. □

**Remark 5.5.** Let $\eta$ be a convex entropy for the system of conservation laws (5.1), with entropy flux $q$. If $u^-$, $u^+$ are sufficiently close, by Theorem 5.2 it follows that the function $u$ defined at (5.45) satisfies $\eta(u)_t + q(u)_x \leq 0$ in a distributional sense. In applications, we thus expect the weak solution (5.45) to be the physically relevant one.

## 5.4 An example

The $2 \times 2$ system of conservation laws

$$[u_1]_t + \left[\frac{u_1}{1+u_1+u_2}\right]_x = 0, \quad [u_2]_t + \left[\frac{u_2}{1+u_1+u_2}\right]_x = 0, \qquad u_1, u_2 > 0 \tag{5.47}$$

is motivated by the study of two-component chromatography. By writing (5.47) in the quasilinear form (5.3), the eigenvalues and (normalized) eigenvectors of $A(u)$ are found to be

$$\lambda_1(u) = \frac{1}{(1+u_1+u_2)^2}, \qquad \lambda_2(u) = \frac{1}{1+u_1+u_2},$$

$$r_1(u) = \frac{1}{\sqrt{u_1^2+u_2^2}} \cdot (-u_1, -u_2), \qquad r_2(u) = \frac{1}{\sqrt{2}} \cdot (1, -1).$$

The first characteristic field is genuinely non-linear, the second is linearly degenerate. In this example, the shock and rarefaction curves $S_i$, $R_i$ always coincide, for $i = 1, 2$. Their computation is straightforward, because they are straight lines (Fig. 5.3):

$$R_1(\sigma)(u) = u + \sigma r_1(u), \qquad R_2(\sigma)(u) = u + \sigma r_2(u). \tag{5.48}$$

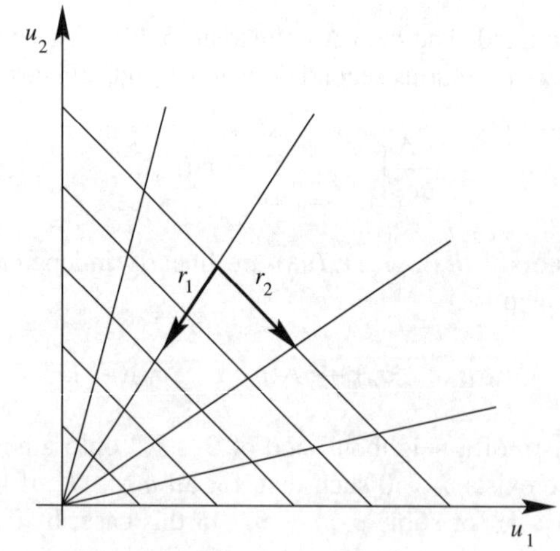


**Figure 5.3**

Observe that the integral curves of the vector field $r_1$ are precisely the rays through the origin, while the integral curves of $r_2$ are the lines with slope $-1$. Now let two states $u^- = (u_1^-, u_2^-)$, $u^+ = (u_1^+, u_2^+)$ be given. To solve the Riemann problem (5.47)–(5.2), we first compute an intermediate state $u^*$ such that $u^* = R_1(\sigma_1)(u^-)$, $u^+ = R_2(\sigma_2)(u^*)$ for some $\sigma_1, \sigma_2$. By (5.48), the components of $u^*$ satisfy

$$u_1^* + u_2^* = u_1^+ + u_2^+, \qquad u_1^* u_2^- = u_1^- u_2^*.$$

The solution of the Riemann problem thus takes two different forms, depending on the sign of $\sigma_1 = \sqrt{(u_1^-)^2 + (u_2^-)^2} - \sqrt{(u_1^*)^2 + (u_2^*)^2}$.

CASE 1: If $\sigma_1 > 0$ then the solution consists of a centred rarefaction wave of the first family and a contact discontinuity of the second family:

$$u(t,x) = \begin{cases} u^- & \text{if } x/t < \lambda_1(u^-), \\ su^* + (1-s)u^- & \text{if } x/t = \lambda_1(su^* + (1-s)u^-), \quad s \in [0,1], \\ u^* & \text{if } \lambda_1(u^*) < x/t < \lambda_2(u^+), \\ u^+ & \text{if } x/t \geq \lambda_2(u^+). \end{cases}$$

CASE 2: If $\sigma_1 \leq 0$ then the solution contains a compressive shock of the first family (which vanishes if $\sigma_1 = 0$) and a contact discontinuity of the second family:

$$u(t,x) = \begin{cases} u^- & \text{if } x/t < \lambda_1(u^-, u^*), \\ u^* & \text{if } \lambda_1(u^-, u^*) \leq x/t < \lambda_2(u^+), \\ u^+ & \text{if } x/t \geq \lambda_2(u^+). \end{cases} \tag{5.49}$$

Observe that $\lambda_2(u^*) = \lambda_2(u^+) = (1 + u_1^+ + u_2^+)^{-1}$, because the second characteristic field is linearly degenerate. In this special case, since the integral curves of $r_1$ are straight lines, the shock speed in (5.49) can be computed as

$$\begin{aligned} \lambda_1(u^-, u^*) &= \int_0^1 \lambda_1(su^* + (1-s)u^-)\, ds \\ &= \int_0^1 [1 + s(u_1^* + u_2^*) + (1-s)(u_1^- + u_2^-)]^{-2}\, ds \\ &= \frac{1}{(1 + u_1^* + u_2^*)(1 + u_1^- + u_2^-)}. \end{aligned}$$

## 5.5 Isentropic gas dynamics

A model for isentropic gas dynamics (in Lagrangian coordinates) is provided by the following $2 \times 2$ hyperbolic system:

$$v_t - u_x = 0, \qquad u_t + p(v)_x = 0. \tag{5.50}$$

Here $v > 0$ is the specific volume, i.e. $v = \rho^{-1}$ where $\rho$ is the density and $u$ is the velocity. The function $p = p(v)$ gives the pressure in terms of the specific volume. It is thus natural to assume that

$$p > 0, \qquad p' < 0, \qquad p'' > 0. \tag{5.51}$$

A typical choice, valid for most gases, is

$$p(v) = \frac{k}{v^\gamma}, \qquad 1 < \gamma < 3.$$

Here $\gamma$ is called the *adiabatic gas constant.*

By introducing the vectors

$$U \doteq (v, u) \qquad F(U) \doteq (-u,\ p(v)),$$

the system (5.50) can be written in the standard form

$$U_t + F(U)_x = 0. \tag{5.52}$$

If $p'(v) < 0$ for all $v$, then the system is strictly hyperbolic. Indeed, its Jacobian matrix

$$A(U) = \begin{pmatrix} 0 & -1 \\ p'(v) & 0 \end{pmatrix}$$

has the two real distinct eigenvalues

$$\lambda_1 = -\sqrt{-p'(v)}, \qquad \lambda_2 = \sqrt{-p'(v)}, \tag{5.53}$$

with corresponding (unnormalized) eigenvectors

$$r_1 = \left(1,\ \sqrt{-p'(v)}\right), \qquad r_2 = \left(-1,\ \sqrt{-p'(v)}\right). \tag{5.54}$$

We now study the Riemann problem for the system (5.52), with initial data

$$U(0, x) = \begin{cases} U^- = (v^-, u^-) & \text{if } x < 0, \\ U^+ = (v^+, u^+) & \text{if } x > 0, \end{cases} \tag{5.55}$$

assuming, of course, that $v^-, v^+ > 0$.

By (5.54), the 1-rarefaction curve through $U^-$ (Fig. 5.4) is obtained by solving the Cauchy problem

$$\frac{du}{dv} = \sqrt{-p'(v)}, \qquad u(v^-) = u^-.$$

This yields the curve

$$\mathbf{R}_1 = \left\{(v, u);\ u - u^- = \int_{v^-}^{v} \sqrt{-p'(y)}\, dy\right\}. \tag{5.56}$$

Similarly, the 2-rarefaction curve through the point $U^-$ is

$$\mathbf{R}_2 = \left\{(v, u);\ u - u^- = -\int_{v^-}^{v} \sqrt{-p'(y)}\, dy\right\}. \tag{5.57}$$

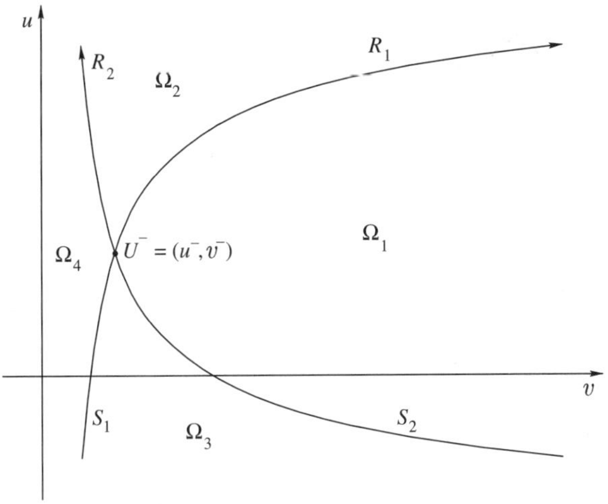


**Figure 5.4**

Next, the shock curves $S_1$, $S_2$ through $U^-$ are derived from the Rankine–Hugoniot conditions

$$\lambda(v - v^-) = -(u - u^-), \qquad \lambda(u - u^-) = p(v) - p(v^-). \tag{5.58}$$

Using the first equation in (5.58) to eliminate $\lambda$, these shock curves are computed as

$$\mathbf{S}_1 = \left\{ (v, u);\ -(u - u^-)^2 = (v - v^-)(p(v) - p(v^-)),\ \lambda \doteq -\frac{u - u^-}{v - v^-} < 0 \right\}, \tag{5.59}$$

$$\mathbf{S}_2 = \left\{ (v, u);\ -(u - u^-)^2 = (v - v^-)(p(v) - p(v^-)),\ \lambda \doteq -\frac{u - u^-}{v - v^-} > 0 \right\}. \tag{5.60}$$

Recalling (5.53)–(5.54) and the assumptions (5.51), we now compute the directional derivatives

$$r_1 \bullet \lambda_1 = r_2 \bullet \lambda_2 = \frac{p''(v)}{2\sqrt{-p'(v)}} > 0. \tag{5.61}$$

From (5.61) it is clear that the Riemann problem (5.55) admits a solution in the form of a centred rarefaction wave in the two cases $U^+ \in \mathbf{R}_1$, $v^+ > v^-$, or else $U^+ \in \mathbf{R}_2$, $v^+ < v^-$. On the other hand, a shock connecting $U^-$ with $U^+$ will be admissible provided that either $U^+ \in \mathbf{S}_1$ and $v^+ < v^-$, or else $U^+ \in \mathbf{S}_2$ and $v^+ > v^-$.

Taking the above admissibility conditions into account, we thus obtain four parametrized curves (Fig. 5.5) originating from the point $U^- = (v^-, u^-)$, i.e. the two rarefaction curves

$$\sigma \mapsto R_1(\sigma)(U^-), \quad \sigma \mapsto R_2(\sigma)(U^-) \qquad \sigma \geq 0,$$

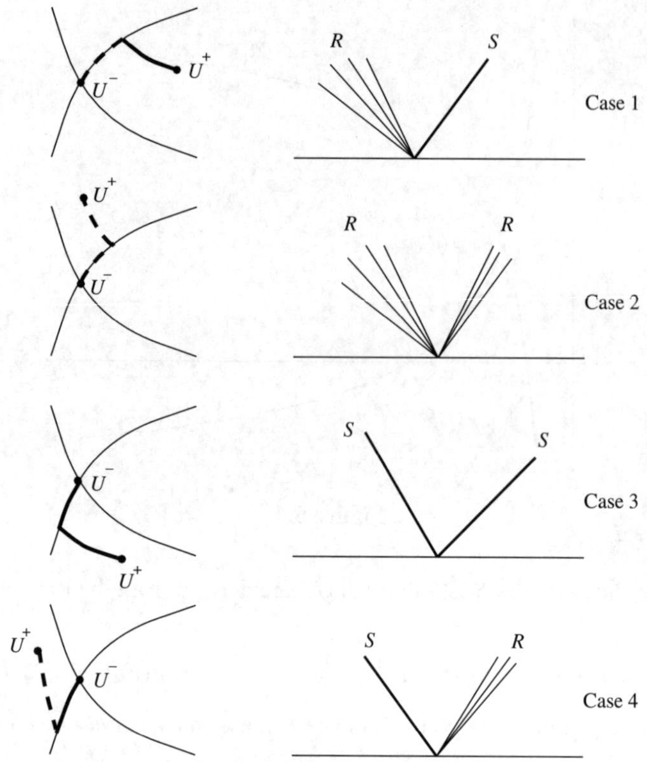


**Figure 5.5**

and the two shock curves

$$\sigma \mapsto S_1(\sigma)(U^-), \quad \sigma \mapsto S_2(\sigma)(U^-) \qquad \sigma \leq 0.$$

In turn, these curves divide a neighbourhood of $U^-$ into four regions (see Fig. 5.4):

$$\Omega_1, \text{ bordering on } R_1, S_2, \quad \Omega_2, \text{ bordering on } R_1, R_2,$$
$$\Omega_3, \text{ bordering on } S_1, S_2, \quad \Omega_4, \text{ bordering on } S_1, R_2$$

For $U^+$ sufficiently close to $U^-$, the structure of the general solution to the Riemann problem (5.52)–(5.55) is now determined by the location of the state $U^+$, with respect to the curves $R_i$, $S_i$ (see Fig. 5.5).

CASE 1: $U^+ \in \Omega_1$. The solution consists of a 1-rarefaction and a 2-shock wave.

CASE 2: $U^+ \in \Omega_2$. The solution consists of two centred rarefaction waves.

CASE 3: $U^+ \in \Omega_3$. The solution consists of two shocks.

CASE 4: $U^+ \in \Omega_4$. The solution consists of a 1-shock and a 2-rarefaction wave.

## Problems

(1) Consider the system (5.1). For a given $i \in \{1, \ldots, n\}$, assume that $r_i \bullet r_i(u) \equiv 0$ for all $u \in \Omega$. Show that, for every $u_0 \in \Omega$, the $i$-shock and the $i$-rarefaction curves in (5.35) are straight lines and coincide. If $u, u'$ lie on the same curve of the $i$-th characteristic field, i.e. $u' = u + \sigma r_i(u)$ for some $\sigma$, prove that the $i$-th eigenvector of the averaged matrix $A(u, u')$ is

$$\lambda_i(u, u') = \int_0^1 \lambda_i(\theta u' + (1-\theta)u)\, d\theta.$$

(2) For the system (5.1), assume that the $i$-th field is genuinely non-linear and normalize the $i$-th eigenvectors so that $r_i \bullet \lambda_i \equiv 1$. At a given point $u_0$, assume that the vectors $r_i(u_0)$ and $r_i \bullet r_i(u_0)$ are linearly independent. Show that the $i$-shock and $i$-rarefaction curves through $u_0$ do not coincide.

Hint: call $\sigma \mapsto R_i(\sigma)$ the $i$-rarefaction curve through the point $u_0$, parametrized as in (5.37). Consider the wedge product

$$\chi(\sigma) \doteq (f(R_i(\sigma)) - f(u_0)) \wedge (R_i(\sigma) - u_0).$$

Using the identity $(d/d\sigma) f(R_i(\sigma)) = (\lambda_i(u_0) + \sigma)\dot{R}_i(\sigma)$ and the skew symmetry of the wedge product, prove that

$$\left.\frac{d^4\chi(\sigma)}{d\sigma^4}\right|_{\sigma=0} = 2\dot{R}_i(0) \wedge \ddot{R}_i(0) = 2r_i(u_0) \wedge r_i \bullet r_i(u_0).$$

(3) In addition to (5.1), consider a second system

$$u_t + \tilde{f}(u)_x = 0. \tag{5.62}$$

Assume that both systems are strictly hyperbolic, with each characteristic field either linearly degenerate or genuinely non-linear. Given two states $u^-, u^+ \in \mathbb{R}^n$, let $u$, $\tilde{u}$ be respectively the solutions of (5.1) and (5.62), both with the same initial data (5.2). Prove the estimate

$$\frac{1}{t}\int_{-\infty}^{\infty} |u(t,x) - \tilde{u}(t,x)|\, dx = \mathcal{O}(1) \cdot |u^+ - u^-| \cdot \|f - \tilde{f}\|_{C^1} \quad \text{for all } t > 0.$$

# 6

# The single conservation law

This chapter is concerned with the Cauchy problem for a scalar conservation law

$$u_t + f(u)_x = 0, \tag{6.1}$$

$$u(0, \cdot) = \bar{u}, \tag{6.2}$$

assuming that $f : \mathbb{R} \mapsto \mathbb{R}$ is locally Lipschitz continuous and that $\bar{u} \in \mathbf{L}^1_{\text{loc}}$. Recalling the analysis in the last section of Chapter 4, we define an *entropy solution* of (6.1)–(6.2) as a continuous map $u : [0, \infty) \mapsto \mathbf{L}^1_{\text{loc}}(\mathbb{R})$ which satisfies (6.2) together with

$$\iint \{|u - k|\phi_t + (f(u) - f(k))\text{sgn}(u - k)\phi_x\}\, dx\, dt \geq 0, \tag{6.3}$$

for every $k \in \mathbb{R}$ and every non-negative function $\phi \in \mathcal{C}^1_c(\mathbb{R}^2)$, whose compact support is contained in the half plane where $t > 0$. In (6.3) we implicitly assume that both $u$ and $f(u)$ are locally integrable on the half plane $[0, \infty[\, \times \mathbb{R}$. The inequality (6.3) means that

$$\eta(u)_t + q(u)_x \leq 0 \tag{6.4}$$

for every entropy of the form $\eta(u) = |u - k|$, with entropy flux $q(u) \doteq (f(u) - f(k))\text{sgn}(u - k)$. If the function $u$ is bounded, choosing $k < \inf u(t, x)$, it follows from (6.3) that

$$\iint \{u\phi_t + f(u)\phi_x\}\, dx\, dt \geq 0$$

for every $\phi \geq 0$ with support in the half plane where $t > 0$. Choosing $k > \sup u(t, x)$ we obtain the opposite inequality. Hence (6.3) implies that $u$ is a distributional solution of (6.1).

The existence of solutions to (6.1)–(6.2) will be proved by the method of *wave-front tracking*. For given initial data $\bar{u} \in \mathbf{L}^1$, we will construct a sequence $(u_\nu)_{\nu \geq 1}$ of piecewise constant approximate solutions, with $u_\nu(0, \cdot) \to \bar{u}$. As $\nu \to \infty$, a compactness argument will yield a subsequence $(u_\mu)_{\mu \geq 1}$ converging in $\mathbf{L}^1_{\text{loc}}$ to an entropy solution.

The uniqueness and continuous dependence will then be proved by showing that, for any two bounded entropy solutions $u, v$ of (6.1), one has

$$\int_{-\infty}^{\infty} |u(t, x) - v(t, x)|\, dx \leq \int_{-\infty}^{\infty} |u(0, x) - v(0, x)|\, dx$$

for every $t \geq 0$. In other words, the flow generated by a scalar conservation law is contractive w.r.t. the $\mathbf{L}^1$ distance.

## 6.1 Piecewise constant approximations

Fix an integer $\nu \geq 1$ and let $f_\nu$ be the piecewise affine function which coincides with $f$ at all nodes $2^{-\nu}j$ with $j$ integer, i.e.

$$f_\nu(s) = \frac{s - 2^{-\nu}j}{2^{-\nu}} \cdot f(2^{-\nu}(j+1)) + \frac{2^{-\nu}(j+1) - s}{2^{-\nu}} \cdot f(2^{-\nu}j)$$
$$s \in [2^{-\nu}j, 2^{-\nu}(j+1)]. \tag{6.5}$$

Let $\bar{u}$ be a piecewise constant function with compact support, taking values inside the discrete set $2^{-\nu}\mathbb{Z} \doteq \{2^{-\nu}j;\ j \text{ integer}\}$. We will show that the Cauchy problem

$$u_t + [f_\nu(u)]_x = 0 \tag{6.6}$$

with initial data $\bar{u}$ admits a piecewise constant entropy-admissible solution $u = u(t,x)$, still taking values within the discrete set $2^{-\nu}\mathbb{Z}$. As a preliminary, consider a Riemann problem for (6.6), with initial data

$$u(0,x) = \begin{cases} u^- & \text{if } x < 0, \\ u^+ & \text{if } x > 0, \end{cases} \qquad u^-, u^+ \in 2^{-\nu}\mathbb{Z}. \tag{6.7}$$

The functions $f^*$, $f_*$ constructed below are illustrated in Fig. 6.1.

CASE 1: $u^- < u^+$. Let $f_*$ be the largest convex function such that

$$f_*(s) \leq f_\nu(s) \quad \text{for all } s \in [u^-, u^+].$$

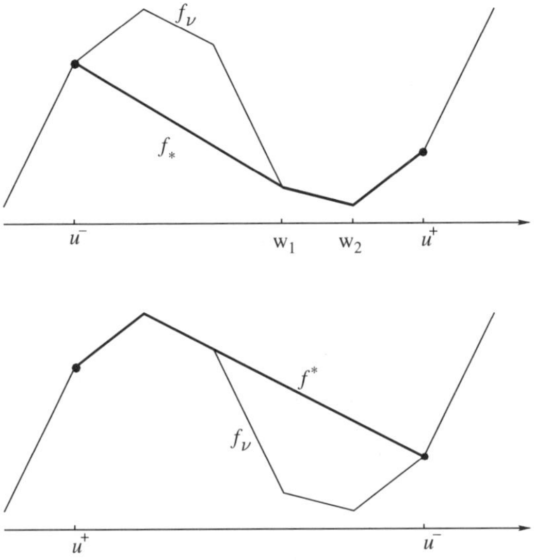


**Figure 6.1**

Observe that $f_*$ is piecewise linear, being the convex hull of a piecewise linear function. By convexity, its derivative $f_*'$ is a piecewise constant non-decreasing function, say with jumps at the points $w_0 \doteq u^- < w_1 < \cdots < w_q \doteq u^+$. We define the increasing sequence of shock speeds as

$$\lambda_\ell = \frac{f_\nu(w_\ell) - f_\nu(w_{\ell-1})}{w_\ell - w_{\ell-1}} \qquad \ell = 1, \ldots, q. \tag{6.8}$$

We claim that the function

$$\omega(t,x) \doteq \begin{cases} u^- & \text{if } x < t\lambda_1, \\ w_\ell & \text{if } t\lambda_\ell < x < t\lambda_{\ell+1} \ (1 \le \ell \le q-1), \\ u^+ & \text{if } t\lambda_q < x, \end{cases} \tag{6.9}$$

provides a weak, entropy-admissible solution of the Riemann problem (6.6)–(6.7). Indeed, fix any constant $k$ and any $\mathcal{C}^1$ function $\phi \ge 0$ with compact support contained in the half plane where $t > 0$. We define the characteristic function

$$\chi_{[w_{\ell-1}, w_\ell]}(k) \doteq \begin{cases} 1 & \text{if } k \in [w_{\ell-1}, w_\ell], \\ 0 & \text{if } k \notin [w_{\ell-1}, w_\ell]. \end{cases}$$

From the above construction it follows that

$$\begin{aligned} &\iint \{|\omega - k|\phi_t + (f_\nu(\omega) - f_\nu(k))\operatorname{sgn}(\omega - k)\phi_x\}\, dx\, dt \\ &\quad = \sum_{\ell=1}^{q} \int \{(|w_\ell - k| - |w_{\ell-1} - k|)\lambda_\ell - (f_\nu(w_\ell) - f_\nu(k))\operatorname{sgn}(w_\ell - k) \\ &\qquad + (f_\nu(w_{\ell-1}) - f_\nu(k))\operatorname{sgn}(w_{\ell-1} - k)\}\phi(t, t\lambda_\ell)\, dt \\ &\quad = \sum_{\ell=1}^{q} \int [(w_\ell + w_{\ell-1} - 2k)\lambda_\ell + 2f_\nu(k) - f_\nu(w_\ell) - f_\nu(w_{\ell-1})] \\ &\qquad \cdot \chi_{[w_{\ell-1}, w_\ell]}(k) \cdot \phi(t, t\lambda_\ell)\, dt \\ &\quad \ge 0. \end{aligned}$$

Indeed, for $k \in [w_{\ell-1}, w_\ell]$ the definition of $f_*$ implies

$$2f_\nu(k) \ge 2f_*(k) = [f_\nu(w_\ell) + (k - w_\ell)\lambda_\ell] + [f_\nu(w_{\ell-1}) + (k - w_{\ell-1})\lambda_\ell].$$

CASE 2: $u^- > u^+$. Let $f^*$ be the smallest concave function such that

$$f^*(s) \ge f_\nu(s) \quad \text{for all } s \in [u^+, u^-].$$

The derivative of $f^*$ is then a piecewise constant, non-increasing function, say with jumps at the points $w_0 \doteq u^+ < w_1 < \cdots < w_q \doteq u^-$. Letting the shock speeds $\lambda_\ell$ be as in (6.8), the function

$$\omega(t,x) \doteq \begin{cases} u^- & \text{if } x < t\lambda_q, \\ w_\ell & \text{if } t\lambda_{\ell+1} < x < t\lambda_\ell \ (1 \le \ell \le q-1), \\ u^+ & \text{if } t\lambda_1 < x \end{cases}$$

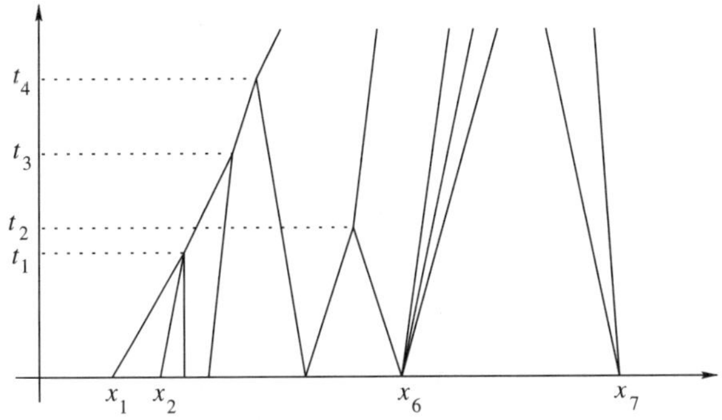


**Figure 6.2**

again provides an entropy solution to the Riemann problem (6.6)–(6.7). Observe that, in the above construction, all values $w_\ell$ lie within the set $2^{-\nu}\mathbb{Z}$.

Next, consider the more general Cauchy problem (6.6)–(6.2), still assuming that the initial condition $\bar{u}$ is piecewise constant, taking values within the set $2^{-\nu}\mathbb{Z}$. Let $x_1 < \cdots < x_N$ be the points where $\bar{u}$ has a jump. At each $x_i$, consider the left and right limits $\bar{u}(x_i-), \bar{u}(x_i+) \in 2^{-\nu}\mathbb{Z}$. Solving the corresponding Riemann problems, we thus obtain a local solution $u = u(t, x)$, defined for $t > 0$ sufficiently small. This solution can be prolonged up to a first time $t_1 > 0$ where two or more lines of discontinuity (emerging from different Riemann problems at $t = 0$) cross each other (Fig. 6.2). Since the values of $u(t, \cdot)$ always remain within the set $2^{-\nu}\mathbb{Z}$, we can again solve the new Riemann problems generated by the interactions, according to the above procedure. The solution is then prolonged up to a time $t_2 > t_1$ where a second set of wave-front interactions take place, and so on.

We claim that the total number of interactions is finite, and hence that the solution can be prolonged for all $t \geq 0$. Indeed, let

$$\xi_1(t) < \cdots < \xi_m(t) \quad (t < \tau) \tag{6.10}$$

be the locations of $m$ discontinuities, which interact all together at some time $\tau$. For $t < \tau$, let $u_0, u_1, \ldots, u_m$ be the constant values taken by $u$ and consider the jumps

$$u_i - u_{i-1} = u(t, \xi_i(t)+) - u(t, \xi_i(t)-) \quad i = 1, \ldots, m. \tag{6.11}$$

Two cases can occur.

CASE 1: All jumps in (6.11) have the same sign. In this case (Fig. 6.3), we claim that the Riemann problem determined by the interaction is solved by a single jump, connecting $u_0$ with $u_m$. To fix the ideas, assume $u_0 < u_1 < \cdots < u_m$, the opposite case being entirely similar. By construction, all the incoming fronts are entropy admissible so that

$$\dot{\xi}_i = \frac{f_\nu(u_i) - f_\nu(u_{i-1})}{u_i - u_{i-1}},$$

$$f_\nu(s) \geq \frac{s - u_{i-1}}{u_i - u_{i-1}} \cdot f_\nu(u_i) + \frac{u_i - s}{u_i - u_{i-1}} \cdot f_\nu(u_{i-1}) \quad \text{for all } s \in [u_{i-1}, u_i].$$

Moreover, since all fronts $\xi_i$ meet at the same point, (6.10) clearly implies $\dot\xi_1 > \cdots > \dot\xi_m$. From the above relations we deduce that

$$f_\nu(s) \geq \frac{s - u_0}{u_m - u_0} \cdot f_\nu(u_m) + \frac{u_m - s}{u_m - u_0} \cdot f_\nu(u_0) \quad \text{for all } s \in [u_0,\ u_m].$$

Therefore, the single jump $(u_0, u_m)$, travelling with speed

$$\dot\xi = \frac{f_\nu(u_m) - f_\nu(u_0)}{u_m - u_0},$$

is entropy admissible, proving our claim.

We conclude that, in this case, the total variation of $u(t, \cdot)$ does not change as a consequence of the interaction. Moreover, the number of lines where $u$ is discontinuous decreases at least by 1.

CASE 2: At least two of the jumps in (6.11) have opposite signs. In this case, the total number of wave-fronts may increase through the interaction. However, since the total strength of the outgoing fronts is given by $|u_m - u_0|$, the total variation of the solution must decrease by at least $2 \cdot 2^{-\nu}$, owing to a cancellation effect.

Figure 6.4 shows an example with two incoming fronts (i.e. $m = 2$) having opposite signs. The Riemann problem determined by the interaction is solved by three outgoing

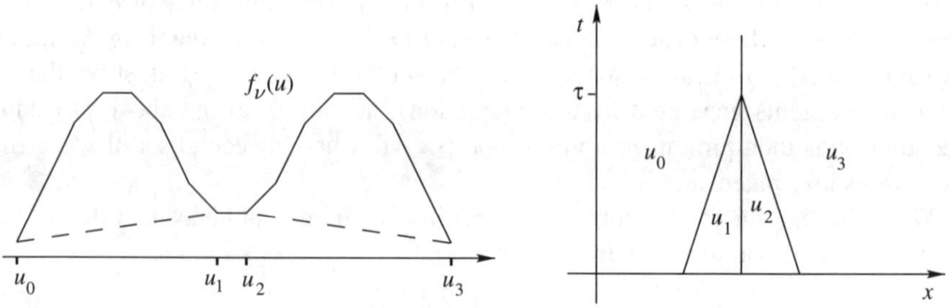


**Figure 6.3**

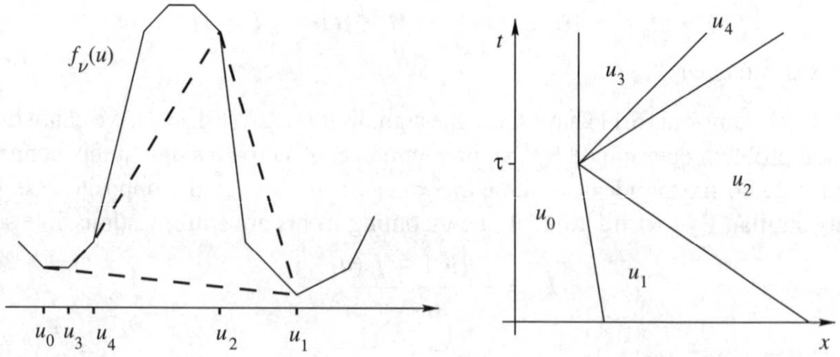


**Figure 6.4**

fronts, connecting the states $u_0 < u_3 < u_4 < u_2$. Observe that this configuration is possible owing to the particular shape of the function $f$, with two inflection points.

Since the total variation of $u(t, \cdot)$ is bounded when $t = 0$ and never increases, CASE 2 can occur only finitely many times, and hence also CASE 1. This proves that the total number of interactions is finite. The above method of wave-front tracking thus defines a piecewise constant solution to (6.6)–(6.2), with jumps occurring along a finite number of straight lines in the $t$–$x$ plane.

## 6.2 Global existence of *BV* solutions

Relying on a compactness argument, we prove here an intermediate result concerning the global existence of entropy weak solutions, within a class of functions with bounded variation.

**Theorem 6.1.** *Let $f$ be locally Lipschitz continuous and let $\bar{u} \in \mathbf{L}^1$ have bounded variation. Then the Cauchy problem (6.1)–(6.2) admits an entropy weak solution $u = u(t, x)$, defined for all $t \geq 0$, with*

$$\text{Tot. Var.}\{u(t,\cdot)\} \leq \text{Tot. Var.}\{\bar{u}\}, \qquad \|u(t,\cdot)\|_{\mathbf{L}^\infty} \leq \|\bar{u}\|_{\mathbf{L}^\infty} \qquad \textit{for all } t \geq 0. \tag{6.12}$$

*Proof.* Call $M \doteq \|\bar{u}\|_{\mathbf{L}^\infty}$. Recalling Lemma 2.2, we construct a sequence $(\bar{u}_\nu)_{\nu\geq 1}$ of piecewise constant functions such that

(i) $\bar{u}_\nu(x) \in 2^{-\nu}\mathbb{Z}$, for all $x$,
(ii) $\|\bar{u}_\nu - \bar{u}\|_{\mathbf{L}^1} \to 0$,
(iii) Tot. Var. $\{\bar{u}_\nu\} \leq$ Tot. Var. $\{\bar{u}\}$,
(iv) $\|\bar{u}_\nu\|_{\mathbf{L}^\infty} \leq M$.

For each $\nu$, let $u_\nu = u_\nu(t, x)$ be the piecewise constant entropy solution of the conservation law (6.6) with initial data $u_\nu(0, \cdot) = \bar{u}_\nu$, constructed by the front tracking algorithm in the previous section. Observe that from (iii) and (iv) it follows that

$$\text{Tot. Var.}\{u_\nu(t,\cdot)\} \leq \text{Tot. Var.}\{\bar{u}\}, \qquad |u_\nu(t,x)| \leq M, \tag{6.13}$$

for all $\nu, t, x$. Let $L$ be a Lipschitz constant such that

$$|f(w) - f(w')| \leq L|w - w'| \quad \text{for all } w, w' \in [-M, M].$$

Clearly, $L$ provides a Lipschitz constant also for all functions $f_\nu$, on the interval $[-M, M]$. By (6.8), this implies that the speed of all discontinuities in $u_\nu(t, \cdot)$ is bounded by $L$. Using the bound (6.13) on the total variation, for every $t, t' \geq 0$ one obtains

$$\|u_\nu(t,\cdot) - u_\nu(t',\cdot)\|_{\mathbf{L}^1} \leq L|t - t'| \cdot \text{Tot. Var.}\{\bar{u}\}. \tag{6.14}$$

We can thus apply Theorem 2.4 and deduce the existence of a subsequence $(u_\mu)_{\mu\geq 1}$ which converges to some function $u$ in $\mathbf{L}^1_{\text{loc}}([0, \infty[\, \times\mathbb{R})$. Clearly (6.13) implies (6.12). Observing that the convergence $f_\mu \to f$ is uniform on the interval $[-M, M]$

and recalling that each $u_\mu$ is an entropy solution of (6.6) (with $\nu$ replaced by $\mu$), we obtain

$$\begin{aligned}
&\iint \{|u-k|\phi_t + (f(u)-f(k))\text{sgn}(u-k)\phi_x\}\,dx\,dt \\
&\quad = \lim_{\mu\to\infty} \iint \{|u_\mu - k|\phi_t + (f_\mu(u_\mu) - f_\mu(k))\text{sgn}(u_\mu - k)\phi_x\}\,dx\,dt \\
&\quad \geq 0,
\end{aligned}$$

for every $C^1$ function $\phi \geq 0$ with compact support contained in the half plane where $t > 0$. This proves that $u$ is an entropy weak solution of (6.1). Finally, (6.14) and property (ii) of the approximating sequence imply that the initial condition (6.2) is attained. □

## 6.3 Uniqueness

The goal of this section is to prove the classical theorem of Kruzhkov, providing an estimate of the $\mathbf{L}^1$ distance between any two bounded entropy-admissible solutions of (6.1). In particular, we will show that the entropy solution of the Cauchy problem is unique, within a class of $\mathbf{L}^\infty$ functions.

**Theorem 6.2 (Kruzhkov).** *Let $f : \mathbb{R} \mapsto \mathbb{R}$ be locally Lipschitz continuous. Let $u, v$ be entropy-admissible solutions of (6.1) defined for $t \geq 0$, and let $M, L$ be constants such that*

$$|u(t,x)| \leq M, \quad |v(t,x)| \leq M \qquad \textit{for all } t, x, \tag{6.15}$$

$$|f(w) - f(w')| \leq L|w - w'| \quad \textit{for all } w, w' \in [-M, M]. \tag{6.16}$$

*Then, for every $R > 0$ and $\tau \geq \tau_0 \geq 0$, one has*

$$\int_{|x|\leq R} |u(\tau,x) - v(\tau,x)|\,dx \leq \int_{|x|\leq R+L(\tau-\tau_0)} |u(\tau_0,x) - v(\tau_0,x)|\,dx. \tag{6.17}$$

*Proof.* To help the reader, we first give an intuitive sketch of the main arguments. Consider the trapezoid (Fig. 6.5)

$$\Omega \doteq \{(t,x);\ \tau_0 \leq t \leq \tau,\ |x| \leq R + L(\tau - t)\}. \tag{6.18}$$

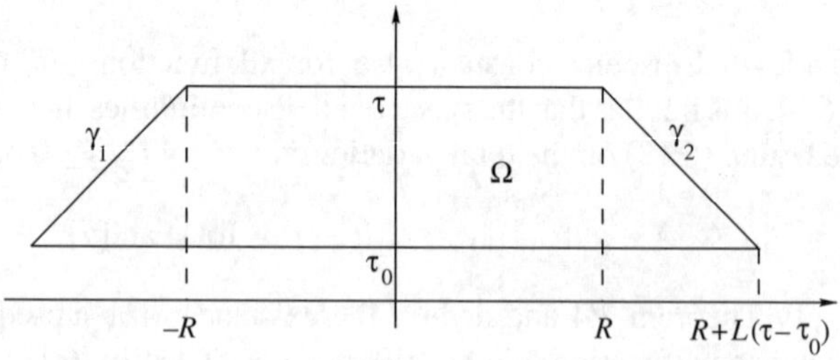


**Figure 6.5**

If $u$ is an entropy solution, we can apply the divergence theorem to the vector field $\Phi \doteq (\eta(u),\ q(u))$ on the domain $\Omega$. Using the inequality (6.4), we formally obtain

$$
\begin{aligned}
0 &\geq \iint_{\Omega} \{\eta(u)_t + q(u)_x\}\,dx\,dt \\
&= \int_{-R}^{R} \eta(u(\tau, x))\,dx - \int_{-R-L(\tau-\tau_0)}^{R+L(\tau-\tau_0)} \eta(u(\tau_0, x))\,dx \\
&\quad + \int_{\tau_0}^{\tau} \{L\,\eta(u(t, \gamma_1(t))) - q(u(t, \gamma_1(t)))\}\,dt + \int_{\tau_0}^{\tau} \{L\,\eta(u(t, \gamma_2(t))) + q(u(t, \gamma_2(t)))\}\,dt.
\end{aligned}
\tag{6.19}
$$

As in Fig. 6.5, the lines

$$\gamma_1(t) \doteq -R - L(\tau - t), \qquad \gamma_2(t) \doteq R + L(\tau - t)$$

represent the two sides of $\Omega$. Observe that (6.19) is valid for every entropy $\eta(u) = |u-k|$, with entropy flux $q(u) \doteq (f(u) - f(k))\text{sgn}(u - k)$. By the assumption (6.16), $f$ is Lipschitz continuous with constant $L$. As a consequence, the last two integrals on the right hand side of (6.19) are both $\geq 0$. From (6.19) we thus obtain the inequality

$$\int_{|x|\leq R} |u(\tau, x) - k|\,dx \leq \int_{|x|\leq R+L(\tau-\tau_0)} |u(\tau_0, x) - k|\,dx, \tag{6.20}$$

valid for every $k \in \mathbb{R}$. Observe that (6.20) gives precisely the estimate (6.17) that we are looking for, in the special case where $v(t, x) \equiv k$ is a constant function.

Motivated by the previous analysis, the proof of the theorem will thus consist of two parts:

(i) Show that the inequality in (6.3) remains valid if the constant $k$ is replaced by any entropy solution $v = v(t, x)$.

(ii) Make rigorous the formal derivation at (6.19).

To achieve (i), we first consider two entropy solutions $u = u(s, x)$ and $v = v(t, y)$ acting on distinct independent variables. The corresponding entropy inequalities can then be written on the product space $\mathbb{R}^2 \times \mathbb{R}^2$, with variables $(s, x, t, y)$. At this stage, the trick is to use test functions $\phi = \phi(s, x, t, y)$ which concentrate most of the mass along the diagonal where $s = t$ and $x = y$. By taking the limit over a suitable sequence of these test functions we shall obtain (6.27), providing the desired extension of (6.3). To achieve (ii), we shall use (6.27) in connection with test functions which approximate the characteristic function of the domain $\Omega$. We can now begin with the actual details of the proof.

1. Let $u, v$ be entropy solutions of (6.1). Given any constants $k, k' \in \mathbb{R}$ and any smooth function $\phi = \phi(s, x, t, y) \geq 0$ with compact support contained in the set where $s, t > 0$,

by assumption one has

$$\iint \{|u(s,x)-k|\phi_s(s,x,t,y)+\operatorname{sgn}(u(s,x)-k)(f(u(s,x)) - f(k))\phi_x(s,x,t,y)\}\,dx\,ds \geq 0, \tag{6.21}$$

$$\iint \{|v(t,y)-k'|\phi_t(s,x,t,y)+\operatorname{sgn}(v(t,y)-k')(f(v(t,y)) - f(k'))\phi_y(s,x,t,y)\}\,dy\,dt \geq 0. \tag{6.22}$$

Set $k = v(t, y)$ in (6.21) and integrate w.r.t. $t, y$. Then set $k' = u(s, x)$ in (6.22) and integrate w.r.t. $s, x$. Adding the two results, one obtains

$$\iiiint \{|u(s,x)-v(t,y)|(\phi_s+\phi_t)(s,x,t,y)+[f(u(s,x))-f(v(t,y))] \cdot [(\phi_x+\phi_y)(s,x,t,y)]\operatorname{sgn}(u(s,x)-v(t,y))\}\,dx\,dy\,ds\,dt \geq 0. \tag{6.23}$$

2. Now choose a sequence of functions $(\delta_h)_{h\geq 1}$, approximating the Dirac mass at the origin. More precisely, let $\delta : \mathbb{R} \mapsto [0, 1]$ be a $\mathcal{C}^\infty$ function such that

$$\int_{-\infty}^{\infty} \delta(z)\,dz = 1, \quad \delta(z) = 0 \qquad \text{for all } z \notin [-1, 1],$$

and define

$$\delta_h(z) = h\delta(hz), \qquad \alpha_h(z) = \int_{-\infty}^{z} \delta_h(s)\,ds. \tag{6.24}$$

Consider any non-negative smooth function $\psi = \psi(T, X)$ whose support is a compact subset of the open half plane where $T > 0$, and define

$$\phi(s,x,t,y) = \psi\left(\frac{s+t}{2}, \frac{x+y}{2}\right)\delta_h\left(\frac{s-t}{2}\right)\delta_h\left(\frac{x-y}{2}\right).$$

A direct computation yields

$$(\phi_s+\phi_t)(s,x,t,y) = \psi_T\left(\frac{s+t}{2}, \frac{x+y}{2}\right)\delta_h\left(\frac{s-t}{2}\right)\delta_h\left(\frac{x-y}{2}\right),$$

$$(\phi_x+\phi_y)(s,x,t,y) = \psi_X\left(\frac{s+t}{2}, \frac{x+y}{2}\right)\delta_h\left(\frac{s-t}{2}\right)\delta_h\left(\frac{x-y}{2}\right).$$

For $h$ sufficiently large, the support of $\phi$ is contained in the set where $s > 0, t > 0$. From (6.23) it thus follows that

$$\iiiint \delta_h\left(\frac{s-t}{2}\right)\delta_h\left(\frac{x-y}{2}\right)\left\{|u(s,x)-v(t,y)|\psi_T\left(\frac{s+t}{2}, \frac{x+y}{2}\right) + [f(u(s,x))-f(v(t,y))]\operatorname{sgn}\big(u(s,x)-v(t,y)\big)\psi_X\left(\frac{s+t}{2}, \frac{x+y}{2}\right)\right\}dx\,dy\,ds\,dt \geq 0. \tag{6.25}$$

3. We now compute the limit of the left hand side of (6.25) as $h \to \infty$. Using the variables

$$T = \frac{s+t}{2}, \qquad S = \frac{s-t}{2}, \qquad X = \frac{x+y}{2}, \qquad Y = \frac{x-y}{2},$$

the inequality (6.25) becomes

$$\begin{aligned}\iiiint &\{|u(T+S, X+Y) - v(T-S, X-Y)|\psi_T(T,X) \\ &\quad + [f(u(T+S, X+Y)) - f(v(T-S, X-Y))] \\ &\quad\quad \cdot \operatorname{sgn}(u(T+S, X+Y) - v(T-S, X-Y)) \cdot \psi_X(T,X)\} \\ &\quad \cdot \delta_h(S)\delta_h(Y)\, dX\, dY\, dS\, dT \geq 0. \end{aligned} \tag{6.26}$$

Letting $h \to \infty$ in (6.26) and renaming the variables $T, X$, we thus obtain

$$\begin{aligned}\iint &\{|u(t,x) - v(t,x)|\psi_t(t,x) + [f(u(t,x)) - f(v(t,x))] \\ &\quad \cdot \operatorname{sgn}(u(t,x) - v(t,x))\psi_x(t,x)\}\, dx\, dt \geq 0 \end{aligned} \tag{6.27}$$

for every $C^1$ function $\psi$ with compact support contained in the half plane where $t > 0$.

4. Now let $0 < \tau_0 < \tau$ and $R > 0$ be given. We construct a smooth approximation $\psi$ to the characteristic function of the trapezoid $\Omega$ in (6.18), by setting

$$\psi(t,x) \doteq [\alpha_h(t-\tau_0) - \alpha_h(t-\tau)] \cdot [1 - \alpha_h(|x| - R + L(\tau - t))].$$

Recall that $\alpha_h$ was defined at (6.24), so that $\alpha_h' = \delta_h \geq 0$. Using (6.27) with this particular test function $\psi$, one obtains

$$\begin{aligned}&\iint |u(t,x) - v(t,x)|\, [\delta_h(t-\tau_0) - \delta_h(t-\tau)] \cdot [1 - \alpha_h(|x| - R - L(\tau - t))]\, dx\, dt \\ &\quad \geq \iint \left\{ \frac{x}{|x|}[f(u(t,x)) - f(v(t,x))]\operatorname{sgn}(u(t,x) - v(t,x)) + L|u(t,x) - v(t,x)| \right\} \\ &\quad\quad \cdot [\alpha_h(t-\tau_0) - \alpha_h(t-\tau)]\, \delta_h(|x| - R - L(\tau - t))\, dx\, dt. \end{aligned} \tag{6.28}$$

By (6.15) and (6.16) we have $|f(u) - f(v)| \leq L|u - v|$. Moreover, (6.24) yields $\alpha_h(t-\tau_0) - \alpha_h(t-\tau) \geq 0$, $\alpha_h' = \delta_h \geq 0$. Hence

$$\iint |u(t,x) - v(t,x)|\, [\delta_h(t-\tau_0) - \delta_h(t-\tau)] \cdot [1 - \alpha_h(|x| - R - L(\tau - t))]\, dx\, dt \geq 0. \tag{6.29}$$

Recalling that the maps $t \mapsto u(t,\cdot)$, $t \mapsto v(t,\cdot)$ are both continuous from $[0, \infty[$ into $\mathbf{L}^1_{\mathrm{loc}}$, we now let $h \to \infty$ in (6.29) and obtain (6.17), in the case where $0 < \tau_0 < \tau$. By continuity, (6.17) still holds if $\tau_0 = \tau$ or if $\tau_0 = 0$. □

**Corollary 6.1 (Uniqueness in $\mathbf{L}^\infty$).** *Let $f : \mathbb{R} \mapsto \mathbb{R}$ be locally Lipschitz continuous. If $u, v$ are bounded entropy solutions of (6.1) such that $\|u(0,\cdot) - v(0,\cdot)\|_{\mathbf{L}^1} < \infty$, then for every $t > 0$ we have*

$$\int_{-\infty}^{\infty} |u(t,x) - v(t,x)|\,dx \le \int_{-\infty}^{\infty} |u(0,x) - v(0,x)|\,dx. \tag{6.30}$$

*For all initial data $\bar{u} \in \mathbf{L}^\infty$, the Cauchy problem (6.1)–(6.2) has at most one bounded entropy solution.*

*Proof.* By assumption, there exist constants $M, L$ for which (6.15) and (6.16) hold. For every $R, t \ge 0$, by (6.17) one has

$$\int_{|x|\le R} |u(t,x) - v(t,x)|\,dx \le \int_{|x|\le R+Lt} |u(0,x) - v(0,x)|\,dx. \tag{6.31}$$

Letting $R \to \infty$ in (6.31) we obtain (6.30), and hence the uniqueness of the solution. □

## 6.4 A contractive semigroup

By Theorem 6.1, a weak solution of (6.1) exists for every initial condition $\bar{u} \in \mathbf{L}^1$ with bounded variation. Since the $\mathbf{L}^1$ distance between any two such solutions does not increase in time, the solution operator can be extended by continuity to a much larger family of initial conditions. In particular, this yields the existence of a unique entropy solution to the Cauchy problem (6.1)–(6.2) for all initial data $\bar{u} \in \mathbf{L}^1 \cap \mathbf{L}^\infty$. We recall that weak solutions are defined up to equivalence in $\mathbf{L}^1_{\mathrm{loc}}$. The results stated below are thus understood to be valid after possibly changing the values of the solutions on a set of measure zero.

**Theorem 6.3.** *Let $f : \mathbb{R} \mapsto \mathbb{R}$ be locally Lipschitz continuous. Then there exists a continuous semigroup $S : [0,\infty) \times \mathbf{L}^1 \mapsto \mathbf{L}^1$ with the following properties*

*(i)* $S_0\bar{u} = \bar{u}$, $S_s(S_t\bar{u}) = S_{s+t}\bar{u}$.

*(ii)* $\|S_t\bar{u} - S_t\bar{v}\|_{\mathbf{L}^1} \le \|\bar{u} - \bar{v}\|_{\mathbf{L}^1}$.

*(iii) For each $\bar{u} \in \mathbf{L}^1 \cap \mathbf{L}^\infty$, the trajectory $t \mapsto S_t\bar{u}$ yields the unique bounded, entropy-admissible, weak solution of the corresponding Cauchy problem (6.1)–(6.2).*

*(iv) If $\bar{u}(x) \le \bar{v}(x)$ for all $x \in \mathbb{R}$, then $S_t\bar{u}(x) \le S_t\bar{v}(x)$ for every $x \in \mathbb{R}$, $t \ge 0$.*

*Proof.* For all initial data $\bar{w} \in \mathbf{L}^1 \cap BV$, let $S_t\bar{w}$ be the value at time $t$ of the entropy solution to (6.1) with initial condition $\bar{w}$. The existence and uniqueness of such a solution are guaranteed by Theorems 6.1 and 6.2. By continuity, we can now extend the domain of the semigroup $S$ to the entire space $\mathbf{L}^1$ by setting

$$S_t\bar{u} = \lim_{\substack{\bar{w}\to\bar{u}\\ \bar{w}\in BV}} S_t\bar{w}, \tag{6.32}$$

the convergence taking place in the $\mathbf{L}^1$ norm. Because of (6.30), the limit in (6.32) is well defined and satisfies (i) and (ii).

To prove (iii), let $\bar{u} \in \mathbf{L}^1 \cap \mathbf{L}^\infty$ be given, say with $|\bar{u}(x)| \leq M$ for all $x$. Consider a sequence of functions $\bar{u}_\nu \in BV$ with $\|\bar{u}_\nu\|_{\mathbf{L}^\infty} \leq M$, $\bar{u}_\nu \to \bar{u}$ in $\mathbf{L}^1$. Observe that the corresponding solutions $u_\nu(t, \cdot) = S_t \bar{u}_\nu$ all take values inside the interval $[-M, M]$. By the contractivity of the semigroup, for each $\nu$ and every $t \geq 0$ we now have

$$\|S_t\bar{u}_\nu - S_t\bar{u}\|_{\mathbf{L}^1} \leq \|\bar{u}_\nu - \bar{u}\|_{\mathbf{L}^1}, \qquad \|f(S_t\bar{u}_\nu) - f(S_t\bar{u})\|_{\mathbf{L}^1} \leq L \cdot \|\bar{u}_\nu - \bar{u}\|_{\mathbf{L}^1}, \tag{6.33}$$

where $L$ is a Lipschitz constant for $f$ on $[-M, M]$. Fix any $k \in \mathbb{R}$ and any $C^1$ function $\phi \geq 0$ with compact support contained in the half plane where $t > 0$. Using (6.33) we obtain

$$\begin{aligned} &\iint \{|S_t\bar{u} - k|\phi_t + (f(S_t\bar{u}) - f(k))\,\mathrm{sgn}(S_t\bar{u} - k)\phi_x\}\,dx\,dt \\ &\quad = \lim_{\nu\to\infty} \iint \{|S_t\bar{u}_\nu - k|\phi_t + (f(S_t\bar{u}_\nu) - f(k))\,\mathrm{sgn}(S_t\bar{u}_\nu - k)\phi_x\}\,dx\,dt \\ &\quad \geq 0, \end{aligned}$$

showing that each trajectory of the semigroup is an entropy-admissible solution of (6.1).

Concerning (iv), by continuity it suffices to consider the case where both $\bar{u}$ and $\bar{v}$ have bounded variation. In this case, the corresponding solutions of (6.1)–(6.2) can be obtained as limits of the piecewise constant approximations constructed in Section 6.1. The proof is thus reduced to showing that for any given $\nu \geq 1$, if $u, v$ are piecewise constant solutions of (6.6) and $u(0, x) \leq v(0, x)$ for all $x$, then

$$u(t, x) \leq v(t, x) \quad \text{for all } t \geq 0, \quad x \in \mathbb{R}. \tag{6.34}$$

If (6.34) fails, by continuity there exists a largest time $\tau$ such that $u(t, x) \leq v(t, x)$ for all $x \in \mathbb{R}$ and $t \leq \tau$. Since $u$ are piecewise constant, on a small time interval $[\tau, \tau + \delta]$ it is obtained by piecing together the corresponding solutions of the Riemann problems at every point of jump of $u(\tau, \cdot)$. The same of course holds for $v$. To derive a contradiction, it thus suffices to prove a comparison result for solutions of two Riemann problems.

We thus consider two Riemann data $(u^-, u^+)$, $(v^-, v^+)$ for the conservation law (6.6). If $\max\{u^-, u^+\} \leq \min\{v^-, v^+\}$, it is obvious that the corresponding solutions satisfy $u(t, x) \leq v(t, x)$ for every $x \in \mathbb{R}$, $t > 0$. We thus need to consider two non-trivial cases:

CASE 1: $u^- \leq v^- < u^+ \leq v^+$.

CASE 2: $u^+ \leq v^+ < u^- \leq v^-$.

In the first case, we observe that the piecewise constant solution $u = u(t, x)$ constructed at (6.9) can be characterized as follows:

$$u(t, x) = w \quad \text{iff } f_\nu(w) - \frac{x}{t} \cdot w = \min_{s \in [u^-, u^+]} \left\{ f_\nu(s) - \frac{x}{t} \cdot s \right\}. \tag{6.35}$$

In other words, $u(t, x) = w$ iff the line with slope $\lambda = x/t$ supports the graph of $f_\nu$ (restricted to the interval $[u^-, u^+]$) at the point $(w, f(w))$. From (6.35) and the analogous property of $v$ it follows that

$$u(t,x) = \arg\min_{s\in[u^-,u^+]} \left\{ f_v(s) - \frac{x}{t}\cdot s \right\} \le \arg\min_{s\in[v^-,v^+]} \left\{ f_v(s) - \frac{x}{t}\cdot s \right\} = v(t,x).$$

Similarly, in the second case we have

$$u(t,x) = \arg\max_{s\in[u^+,u^-]} \left\{ f_v(s) - \frac{x}{t}\cdot s \right\} \le \arg\max_{s\in[v^+,v^-]} \left\{ f_v(s) - \frac{x}{t}\cdot s \right\} = v(t,x).$$

In both cases we have obtained a contradiction with the maximality of $\tau$. This establishes (6.34), completing the proof of the theorem. □

## Problems

(1) Compute the unique entropy solution of the Riemann problem

$$u_t + (u^3 - 3u)_x = 0, \qquad u(0,x) = \begin{cases} -2 & \text{if } x < 0, \\ 2 & \text{if } x > 0. \end{cases}$$

(2) Let $f : \mathbb{R} \mapsto \mathbb{R}$ be globally Lipschitz continuous with constant $L$, and let $S$ be the semigroup generated by (6.1). If $\bar{u} \in \mathbf{L}^1$ has support contained inside $[a, b]$, prove that the support of $S_t\bar{u}$ is contained in the interval $[a - Lt, b + Lt]$.

(3) Let $f$ be locally Lipschitz continuous and consider the Riemann problem

$$u_t + f(u)_x = 0, \qquad u(0,x) = \begin{cases} u^- & \text{if } x < 0, \\ u^+ & \text{if } x > 0, \end{cases} \tag{6.36}$$

with $u^- < u^+$. Prove that the unique entropy solution of (6.36) has the form $u(t,x) = \psi(x/t)$, where

$$\psi(\lambda) \doteq \arg\min_{s\in[u^-,u^+]} f(s) - \lambda s.$$

Otherwise stated, $u(t,x) = w \in [u^-, u^+]$ iff the line with slope $x/t$ supports the epigraph of $f$, restricted to the interval $[u^-, u^+]$, at the point $(w, f(w))$.
Hint: consider first the case where $f$ is piecewise affine.

(4) Consider initial data $\bar{u} \in \mathbf{L}^\infty$ which is periodic with period $p$. Let $u = u(t,x)$ be an entropy solution of the Cauchy problem (6.1)–(6.2). For every $t \ge 0$, prove that $u(t,\cdot)$ is periodic with period $p$. Moreover,

$$\int_0^p u(t,x)\,dx = \int_0^p \bar{u}(x)\,dx.$$

Find an example where, for $t > 0$,

$$\int_0^p |u(t,x)|\,dx < \int_0^p |\bar{u}(x)|\,dx.$$

(5) Let $f \in \mathcal{C}^2$ satisfy

$$f(0) = f'(0) = 0, \qquad f''(u) \geq c > 0 \quad \text{for all } u.$$

Consider the initial condition

$$\bar{u}(x) = \begin{cases} k(x-a) & \text{if } x \in [a,b], \\ 0 & \text{if } x \notin [a,b], \end{cases}$$

for some constant $k > 0$. Prove that the entropy solution of (6.1)–(6.2) is given by

$$u(t,x) = \begin{cases} \bar{u}(y(t,x)) & \text{if } x \in [a,b(t)], \\ 0 & \text{if } x \notin [a,b(t)], \end{cases}$$

where $y = y(t,x)$ and $b = b(t)$ are implicitly defined by

$$y + tf'(\bar{u}(y)) = x, \quad y \in [a,b],$$

$$\int_a^{b(t)} u(t,x)\,dx = \int_a^{b(t)} \bar{u}(y(t,x))\,dx = \int_a^b \bar{u}(x)\,dx = \frac{k(b-a)}{2}.$$

In addition, show that

$$\frac{u(t,x') - u(t,x)}{x'-x} \leq \frac{1}{ct} \quad x' > x, \ t > 0. \tag{6.37}$$

Finally, prove the decay estimate

$$\|u(t,\cdot)\|_{\mathbf{L}^\infty} \leq \sqrt{\frac{2\|\bar{u}\|_{\mathbf{L}^1}}{ct}} \quad t > 0. \tag{6.38}$$

Hint: outside the shock at $x = b(t)$, the function $u$ satisfies

$$(u_x)_t + f'(u)(u_x)_x = -f''(u)u_x^2 \leq -cu_x^2. \tag{6.39}$$

Integrate (6.39) along the characteristics $x(t) = x_0 + tf'(u(x_0))$ to obtain $u_x(t,x) \leq (ct)^{-1}$. To prove (6.38) observe that, if $u(t,x') = h > 0$ at some point $x'$, then (6.37) implies

$$\frac{cth^2}{2} \leq \|u(t,\cdot)\|_{\mathbf{L}^1} \leq \|\bar{u}\|_{\mathbf{L}^1}. \tag{6.40}$$

(6) Let $f \in \mathcal{C}^2$ satisfy $f''(u) \geq c > 0$ for all $u \in \mathbb{R}$. Fix an integer $\nu \geq 1$ and define the piecewise constant approximation $f_\nu$ as in (6.5). Let $u_\nu = u_\nu(t,x)$ be a piecewise constant solution of (6.6), taking values in the discrete set $2^{-\nu}\mathbb{Z}$. Lines $x = x_\alpha(t)$ where $u_\nu$ has an upward jump, i.e. $u_\nu(t,x_\alpha-) < u_\nu(t,x_\alpha+)$, are called *rarefaction fronts*. Lines where $u_\nu$ has a downward jump are called *shock fronts*.

(i) Show that all rarefaction fronts have strength $u_\nu(t,x_\alpha+) - u_\nu(t,x_\alpha-) = 2^{-\nu}$.

(ii) Show that, if two or more fronts collide, at least one of them is a shock. From the interaction, a single shock emerges (unless all fronts completely cancel each other).

(iii) If two adjacent fronts $x_\alpha, x_{\alpha+1}$ are both rarefactions, prove that their speeds satisfy $\dot{x}_{\alpha+1} - \dot{x}_\alpha \geq c \cdot 2^{-\nu}$.

(iv) From (iii) deduce the inequality

$$u_\nu(t, y) - u_\nu(t, x) \leq 2^{-\nu} + \frac{y - x}{ct} \qquad x < y, \quad t > 0.$$

(v) Letting $\nu \to \infty$ in (6.41), prove that every entropy solution $u$ of (6.1) satisfies

$$u(t, y) - u(t, x) \leq \frac{y - x}{ct} \qquad x < y, \quad t > 0.$$

(vi) Using (6.42), show that the decay estimate (6.38) holds for every solution $u$ of (6.1), provided that $f'' \geq c > 0$.

# 7

# The Cauchy problem for systems

This chapter is concerned with the global existence of solutions to the Cauchy problem

$$u_t + f(u)_x = 0, \tag{7.1}$$

$$u(0, x) = \bar{u}(x), \tag{7.2}$$

under the assumptions

(♣) The $n \times n$ system of conservation laws (7.1) is strictly hyperbolic, with smooth coefficients, defined for $u$ in an open set $\Omega \subseteq \mathbb{R}^n$. Each characteristic field is either genuinely non-linear or linearly degenerate.

By possibly performing a translation in the $u$-coordinates, it is not restrictive to assume that $\Omega$ contains the origin. Given an initial condition $\bar{u}$ with sufficiently small total variation, we will construct a weak, entropy-admissible solution $u$, defined for all $t \geq 0$. We recall that a function $u : [0, T] \times \mathbb{R} \mapsto \mathbb{R}^n$ is a weak solution to the Cauchy problem (7.1)–(7.2) if the map $t \mapsto u(t, \cdot)$ is continuous with values in $\mathbf{L}^1_{\text{loc}}$, the initial condition (7.2) is satisfied and, for every $\mathcal{C}^1$ function $\phi$ with compact support contained in the open strip $]0, T[\, \times \mathbb{R}$, one has

$$\int_0^T \int_{-\infty}^{\infty} \left\{ \phi_t(t, x) u(t, x) + \phi_x(t, x) f(u(t, x)) \right\} dx\, dt = 0. \tag{7.3}$$

Given a convex entropy $\eta$ for the system (7.1), with entropy flux $q$, we say that the solution $u$ is $\eta$-admissible if it satisfies the entropy inequality

$$\eta(u)_t + q(u)_x \leq 0$$

in the distributional sense. For every non-negative $\mathcal{C}^1$ function $\phi$ with compact support contained in the strip $]0, T[\, \times \mathbb{R}$, we thus require

$$\int_0^T \int_{-\infty}^{\infty} \left\{ \phi_t(t, x) \eta(u(t, x)) + \phi_x(t, x) q(u(t, x)) \right\} dx\, dt \geq 0. \tag{7.4}$$

Most of this chapter is devoted to the proof of the following basic existence theorem.

**Theorem 7.1 (Global existence of entropy solutions).** *Under the basic assumptions (♣), there exists a constant $\delta_0 > 0$ with the following property. For every initial condition $\bar{u} \in \mathbf{L}^1$ with*

$$\text{Tot. Var.}\,\{\bar{u}\} \leq \delta_0, \tag{7.5}$$

*the Cauchy problem (7.1)–(7.2) has a weak solution $u = u(t, x)$, defined for all $t \geq 0$. In addition, if the system (1.1) admits a convex entropy $\eta$, then one can find a solution $u$ which is also $\eta$-admissible.*

Theorem 7.1 was first proved in a fundamental paper of Glimm, constructing approximate solutions by means of a random restarting procedure. An alternative method for constructing approximate solutions, based on wave-front tracking, will be described in the next section.

## 7.1 Wave-front tracking approximations

Roughly speaking, an $\varepsilon$-approximate front tracking solution of the system of conservation laws (7.1) is a piecewise constant function $u = u(t, x)$, whose jumps are located along finitely many straight lines $x = x_\alpha(t)$ in the $t$–$x$ plane and approximately satisfy the Rankine–Hugoniot conditions. For each time $t > 0$, one should thus have an estimate of the form

$$\sum_\alpha \Big| [f(u(t, x_\alpha+)) - f(u(t, x_\alpha-))] - \dot{x}_\alpha [u(t, x_\alpha+) - u(t, x_\alpha-)] \Big| = \mathcal{O}(1) \cdot \varepsilon.$$

If $\eta$ is a convex entropy with flux $q$, in view of (4.37), at each time $t > 0$ one should also have

$$\sum_\alpha \Big\{ [q(u(t, x_\alpha+)) - q(u(t, x_\alpha-))] - \dot{x}_\alpha [\eta(u(t, x_\alpha+)) - \eta(u(t, x_\alpha-))] \Big\} \leq \mathcal{O}(1) \cdot \varepsilon.$$

Since we want to use these approximations for a detailed analysis of solutions, it is convenient to require a number of additional properties, described below.

For convenience, we recall the main notation introduced in Chapter 5. For $k \in \{1, \ldots, n\}$, $\lambda_k(u)$ is the $k$-th eigenvalue of the matrix $A(u)$. If the states $u^-, u^+$ are connected by a $k$-shock, by $\lambda_k(u^-, u^+)$ we denote the Rankine–Hugoniot speed of this shock. The $k$-shock and the $k$-rarefaction curves through a state $\omega$ are written as

$$\sigma \mapsto S_k(\sigma)(\omega), \qquad \sigma \mapsto R_k(\sigma)(\omega).$$

If the $k$-th field is linearly degenerate, the above curves are parametrized by arc-length. In the genuinely non-linear case, the parametrization is chosen so that

$$\lambda_k\left(S_k(\sigma)(\omega)\right) = \lambda_k\left(R_k(\sigma)(\omega)\right) = \lambda_k(\omega) + \sigma.$$

**Definition 7.1.** Given $\varepsilon > 0$, we say that a continuous map $u : [0, \infty[ \mapsto \mathbf{L}^1_{\mathrm{loc}}(\mathbb{R};\ \mathbb{R}^n)$ is an *$\varepsilon$-approximate front tracking solution* of (7.1) if the following four conditions hold:

1. As a function of two variables, $u = u(t, x)$ is piecewise constant, with discontinuities occurring along finitely many straight lines in the $t$–$x$ plane. Jumps can be of three types: shocks (or contact discontinuities), rarefactions and non-physical waves, denoted as $\mathcal{J} = \mathcal{S} \cup \mathcal{R} \cup \mathcal{NP}$.
2. Along each shock (or contact discontinuity) $x = x_\alpha(t)$, $\alpha \in \mathcal{S}$, the values $u^- \doteq u(t, x_\alpha-)$ and $u^+ \doteq u(t, x_\alpha+)$ are related by
$$u^+ = S_{k_\alpha}(\sigma_\alpha)(u^-), \tag{7.6}$$
for some $k_\alpha \in \{1, \dots, n\}$ and some wave size $\sigma_\alpha$. If the $k_\alpha$-th family is genuinely non-linear, then the admissibility condition $\sigma_\alpha < 0$ also holds. Moreover, the speed of the shock front satisfies
$$|\dot{x}_\alpha - \lambda_{k_\alpha}(u^+, u^-)| \le \varepsilon. \tag{7.7}$$
3. Along each rarefaction front $x = x_\alpha(t)$, $\alpha \in \mathcal{R}$, one has
$$u^+ = R_{k_\alpha}(\sigma_\alpha)(u^-), \qquad \sigma_\alpha \in ]0, \varepsilon] \tag{7.8}$$
for some genuinely non-linear family $k_\alpha$. Moreover,
$$|\dot{x}_\alpha - \lambda_{k_\alpha}(u^+)| \le \varepsilon. \tag{7.9}$$
4. All non-physical fronts $x = x_\alpha(t)$, $\alpha \in \mathcal{NP}$ have the same speed:
$$\dot{x}_\alpha \equiv \hat{\lambda}, \tag{7.10}$$
where $\hat{\lambda}$ is a fixed constant strictly greater than all characteristic speeds. The total strength of all non-physical fronts in $u(t, \cdot)$ remains uniformly small, i.e.
$$\sum_{\alpha \in \mathcal{NP}} |u(t, x_\alpha+) - u(t, x_\alpha-)| \le \varepsilon \quad \text{for all } t \ge 0. \tag{7.11}$$

If, in addition, the initial value of $u$ satisfies
$$\|u(0, \cdot) - \bar{u}\|_{\mathbf{L}^1} < \varepsilon, \tag{7.12}$$
then we say that $u$ is an *$\varepsilon$-approximate front tracking solution to the Cauchy problem* (7.1)–(7.2).

The following lemma estimates the amount of error in an $\varepsilon$-approximate solution, at a given time $t$. As is customary, by the Landau symbol $\mathcal{O}(1)$ we denote a (possibly vector-valued) function, whose values remain uniformly bounded. The strength of a non-physical front is defined as $|\sigma_\alpha| \doteq |u(t, x_{\alpha+}) - u(t, x_{\alpha-})|$.

**Lemma 7.1.** *Let $u = u(t, x)$ be an $\varepsilon$-approximate front tracking solution of (7.1). Moreover, let $\eta$ be a convex entropy with flux $q$. Fix any time $t$ where no interaction takes place and consider any front, located say at $x_\alpha$ with strength $\sigma_\alpha$. Calling $u^- \doteq u(t, x_\alpha-)$, $u^+ \doteq u(t, x_\alpha+)$, define the errors in the Rankine–Hugoniot equations and in the entropy conditions:*

$$E_\alpha \doteq [f(u^+) - f(u^-)] - \dot{x}_\alpha[u^+ - u^-]$$
$$E'_\alpha \doteq [q(u^+) - q(u^-)] - \dot{x}_\alpha[\eta(u^+) - \eta(u^-)].$$

*(i) At every rarefaction front one has*

$$E_\alpha = \mathcal{O}(1) \cdot \varepsilon|\sigma_\alpha|, \qquad E'_\alpha = \mathcal{O}(1) \cdot \varepsilon|\sigma_\alpha|. \tag{7.13}$$

*(ii) Along every shock front (or contact discontinuity) one has*

$$E_\alpha = \mathcal{O}(1) \cdot \varepsilon|\sigma_\alpha|, \qquad E'_\alpha \leq \mathcal{O}(1) \cdot \varepsilon|\sigma_\alpha|. \tag{7.14}$$

*(iii) Along every non-physical front one has*

$$E_\alpha = \mathcal{O}(1) \cdot |\sigma_\alpha|, \qquad E'_\alpha = \mathcal{O}(1) \cdot |\sigma_\alpha|. \tag{7.15}$$

*Proof.* To prove (i), for a fixed index $k \in \{1, \ldots, n\}$ and a state $\omega$, consider the map

$$\Psi(\sigma, \omega) \doteq \big[f(R_k(\sigma)(\omega)) - f(\omega)\big] - \lambda_k(\omega) \cdot [R_k(\sigma)(\omega) - \omega].$$

Trivially, $\Psi(0, \omega) = 0$. By the analysis in Chapter 5 we have

$$\frac{\partial \Psi}{\partial \sigma}(0, \omega) = Df(\omega) \cdot r_k(\omega) - \lambda_k(\omega) r_k(\omega) = 0.$$

Therefore, Lemma 2.4 and Remark 2.2 yield

$$\Psi(\sigma, \omega) = \mathcal{O}(1) \cdot |\sigma|^2. \tag{7.16}$$

If the front at $x_\alpha$ belongs to the $k_\alpha$-th family, we now use (7.16) with $\omega = u^-$, $k = k_\alpha$. Recalling that $\sigma_\alpha \in [0, \varepsilon]$ and observing that

$$u^+ - u^- = R_{k_\alpha}(\sigma_\alpha)(u^-) - u^- = \mathcal{O}(1) \cdot |\sigma_\alpha| = \mathcal{O}(1) \cdot \varepsilon,$$
$$|\lambda_{k_\alpha}(u^+) - \lambda_{k_\alpha}(u^-)| = \mathcal{O}(1) \cdot |u^+ - u^-| = \mathcal{O}(1) \cdot \varepsilon,$$

by (7.9) we now have

$$E_\alpha = \big\{[f(u^+) - f(u^-)] - \lambda_{k_\alpha}(u^-) \cdot (u^+ - u^-)\big\} + [\lambda_{k_\alpha}(u^-) - \dot{x}_\alpha](u^+ - u^-)$$
$$= \mathcal{O}(1) \cdot |\sigma_\alpha|^2 + \mathcal{O}(1) \cdot \varepsilon|\sigma_\alpha|.$$

The estimate for $E'_\alpha$ is entirely similar, considering the function

$$\Psi(\sigma, \omega) \doteq [q(R_k(\sigma)(\omega)) - q(\omega)] - \lambda_k(\omega) \cdot [\eta(R_k(\sigma)(\omega)) - \eta(\omega)].$$

Concerning (ii), since $u^+ - u^- = \mathcal{O}(1) \cdot \sigma_\alpha$, by (7.6) and (7.7) we obtain

$$E_\alpha = \big\{[f(u^+) - f(u^-)] - \lambda_{k_\alpha}(u^-, u^+) \cdot (u^+ - u^-)\big\} + [\lambda_{k_\alpha}(u^-) - \dot{x}_\alpha](u^+ - u^-)$$
$$= \mathcal{O}(1) \cdot \varepsilon|\sigma_\alpha|.$$

Using (5.34), the estimate for $E'_\alpha$ is entirely analogous:

$$\begin{aligned} E'_\alpha &= \big\{[q(u^+) - q(u^-)] - \lambda_{k_\alpha}(u^-, u^+) \cdot [\eta(u^+) - \eta(u^-)]\big\} \\ &\quad + [\lambda_{k_\alpha}(u^-, u^+) - \dot{x}_\alpha] \cdot [\eta(u^+) - \eta(u^-)] \\ &\le \mathcal{O}(1) \cdot \varepsilon |\sigma_\alpha|. \end{aligned}$$

Finally, the estimates in (iii) are a trivial consequence of the Lipschitz continuity of the functions $f, \eta, q$. □

For a proof of Theorem 7.1, we shall first establish the existence of front tracking approximations, defined for all $t \ge 0$.

**Theorem 7.2 (Global existence of front tracking approximations).** *Under the basic assumptions (♣), there exists a constant $\delta_0 > 0$ with the following property. For every initial condition $\bar{u} \in \mathbf{L}^1$ satisfying (7.5) and for every $\varepsilon > 0$, the Cauchy problem (7.1)–(7.2) admits an $\varepsilon$-approximate front tracking solution, defined for all $t \ge 0$.*

In a second step, we shall show that a suitable sequence of front tracking approximations converges to a limit $u = u(t, x)$ which provides an entropy-admissible weak solution to the Cauchy problem.

## 7.2 A front tracking algorithm

We now describe an algorithm which generates these front tracking approximations. The construction (Fig. 7.1) starts at time $t = 0$ by taking a piecewise constant approximation $u(0, \cdot)$ of $\bar{u}$, satisfying

$$\text{Tot. Var.}\{u(0,\cdot)\} \le \text{Tot. Var.}\{\bar{u}\} \le \delta_0, \qquad \|u(0,\cdot) - \bar{u}\|_{\mathbf{L}^1} < \varepsilon. \tag{7.17}$$

Let $x_1 < \cdots < x_N$ be the points where $u(0, \cdot)$ is discontinuous. For each $\alpha = 1, \ldots, N$, the Riemann problem generated by the jump $(u(0, x_\alpha-),\ u(0, x_\alpha+))$ is approximately

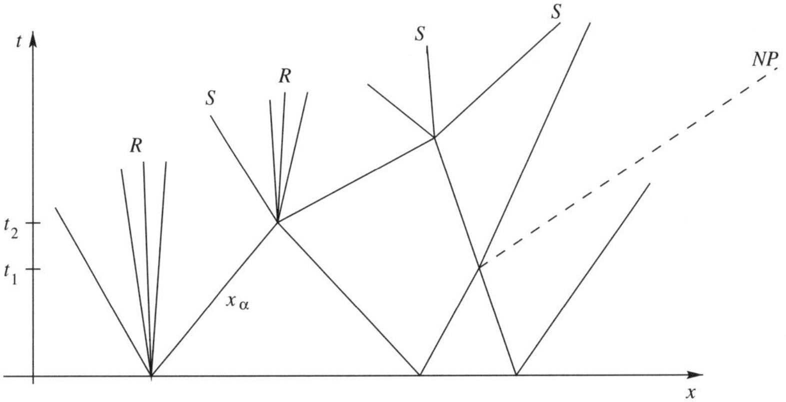


**Figure 7.1**

solved on a forward neighbourhood of $(0, x_\alpha)$ in the $t$–$x$ plane by a function of the form $u(t, x) = \varphi((x - x_\alpha)/t)$, with $\varphi : \mathbb{R} \mapsto \mathbb{R}^n$ piecewise constant. More precisely, if the exact solution of the Riemann problem contains only shocks and contact discontinuities, then we let $u$ coincide with the exact solution, which is already piecewise constant. On the other hand, if centred rarefaction waves are present, they are approximated by a *centred rarefaction fan* containing several small jumps travelling at a speed close to the characteristic speed.

The approximate solution $u$ can then be prolonged until a time $t_1$ is reached, when the first set of interactions between two or more wave-fronts takes place. Since $u(t_1, \cdot)$ is still a piecewise constant function, the corresponding Riemann problems can again be approximately solved within the class of piecewise constant functions. The solution $u$ is then continued up to a time $t_2$ where the second set of wave interactions takes place, and so on.

For general $n \times n$ systems, the main source of technical difficulty stems from the fact that the number of wave-fronts may approach infinity in finite time, in which case the construction would break down. To see this, observe that at a generic interaction point there will be two incoming fronts, while the number of outgoing fronts is $n$ (if all waves generated by the Riemann problem are shocks or contact discontinuities), or even larger (if rarefaction waves are present). In turn, these outgoing wave-fronts may quickly interact with several other fronts, generating more and more lines of discontinuity (Fig. 7.2).

One thus needs to modify the algorithm to ensure that the number of fronts will not become infinite within finite time. We shall use two different procedures for solving a Riemann problem within the class of piecewise constant functions: an accurate Riemann solver (Fig. 7.4), which introduces several new wave-fronts, and a simplified Riemann solver (Fig. 7.5), which involves a minimum number of outgoing fronts. In this second case, all new waves are lumped together in a single *non-physical* front, travelling with a fixed speed $\hat{\lambda}$ strictly larger than all characteristic speeds. The main feature of this algorithm is illustrated in Figs 7.2–7.3. If all Riemann problems were solved accurately, the number of wave-fronts could approach infinity within a finite time $\tau$ (Fig. 7.2). However, since the total variation remains small, the new fronts generated by further interactions are very small. When their size becomes smaller than a threshold parameter $\rho > 0$, the simplified Riemann solver is used, which generates one single new

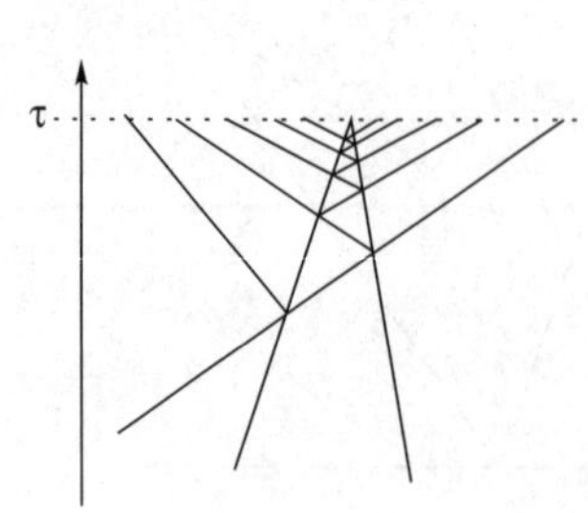


**Figure 7.2**

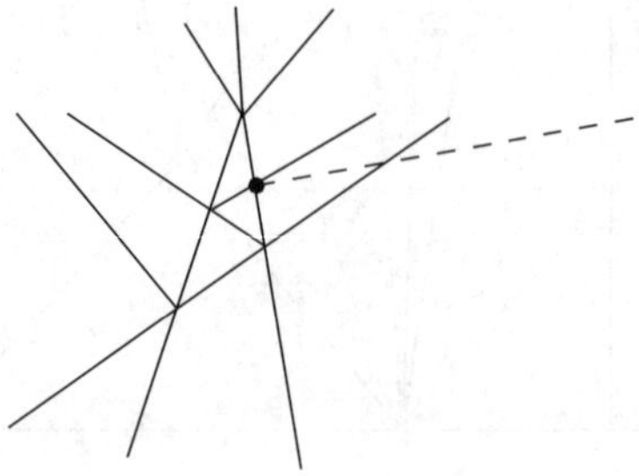

**Figure 7.3**

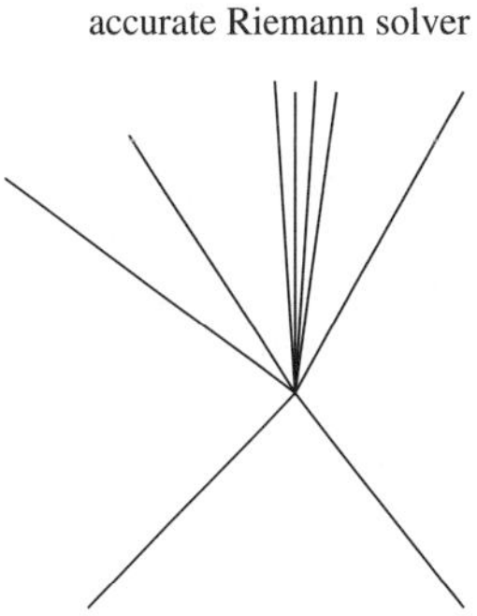


**Figure 7.4**

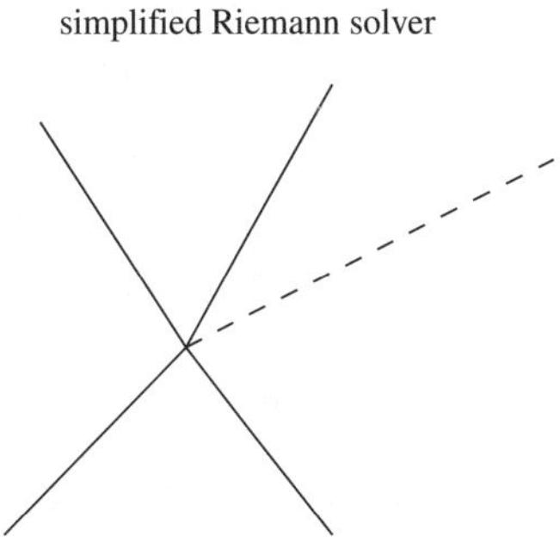


**Figure 7.5**

(non-physical) front, with very small amplitude. The total number of fronts thus remains bounded for all times (Fig. 7.3).

Given a general Riemann problem at a point $(\bar{t}, \bar{x})$,

$$v_t + f(v)_x = 0, \qquad v(\bar{t}, x) = \begin{cases} u^- & \text{if } x < \bar{x}, \\ u^+ & \text{if } x > \bar{x}, \end{cases} \tag{7.18}$$

we now describe two procedures which yield approximate solutions within the class of piecewise constant functions. As in Chapter 5, for $i \in \{1, \ldots, n\}$ and a given state $\omega \in \mathbb{R}^n$, we consider the composed curve

$$\Psi_i(\sigma)(\omega) \doteq \begin{cases} R_i(\sigma)(\omega) & \text{if} \sigma \geq 0, \\ S_i(\sigma)(\omega) & \text{if} \sigma < 0. \end{cases} \tag{7.19}$$

### *Accurate Riemann solver*

Given $u^-, u^+$, we first determine the states $\omega_0, \omega_1, \ldots, \omega_n$ and parameter values $\sigma_1, \ldots, \sigma_n$ such that

$$\omega_0 = u^-, \qquad \omega_n = u^+, \qquad \omega_i = \Psi_i(\sigma_i)(\omega_{i-1}) \quad i = 1, \ldots, n. \tag{7.20}$$

Of course, $\omega_0, \ldots, \omega_n$ are the constant states present in the exact solution of the Riemann problem (7.18). If all jumps $(\omega_{i-1}, \omega_i)$ were shocks or contact discontinuities, then this Riemann problem would have a piecewise constant solution with $\leq n$ lines of discontinuity. In the general case, the exact solution of (7.18) is not piecewise constant, because of the presence of rarefaction waves. These will be approximated by piecewise constant rarefaction fans, inserting additional states $\omega_{i,j}$ as follows.

Let $\delta > 0$ be a small constant, given at the outset of the construction algorithm. If the $i$-th characteristic field is genuinely non-linear and $\sigma_i > 0$, consider the integer

$$p_i \doteq 1 + [\![\sigma_i/\delta]\!], \tag{7.21}$$

where $[\![s]\!]$ denotes the integer part of $s$, i.e. the largest integer $\leq s$. For $j = 1, \ldots, p_i$, define

$$\omega_{i,j} = \Psi_i(j\sigma_i/p_i)(\omega_{i-1}), \qquad x_{i,j}(t) = \bar{x} + (t - \bar{t})\lambda_i(\omega_{i,j}). \tag{7.22}$$

On the other hand, if the $i$-th characteristic field is genuinely non-linear and $\sigma_i \leq 0$, or if the $i$-th characteristic field is linearly degenerate (with $\sigma_i$ arbitrary), define $p_i \doteq 1$ and

$$\omega_{i,1} = \omega_i, \qquad x_{i,1}(t) = \bar{x} + (t - \bar{t})\lambda_i(\omega_{i-1}, \omega_i). \tag{7.23}$$

Here $\lambda_i(\omega_{i-1}, \omega_i)$ is the Rankine–Hugoniot speed of a jump connecting $\omega_{i-1}$ with $\omega_i$ so that

$$\lambda(\omega_{i-1}, \omega_i) \cdot (\omega_i - \omega_{i-1}) = f(\omega_i) - f(\omega_{i-1}). \tag{7.24}$$

As soon as the intermediate states $\omega_{i,j}$ and the locations $x_{i,j}(t)$ of the jumps have been determined by (7.22) and (7.23), we can define an approximate solution to the Riemann problem (7.18) by setting (Fig. 7.6)

$$v(t,x) = \begin{cases} u^- & \text{if } x < x_{1,1}(t), \\ u^+ & \text{if } x > x_{n,p_n}(t), \\ \omega_i (= \omega_{i,p_i}) & \text{if } x_{i,p_i}(t) < x < x_{i+1,1}(t), \\ \omega_{i,j} & \text{if } x_{i,j}(t) < x < x_{i,j+1}(t) \quad (j = 1, \ldots, p_i - 1). \end{cases} \tag{7.25}$$

Observe that the difference between $v$ and the exact self-similar solution of (7.18) is due to the fact that every centred $i$-rarefaction wave is here divided into equal parts and replaced by a rarefaction fan containing $p_i$ wave-fronts. Because of (7.21), the strength of each one of these fronts is $< \delta$.

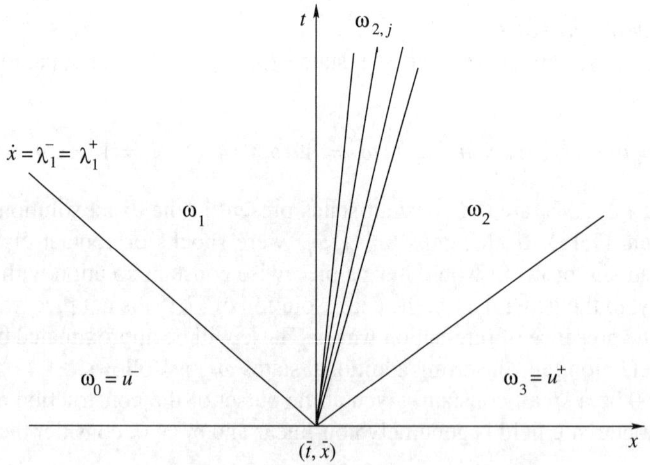


**Figure 7.6**

*Simplified Riemann solver*

CASE 1: Let $j, j' \in \{1, \ldots, n\}$ be the families of the two incoming wave-fronts, with $j \geq j'$. Assume that the left, middle and right states $u_l, u_m, u_r$ before the interaction are related by

$$u_m = \Psi_j(\sigma)(u_l), \qquad u_r = \Psi_{j'}(\sigma')(u_m). \tag{7.26}$$

Define the auxiliary right state

$$\tilde{u}_r = \begin{cases} \Psi_j(\sigma) \circ \Psi_{j'}(\sigma')(u_l) & \text{if } j > j', \\ \Psi_j(\sigma + \sigma')(u_l) & \text{if } j = j'. \end{cases} \tag{7.27}$$

Let $\tilde{v} = \tilde{v}(t, x)$ be the piecewise constant solution of the Riemann problem with data $u_l, \tilde{u}_r$, constructed as in (7.25). Because of (7.27), the piecewise constant function $\tilde{v}$ contains exactly two wave-fronts of sizes $\sigma', \sigma$, if $j > j'$, or a single wave-front of size $\sigma + \sigma'$, if $j = j'$.

Of course, in general one has $\tilde{u}_r \neq u_r$. We let the jump $(\tilde{u}_r, \; u_r)$ travel with a fixed speed $\hat{\lambda}$ strictly bigger than all characteristic speeds. In a forward neighbourhood of the point $(\bar{t}, \bar{x})$, we thus define an approximate solution $v$ as follows (Figs 7.7–7.8):

$$v(t, x) = \begin{cases} \tilde{v}(t, x) & \text{if } x - \bar{x} < (t - \bar{t})\hat{\lambda}, \\ u_r & \text{if } x - \bar{x} > (t - \bar{t})\hat{\lambda}. \end{cases} \tag{7.28}$$

Observe that this simplified Riemann solver introduces a new *non-physical* wave-front, travelling with constant speed $\hat{\lambda}$. In turn, this front may interact with other (physical) fronts. One more case of interaction thus needs to be considered.

CASE 2: A non-physical front hits from the left a wave-front of the $i$-characteristic family (Fig. 7.9), for some $i \in \{1, \ldots, n\}$.

Let $u_l, u_m, u_r$ be the left, middle and right states before the interactions. If

$$u_r = \Psi_i(\sigma)(u_m), \tag{7.29}$$

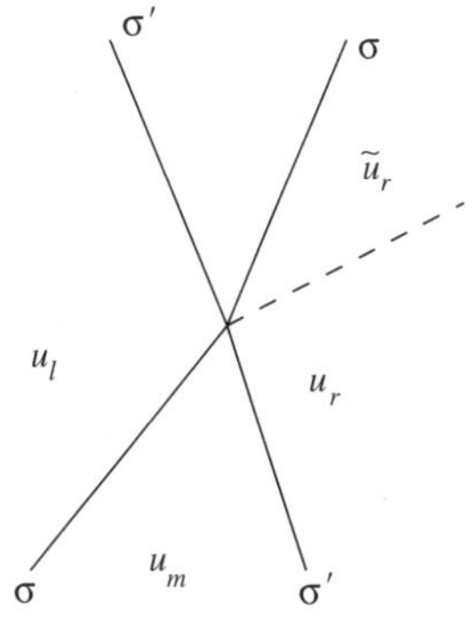


**Figure 7.7**

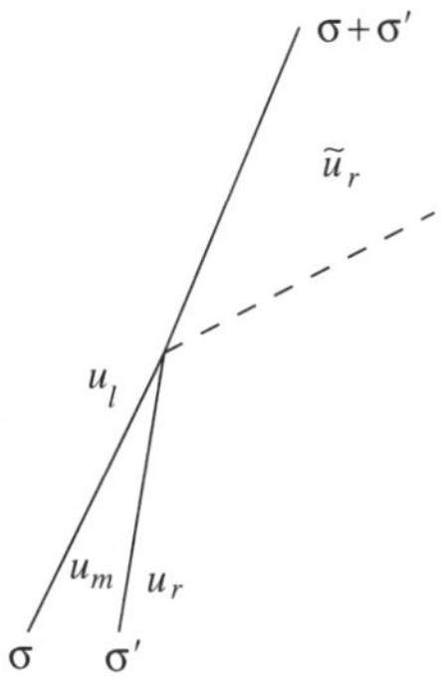


**Figure 7.8**

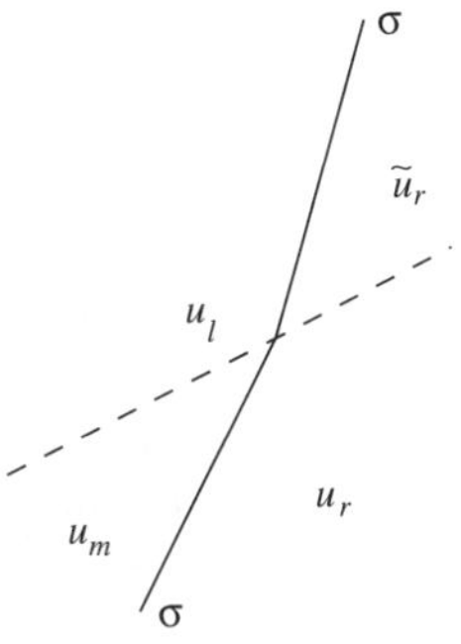


**Figure 7.9**

we define the auxiliary right state

$$\tilde{u}_r = \Psi_i(\sigma)(u_l). \tag{7.30}$$

Call $\tilde{v}$ the solution to the Riemann problem with data $u_l$, $\tilde{u}_r$, constructed as in (7.25). Because of (7.30), $\tilde{v}$ will contain a single wave-front belonging to the $i$-th family, with size $\sigma$. Since $\tilde{u}_r \neq u_r$ in general, we let the jump $(\tilde{u}_r,\ u_r)$ travel with the fixed speed $\hat{\lambda}$. In a forward neighbourhood of the point $(\bar{t}, \bar{x})$, the approximate solution $u$ is thus defined again according to (7.28).

By construction, all non-physical fronts travel with the same speed $\hat{\lambda}$; hence they never interact with each other. The above cases therefore cover all possible interactions between two wave-fronts.

To complete the description of the algorithm, it remains to specify which Riemann solver is used at any given interaction. The choice is made in connection with a threshold parameter $\rho > 0$:

- The accurate method is used at time $t = 0$, and at every interaction between two physical waves (i.e. rarefactions, shocks or contact discontinuities), when the product of the strengths of the incoming waves is $|\sigma\sigma'| \geq \rho$.
- The simplified method is used at every interaction involving a non-physical incoming wave-front, and also at interactions where the product of the strengths of the incoming waves is $|\sigma\sigma'| < \rho$.

We shall also adopt the following provision:

(P) In the accurate Riemann solver, rarefaction fronts of the same family of one of the incoming fronts are never partitioned (even if their strength is $> \delta$).

This guarantees that every wave-front can be uniquely continued forward in time, unless it gets completely cancelled by interacting with another front of the same family and opposite sign.

The construction of an approximate solution thus involves three parameters:

- A fixed speed $\hat{\lambda}$, strictly larger than all characteristic speeds.
- A small constant $\delta > 0$, controlling the maximum strength of rarefaction fronts.
- A threshold parameter $\rho > 0$, determining whether the accurate or the simplified Riemann solver is used.

This completes the definition of our algorithm. To prove Theorem 7.2, we now need to show that for any $\varepsilon > 0$, if the initial data $\bar{u}$ have small total variation, by a suitable choice of the parameters $\delta$, $\rho$ the algorithm will produce an $\varepsilon$-approximate solution defined for all $t \geq 0$. Observe that one can always solve the new Riemann problems generated by wave-front interactions provided that the states $u^-$, $u^+$ are close to each other. A key part of the proof will thus show that the total variation of the approximate solution remains small for all times.

**Remark 7.1.** According to our algorithm, shock fronts travel exactly with Rankine–Hugoniot speed $\dot{x}_\alpha = \lambda_{k_\alpha}(u^+, u^-)$, while rarefaction fronts travel with the characteristic speed of their right state: $\dot{x}_\alpha = \lambda_{k_\alpha}(u^+)$. However, one exception to this rule must be allowed if three or more fronts meet at the same point. To avoid this situation, we must change the speed of one of the incoming fronts. Of course this change of speed can be chosen arbitrarily small.

## 7.3 Global existence of front tracking approximations

The proof of Theorem 7.2 will be given in several steps.

**1. A compact domain.** We start by choosing $\delta_1 > 0$ such that the closed ball $B(0, \delta_1)$ centred at the origin with radius $\delta_1$ is entirely contained in $\Omega$. By possibly reducing the size of $\delta_1$, we can also assume that any Riemann problem with data $u_l, u_r$ has a unique self-similar solution with values inside $\Omega$, provided that $u_l, u_r \in B(0, \delta_1)$.

**2. Interaction estimates.** Whenever two wave-fronts interact, the new Riemann problem is solved in terms of a family of outgoing waves. The next lemma provides an estimate of the difference between the strengths of the corresponding incoming and outgoing fronts.

**Lemma 7.2.** *Consider an interaction between two incoming wave-fronts.*

*(i) Let $\sigma_i', \sigma_j'$ be the sizes of two incoming fronts belonging to the distinct characteristic families $i > j$. Their interaction determines a Riemann problem (Fig. 7.10), whose (exact) solution consists of outgoing waves of sizes $\sigma_1, \ldots, \sigma_n$ say. These are related to the incoming waves by the estimate*

$$|\sigma_i - \sigma_i'| + |\sigma_j - \sigma_j'| + \sum_{k \neq i,j} |\sigma_k| = \mathcal{O}(1) \cdot |\sigma_i' \sigma_j'|. \tag{7.31}$$

*(ii) Let $\sigma, \sigma'$ be the sizes of two incoming fronts, both belonging to the $i$-th characteristic family. As before, calling $\sigma_1, \ldots, \sigma_n$ the sizes of the outgoing waves generated by the corresponding Riemann problem (Fig. 7.11), one has*

$$|\sigma_i - \sigma - \sigma'| + \sum_{k \neq i} |\sigma_k| = \mathcal{O}(1) \cdot |\sigma\sigma'|(|\sigma| + |\sigma'|). \tag{7.32}$$

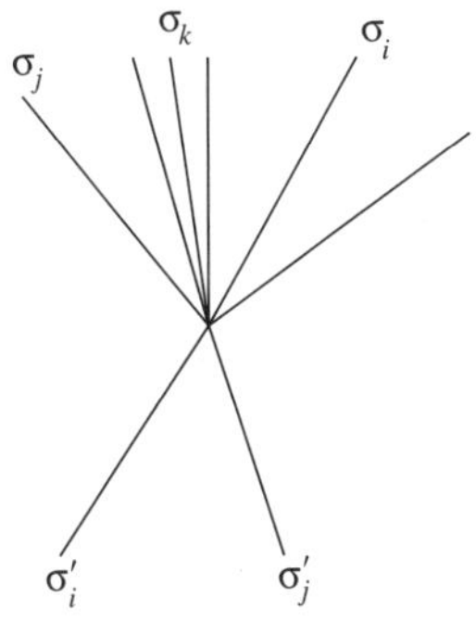


**Figure 7.10**

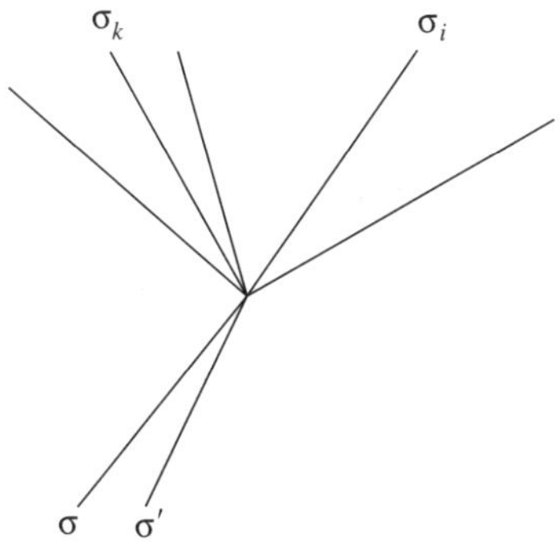


**Figure 7.11**

(iii) *Let $\sigma$, $\sigma'$ be the sizes of two incoming fronts, say of the families $i$, $j$. Let $u_l, u_m, u_r$ be the left, middle and right states before interaction, so that (7.26) holds. Introducing the auxiliary right state $\tilde{u}_r$ as in (7.27), one has (Figs 7.7–7.8)*

$$|\tilde{u}_r - u_r| = \mathcal{O}(1) \cdot |\sigma\sigma'|. \tag{7.33}$$

(iv) *Let a non-physical front connecting the states $u_l, u_m$ interact with an $i$-wave of size $\sigma$, connecting $u_m, u_r$, so that (7.29) holds. Defining the auxiliary right state $\tilde{u}_r$ as in (7.30), one has (Fig. 7.9)*

$$|\tilde{u}_r - u_r| - |u_m - u_l| = \mathcal{O}(1) \cdot |\sigma||u_m - u_l|. \tag{7.34}$$

*All of the above quantities $\mathcal{O}(1)$ remain uniformly bounded, as long as $u_l, u_m, u_r \in B(0, \delta_1)$.*

*Proof of (i).* Call $u_l, u_m, u_r$ the left, middle and right states before the interaction. By the assumptions, we have

$$u_m = \Psi_i(\sigma_i')(u_l), \qquad u_r = \Psi_j(\sigma_j')(u_m). \tag{7.35}$$

Consider the map

$$(u_l, u_r) \mapsto (E_1(u_l, u_r), \ldots, E_n(u_l, u_r)) \doteq (\sigma_1, \ldots, \sigma_n) \tag{7.36}$$

which associates to any two states $u_l, u_r$ the strengths of the waves determined by the corresponding Riemann problem. By the analysis in Chapter 5, this map is well defined provided that $u_l, u_r$ are suitably close to each other. Moreover, by the implicit function theorem, the map (7.36) is twice continuously differentiable with Lipschitz-continuous second derivatives.

For a given left state $u_l$, we now define the composed functions $\Phi_k$, $k = 1, \ldots, n$, by setting

$$\begin{aligned}
\Phi_i(\sigma_i', \sigma_j') &\doteq \sigma_i - \sigma_i' = E_i(u_l,\ \Psi_j(\sigma_j') \circ \Psi_i(\sigma_i')(u_l)) - \sigma_i' \\
\Phi_j(\sigma_i', \sigma_j') &\doteq \sigma_j - \sigma_j' = E_j(u_l,\ \Psi_j(\sigma_j') \circ \Psi_i(\sigma_i')(u_l)) - \sigma_i' \\
\Phi_k(\sigma_i', \sigma_j') &\doteq \sigma_k = E_k(u_l,\ \Psi_j(\sigma_j') \circ \Psi_i(\sigma_i')(u_l)) \quad \text{if } k \neq i, j.
\end{aligned} \tag{7.37}$$

The $\Phi_k$ are $\mathcal{C}^2$ functions of $\sigma_i', \sigma_j'$ with Lipschitz-continuous second derivatives, depending continuously also on the left state $u_l$. Observing that

$$\Phi_k(\sigma_i', 0) = \Phi_k(0, \sigma_j') = 0 \quad \text{for all } \sigma_i', \sigma_j', k = 1, \ldots, n, \tag{7.38}$$

by the first part of Lemma 2.5 we conclude that

$$\Phi_k(\sigma_i', \sigma_j') = \mathcal{O}(1) \cdot |\sigma_i'\sigma_j'| \tag{7.39}$$

for all $k$. This establishes (i).

*Proof of (ii).* In this case, the left, middle and right states are related by

$$u_m = \Psi_i(\sigma)(u_l), \qquad u_r = \Psi_i(\sigma')(u_m). \tag{7.40}$$

As before, let $(E_1, \ldots, E_n)(u_l, u_r)$ be the strengths of the waves in the Riemann problem $(u_l, u_r)$ determined by the interaction. Observe that, by (7.19), in the case where $\sigma, \sigma' \geq 0$ the states $u_l, u_m, u_r$ in (7.40) all lie on a single $i$-rarefaction curve. Hence (5.8) implies

$$E_i = \sigma + \sigma', \qquad E_k = 0 \quad \text{if } k \neq i. \tag{7.41}$$

We now define the functions $\Phi_k$ by setting

$$\begin{aligned} \Phi_i(\sigma, \sigma') &\doteq \sigma_i - \sigma - \sigma' = E_i(u_l,\ \Psi_i(\sigma') \circ \Psi_i(\sigma)(u_l)) - \sigma - \sigma' \\ \Phi_k(\sigma, \sigma') &\doteq \sigma_k = E_k(u_l,\ \Psi_i(\sigma') \circ \Psi_i(\sigma)(u_l)) \quad \text{if } k \neq i, j. \end{aligned} \tag{7.42}$$

As before, the $\Phi_k$ are $\mathcal{C}^2$ functions of $\sigma, \sigma'$ with Lipschitz-continuous second derivatives, depending continuously also on the left state $u_l$. We observe that

$$\Phi_k(\sigma, 0) = \Phi_k(0, \sigma') = 0 \quad k = 1, \ldots, n. \tag{7.43}$$

Moreover, since

$$\Phi_k(\sigma, \sigma') = 0 \quad \text{for all } \sigma, \sigma' \geq 0,\ k = 1, \ldots, n,$$

the continuity of the second derivatives implies

$$\frac{\partial^2 \Phi_k}{\partial \sigma\, \partial \sigma'}(0, 0) = 0 \quad k = 1, \ldots, n. \tag{7.44}$$

Using (7.43)–(7.44), by the second part of Lemma 2.5 we conclude that

$$\Phi_k(\sigma, \sigma') = \mathcal{O}(1) \cdot |\sigma\sigma'|(|\sigma| + |\sigma'|) \tag{7.45}$$

for all $k$. This establishes (ii).

*Proof of (iii).* In the case where $i > j$, consider the map

$$(\sigma, \sigma') \mapsto \Phi(\sigma, \sigma') \doteq \tilde{u}_r - u_r = \Psi_i(\sigma) \circ \Psi_j(\sigma')(u_l) - \Psi_j(\sigma') \circ \Psi_i(\sigma)(u_l). \tag{7.46}$$

This map is $\mathcal{C}^2$ with Lipschitz-continuous second derivatives and satisfies

$$\Phi(\sigma, 0) = \Phi(0, \sigma') = 0 \quad \text{for all } \sigma, \sigma'. \tag{7.47}$$

The estimate (7.33) is thus a consequence of Lemma 2.5.

In the case where $i = j$, we apply the same arguments to the map

$$(\sigma, \sigma') \mapsto \Phi(\sigma, \sigma') \doteq \tilde{u}_r - u_r = \Psi_i(\sigma + \sigma')(u_l) - \Psi_i(\sigma') \circ \Psi_i(\sigma)(u_l). \tag{7.48}$$

*Proof of (iv).* For a given left state $u_l$, consider the map $\Phi : \mathbb{R}^n \times \mathbb{R} \mapsto \mathbb{R}^n$,

$$(v, \sigma) \mapsto \Phi(v, \sigma) \doteq (\Psi_i(\sigma)(u_l) + v) - \Psi_i(\sigma)(u_l + v).$$

Observe that $\Phi$ is well defined in a neighbourhood of the origin, and twice continuously differentiable with Lipschitz-continuous second derivatives. Moreover,

$$\Phi(v, 0) = \Phi(0, \sigma) = 0 \quad \text{for all } v, \sigma. \tag{7.49}$$

Applying Lemma 2.5 once again, we conclude that

$$\Phi(v, \sigma) = \mathcal{O}(1) \cdot |v|\,|\sigma|. \tag{7.50}$$

Choosing $v = u_m - u_l$ we have

$$\Phi(v, \sigma) = (\tilde{u}_r + (u_m - u_l)) - u_r = (\tilde{u}_r - u_r) - (u_l - u_m).$$

Hence (7.34) is a consequence of (7.50).

We now observe that, in the above proof, the various functions $\Phi$, $\Phi_i$ depend continuously on the left state $u_l$, together with their first two derivatives. By Remark 2.2, the quantities $\mathcal{O}(1)$ thus remain uniformly bounded as $u_l$ ranges over a compact set.

**3. Bounds on the total variation.** Let $u = u(t, x)$ be a piecewise constant approximate solution constructed by our algorithm. We seek an estimate of the form

$$\text{Tot. Var.}\{u(t, \cdot\} \leq \delta_1, \tag{7.51}$$

valid for all times $t \geq 0$. Observe that, as long as (7.51) remains valid, at any interaction between two wave-fronts of $u$ the left, middle and right states will all be contained in the ball $B(0, \delta_1)$. Therefore, the function $u$ can be prolonged beyond the interaction time in terms of the accurate or the simplified Riemann solver.

To keep track of the total variation, we introduce two functionals, defined in terms of the strengths of the various wave-fronts. At a fixed time $t$ let $x_\alpha$, $\alpha = 1, \ldots, N$, be the locations of the fronts in $u(t, \cdot)$. Moreover, let $|\sigma_\alpha|$ be the strength of the wave-front at $x_\alpha$. In case of a non-physical front, we simply take

$$|\sigma_\alpha| \doteq |u(t, x_\alpha+) - u(t, x_\alpha-)|. \tag{7.52}$$

In the following, for notational convenience we say that non-physical fronts belong to an $(n+1)$-th characteristic family.

Consider the two functionals

$$V(t) \doteq \sum_\alpha |\sigma_\alpha|, \tag{7.53}$$

measuring the *total strength of waves* in $u(t, \cdot)$, and

$$Q(t) \doteq \sum_{(\alpha,\beta)\in\mathcal{A}} |\sigma_\alpha \sigma_\beta|, \tag{7.54}$$

measuring the *wave interaction potential*. In (7.54), the summation ranges over all couples of approaching wave-fronts. More precisely, we say that two fronts, located at points

$x_\alpha < x_\beta$ and belonging to the characteristic families $k_\alpha, k_\beta \in \{1, \ldots, n+1\}$ respectively, are *approaching* iff $k_\alpha > k_\beta$ or else if $k_\alpha = k_\beta$ and at least one of the waves is a genuinely non-linear shock. Observe that, at this stage, the functionals $V$, $Q$ refer to a particular front tracking approximate solution and are defined only outside interaction times.

Now consider a time $\tau$ where two fronts of strengths $|\sigma|$, $|\sigma'|$ interact. We wish to estimate the change in the values of $V$, $Q$ across time $\tau$. Concerning $V$, the estimates (7.31) (if the accurate Riemann solver is used) or (7.33)–(7.34) (if the simplified Riemann solver is used) yield

$$V(\tau+) - V(\tau-) = \mathcal{O}(1) \cdot |\sigma\sigma'|. \tag{7.55}$$

Concerning $Q$, observe that after time $\tau$ the two fronts $\sigma$, $\sigma'$ are no longer approaching. On the other hand, the new wave-fronts generated by the interaction may approach all other waves. Using the estimates (7.31)–(7.34) we thus obtain

$$Q(\tau+) - Q(\tau-) = -|\sigma\sigma'| + \mathcal{O}(1) \cdot |\sigma\sigma'| \cdot V(\tau-). \tag{7.56}$$

A typical interaction is illustrated in Fig. 7.12. Assume that the incoming fronts belong to distinct families $i > j$ and have sizes $\sigma_i'$, $\sigma_j'$ respectively. Call $\sigma_1, \ldots, \sigma_n$ the sizes of the outgoing waves. By (7.31), since all fronts have small strength, it follows that $\sigma_i$ and $\sigma_i'$ have the same sign. Hence the (non-interacting) fronts $\sigma_\alpha$ that were approaching $\sigma_i'$ before time $\tau$ are precisely the same fronts that approach $\sigma_i$ after time $\tau$. Similarly, the fronts $\sigma_\alpha$ (not involved in the interaction) that were approaching $\sigma_j'$ before time $\tau$ are precisely the same fronts that approach $\sigma_j$ after time $\tau$. In addition, after time $\tau$ the quantity $Q$ contains a number of new terms of the form $|\sigma_k\sigma_\alpha|$, with $k \neq i, j$. By (7.31), the sum of all these terms is bounded by $\mathcal{O}(1) \cdot |\sigma_i'\sigma_j'| V(\tau-)$. The other interaction cases are estimated in a similar way, using (7.32)–(7.34).

If $V$ remains sufficiently small, (7.56) implies

$$Q(\tau+) - Q(\tau-) \leq -\frac{|\sigma\sigma'|}{2}. \tag{7.57}$$

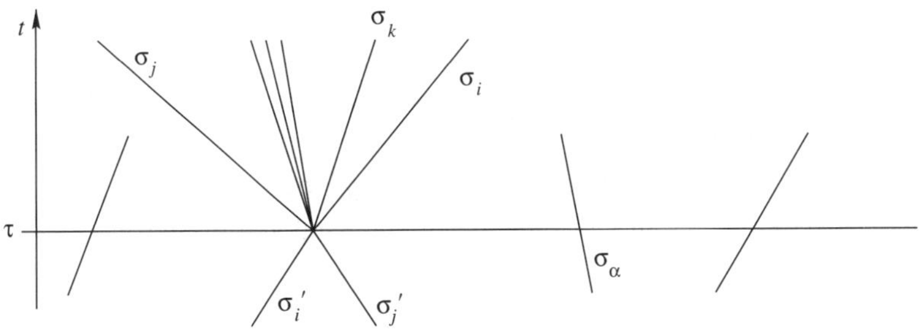


**Figure 7.12**

By (7.57) and (7.55) we can thus choose a constant $C_0$ large enough and $\delta_2 \in ]0, \delta_1]$ so that (7.57) holds and, moreover, the quantity

$$\Upsilon(t) \doteq V(t) + C_0 Q(t) \tag{7.58}$$

decreases at every interaction time $\tau$, provided that $V(\tau-) \leq \delta_2$. Let $C_1$ be a constant such that

$$\frac{1}{C_1} \cdot \text{Tot. Var.}\{u(t, \cdot)\} \leq V(t) \leq C_1 \cdot \text{Tot. Var.}\{u(t, \cdot)\}. \tag{7.59}$$

Let $\delta_3 > 0$ be so small that

$$C_1^2 \delta_3 + C_0 C_1^3 \delta_3^2 \leq \delta_2.$$

If Tot. Var. $\{u(0, \cdot)\} \leq \delta_3$, observing that $Q(t) \leq V^2(t)$ for all $t \geq 0$ we now have

$$\begin{aligned} \text{Tot. Var.}\{u(t, \cdot)\} &\leq C_1[V(t) + C_0 Q(t)] \\ &\leq C_1[V(0) + C_0 Q(0)] \\ &\leq C_1[C_1\delta_3 + C_0(C_1\delta_3)^2] \\ &\leq \delta_2. \end{aligned} \tag{7.60}$$

Indeed, if (7.60) holds before an interaction time $\tau$, then $V(\tau-) \leq \delta_2$. Hence the quantity $\Upsilon$ in (7.58) decreases across the interaction, and (7.60) holds also after time $\tau$. By induction, (7.60) holds for all $t \geq 0$.

**4. Bounds on the number of wave-fronts.** To prove that the total number of wave-fronts remains finite, we recall that the accurate Riemann solver is used when the strengths of the interacting waves satisfy $|\sigma\sigma'| \geq \rho$. This can happen only finitely many times. Indeed, by (7.57) at such times one has

$$Q(\tau+) - Q(\tau-) \leq -\rho/2.$$

Therefore, new physical fronts are introduced only at a number $\leq 2Q(0)/\rho$ of interaction points; hence their total number is finite. In turn, a new non-physical front is generated only when two physical fronts interact. Clearly, any two physical fronts can cross only once. Hence the total number of non-physical fronts is also finite.

The above steps 1–4 show that, for any values of the parameters $\delta, \rho$ that define the algorithm, if the initial data are small enough, then a piecewise constant approximate solution can be constructed for all times $t \geq 0$. The next two steps will show that, given any $\varepsilon > 0$, if the parameters $\delta, \rho$ are chosen small enough, then the function $u$ constructed by the algorithm is an $\varepsilon$-approximate front tracking solution of the Cauchy problem (7.1)–(7.2).

**5. The strength of each rarefaction front is small.** The aim of this step is to prove that, for some constant $C_2$, the strength of every rarefaction front satisfies

$$|\sigma_\alpha| \leq C_2\delta \qquad \alpha \in \mathcal{R}. \tag{7.61}$$

By construction, at a time $t_0$ where a new rarefaction front is introduced by the accurate Riemann solver, its size is $\sigma_\alpha(t_0) \in ]0, \delta]$. Let us see how this size can change at subsequent times. Two rarefaction fronts of the same family never interact. When a rarefaction hits a shock of the same family, its size will decrease owing to a cancellation. When a rarefaction front crosses a non-physical wave, by construction its size remains unchanged. On the other hand, by subsequent interactions with fronts of other families, a rarefaction front may increase its initial strength. To keep track of its size $\sigma_\alpha(t)$ at any time $t > t_0$, consider the quantity

$$V_\alpha(t) = \sum_{\beta \in \mathcal{A}(\alpha)} |\sigma_\beta|, \tag{7.62}$$

where the summation is restricted to the set $\mathcal{A}(\alpha)$ of all wave-fronts which are approaching $\sigma_\alpha$. Consider an interaction time $\tau > t_0$.

CASE 1: The interaction does not involve the front $\sigma_\alpha$. Recalling the interaction estimates (7.31)–(7.34) we then have

$$\Delta\sigma_\alpha(\tau) = 0, \qquad \Delta V_\alpha(\tau) + C_0 \Delta Q(\tau) \le 0. \tag{7.63}$$

CASE 2: At time $\tau$, the rarefaction front $\sigma_\alpha$ interacts with another front of strength $|\sigma_\beta|$. In this case, our interaction estimates imply that for some constant $C_3$ the following holds:

$$\Delta V_\alpha(\tau) = -|\sigma_\beta|, \qquad \Delta Q(\tau) < 0, \qquad \Delta\sigma_\alpha(\tau) \le C_3 |\sigma_\alpha(\tau-)|\,|\sigma_\beta|. \tag{7.64}$$

By (7.63)–(7.64), the map

$$t \mapsto \sigma_\alpha(t) \cdot \exp\{C_3[V_\alpha(t) + C_0 Q(t)]\} \tag{7.65}$$

is non-increasing in time. As a consequence, for all $t > t_0$ we have

$$\begin{aligned} \sigma_\alpha(t) &\le \sigma_\alpha(t_0) \cdot \exp\{C_3[V_\alpha(t_0) + C_0 Q(t_0)]\} \\ &\le \delta \cdot \exp\{C_3[V(0) + C_0 Q(0)]\} \\ &\le \delta e^{C_3 \delta_2}, \end{aligned}$$

where $\delta_2$ is the constant in (7.60). This establishes (7.61), with $C_2 \doteq e^{C_3\delta_2}$. Given $\varepsilon > 0$, we can now choose $\delta$ small enough so that the right hand side of (7.61) is $< \varepsilon$.

**6. The total strength of all non-physical fronts is small.** For this estimate, we first show that, for some constant $C_4$, the strength of each non-physical front satisfies

$$|\sigma_\alpha| \le C_4 \rho \quad \alpha \in \mathcal{NP}. \tag{7.66}$$

Indeed, a new non-physical front is created at a time $t_0$ from the collision of two (physical) waves whenever the product of their strengths is $|\sigma\sigma'| < \rho$. By the intraction estimate (7.33), the initial strength of the non-physical front is

$$\sigma_\alpha(t_0) = \mathcal{O}(1) \cdot |\sigma\sigma'| = \mathcal{O}(1) \cdot \rho. \tag{7.67}$$

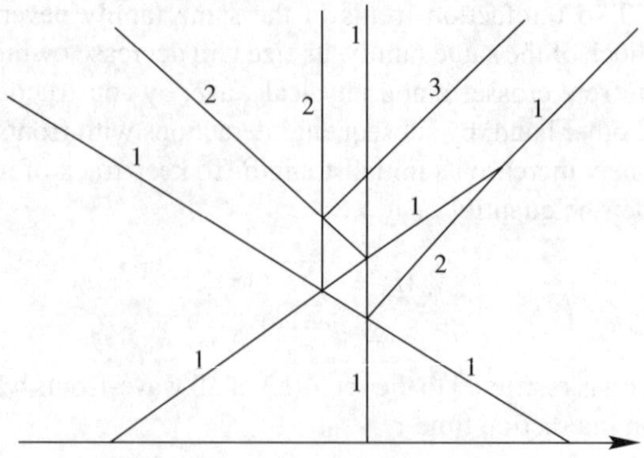


**Figure 7.13**

This initial strength may increase due to subsequent interactions with other fronts. To keep track of its size $\sigma_\alpha(t)$ at any time $t > t_0$, we again consider the quantity (7.62). Using the interaction estimates (7.31)–(7.34), we conclude that for a suitable constant $C_3$, the map (7.65) is non-increasing in time. Repeating the argument in step 5, we thus obtain the bound (7.66).

Next, to each wave-front in $u$ we attach an integer number, counting how many interactions were needed to produce such a front. More precisely, the *generation order* of a front is inductively defined as follows (Fig. 7.13):

- All fronts generated by the Riemann problems at the initial time $t = 0$ have generation order $k = 1$.
- Let two incoming fronts interact, say of the families $i, i' \in \{1, \ldots, n+1\}$, with generation orders $k, k'$. The orders of the outgoing fronts are then defined as follows:

CASE 1: $i \neq i'$. Then

- the outgoing $i$-wave and $i'$-wave have the same orders $k, k'$ as the incoming ones;
- the outgoing fronts of every other family $j \neq i, i'$ have order $\max\{k, k'\} + 1$.

CASE 2: $i = i'$. Then

- the outgoing front of the $i$-th family has order $\min\{k, k'\}$;
- the outgoing fronts of every family $j \neq i$ have order $\max\{k, k'\} + 1$.

For $k \geq 1$, call $V_k(t)$ the sum at time $t$ of the strengths of all waves of order $\geq k$. To obtain an a priori bound on $V_k$, define $Q_k(t) = \sum |\sigma_\alpha \sigma_\beta|$, where the sum extends over all couples of approaching waves in $u(t, \cdot)$, say of order $k_\alpha, k_\beta$, with $\max\{k_\alpha, k_\beta\} \geq k$. Moreover, call $I_k$ the set of times where two waves of order $k_\alpha, k_\beta$ interact, with $\max\{k_\alpha, k_\beta\} = k$. Repeating the arguments used to derive (7.56), but now keeping track of the order of the wave-fronts involved in the interactions, we obtain

$$
\begin{aligned}
\Delta V_k(t) &= 0 \quad t \in I_1 \cup \cdots \cup I_{k-2} \\
\Delta V_k(t) + C_0 \Delta Q_{k-1}(t) &\le 0 \quad t \in I_{k-1} \cup I_k \cup I_{k+1} \ldots \\
\Delta Q_k(t) + C_0 \Delta Q(t) V_k(t-) &\le 0 \quad t \in I_1 \cup \cdots \cup I_{k-2} \\
\Delta Q_k(t) + C_0 \Delta Q_{k-1}(t) V(t-) &\le 0 \quad t \in I_{k-1} \\
\Delta Q_k(t) &\le 0 \quad t \in I_k \cup I_{k+1} \cup \ldots.
\end{aligned}
\tag{7.68$_{1\text{–}5}$}
$$

In other words, $(7.68)_1$ says that the total strength of waves of order $\ge k$ is unaffected by interactions involving wave-fronts of order $\le k-2$. When one of the incoming waves has order $\ge k-1$, by $(7.68)_2$ the growth of $V_k$ is counterbalanced by the decrease of the interaction potential $Q_{k-1}$. The estimates $(7.68)_{3-5}$ provide bounds on the increase of the potentials $Q_k$.

In the following, we write $[s]_+ = \max\{s, 0\}$ and $[s]_- = \max\{-s, 0\}$ for the positive and negative parts of a real number $s$. Observing that $V_1 = V,\ Q_1 = Q$ and $V_k(0+) = Q_k(0+) = 0$ if $k \ge 2$, the estimates (7.68) imply

$$
V_k(t) \le C_0 \sum_{0<\tau\le t} [\Delta Q_{k-1}(\tau)]_- \tag{7.69}
$$

$$
\begin{aligned}
Q_k(t) &\le \sum_{0<\tau\le t} [\Delta Q_k(\tau)]_+ \\
&\le C_0 \sum_{0<\tau\le t} [\Delta Q(\tau)]_- \cdot \sup_t V_k(t) + C_0 \sum_{0<\tau\le t} [\Delta Q_{k-1}(\tau)]_- \cdot \sup_t V(t),
\end{aligned}
\tag{7.70}
$$

both of which are valid for every $t > 0$ and $k \ge 2$. Furthermore,

$$
0 \le Q_k(t) = \sum_{0<\tau\le t} \{[\Delta Q_k(\tau)]_+ - [\Delta Q_k(\tau)]_-\}.
$$

Recalling that the quantity $\Upsilon(t) \doteq V(t) + C_0 Q(t)$ is non-increasing, one has

$$
V_k(t) \le V(t) \le \Upsilon(t) \le \Upsilon(0), \tag{7.71}
$$

$$
\sum_{0<\tau<\infty} [\Delta Q(\tau)]_- \le Q(0) \le \Upsilon(0). \tag{7.72}
$$

Calling

$$
\widetilde{Q}_k \doteq \sum_{t>0} [\Delta Q_k(t)]_+, \qquad \widetilde{V}_k \doteq \sup_{t>0} V_k(t),
$$

from (7.68)–(7.72) we deduce the sequence of inequalities (valid for $k \ge 2$)

$$
\begin{cases}
\widetilde{V}_k \le C_0 \widetilde{Q}_{k-1}, \\
\widetilde{Q}_k \le C_0 \Upsilon(0) \widetilde{V}_k + C_0 \widetilde{Q}_{k-1} \Upsilon(0) \le (C_0^2 + C_0) \Upsilon(0) \widetilde{Q}_{k-1}.
\end{cases}
\tag{7.73}
$$

If Tot. Var. $(u(0+, \cdot))$ is sufficiently small, $\Upsilon(0) \doteq V(0) + C_0 Q(0)$ will satisfy

$$
\gamma \doteq (C_0^2 + C_0) \Upsilon(0) < 1. \tag{7.74}
$$

In this case, for every $t > 0$ and $k \geq 2$, (7.73) and (7.74) yield by induction

$$Q_k(t) \leq \widetilde{Q}_k \leq \delta_2 \gamma^k, \qquad V_k(t) \leq \widetilde{V}_k \leq \delta_2 C_0 \gamma^{k-1}. \tag{7.75}$$

Next, let us make a quick count of the number of wave-fronts in $u$ having a given generation order. Let $N$ be the number of wave-fronts in $u(0+, \cdot)$. At any time $t > 0$, the number of first-order fronts in $u(t, \cdot)$ is thus $\leq N$. From each interaction between fronts of first-order, recalling that rarefaction waves are partitioned into pieces of size $< \delta$, a number $\mathcal{O}(1) \cdot 1/\delta$ of fronts of second-order is generated. The total number of fronts of second-order is thus $\mathcal{O}(1) \cdot N^2/\delta$. By induction, it is clear that the total number of fronts of order $\leq k$ in $u(t, \cdot)$ can be estimated by some polynomial function of $N, \delta^{-1}$, say

$$[\text{number of fronts of order} \leq k] \leq P_k(N, \delta^{-1}). \tag{7.76}$$

The particular form of $P_k$ is of no interest here.

To estimate the total strength of all non-physical waves in $u(t, \cdot)$ we keep track of the fronts having generation order $> k$ and $\leq k$ separately. Using (7.75), (7.76) and (7.66) we deduce that

$$\begin{aligned}
\big[\text{total strength of non-physical fronts in } u(t, \cdot)\big] &= \sum_{\substack{\alpha \in \mathcal{NP} \\ \text{order}(\alpha) > k}} |\sigma_\alpha(t)| + \sum_{\substack{\alpha \in \mathcal{NP} \\ \text{order}(\alpha) \leq k}} |\sigma_\alpha(t)| \\
&\leq \big[\text{total strength of all fronts of order } > k\big] \\
&\quad + \big[\text{number of fronts of order} \leq k\big] \cdot \big[\text{maximum strength of non-physical fronts}\big] \\
&\leq \delta_2 C_0 \gamma^k + P_k(N, \delta^{-1}) \cdot C_4 \rho.
\end{aligned} \tag{7.77}$$

For any given $\varepsilon > 0$, since $\gamma < 1$ we can now choose $k$ large enough so that $\delta_2 C_0 \gamma^k < \varepsilon/2$. We then choose $\rho > 0$ small enough so that the second term on the right hand side of (7.77) is $< \varepsilon/2$. For all $t \geq 0$, this achieves

$$\big[\text{total strength of all non-physical fronts in } u(t, \cdot)\big] < \varepsilon, \tag{7.78}$$

completing the proof of Theorem 7.2. □

**Remark 7.2.** With Definition 7.1 we introduced a rather general concept of front tracking approximate solutions. The proof of Theorem 7.1 will show that every limit of these approximations provides an entropy weak solution to the system of conservation laws. Furthermore, in Chapters 8 and 9 we will prove that all these approximations actually converge to a unique limit, continuously depending on the initial data.

In order to establish various qualitative properties of solutions, however, it is often a technical advantage to work with more special sequences of front tracking approximations, which enjoy a number of additional properties. In this connection, we observe that, within the proof of Theorem 7.2, we constructed a family of front tracking approximations satisfying not only the requirements of Definition 7.1, but also the following properties:

(FT1) At each time $\tau > 0$, at most one interaction takes place, involving exactly two incoming fronts.

(FT2) At each interaction point, if one of the incoming fronts belongs, say, to the $i$-th family, then there is at most one single outgoing front of the $i$-th family.

(FT3) At each time $\tau$ where two fronts interact, the strengths of the outgoing wave-fronts satisfy one of the estimates (7.31)–(7.34), depending on the type of the incoming fronts. In particular, calling $\sigma, \sigma'$ the sizes of the incoming fronts, one has

$$Q(\tau+) - Q(\tau-) \leq -|\sigma\sigma'|/2, \qquad \Upsilon(\tau+) \leq \Upsilon(\tau-).$$

We recall that $Q, \Upsilon$ are the functionals defined at (7.54) and (7.58).

According to Remark 7.1, if we let each shock (or contact discontinuity) travel exactly with Rankine–Hugoniot speed, and each rarefaction front travel exactly with the characteristic speed $\dot{x}_\alpha = \lambda_{k_\alpha}(u^+)$ of its right state, we might have three or more fronts interacting at the same point. This would violate the condition (FT1) and force us to prove more difficult interaction estimates in order to control the total strength of all waves. However, observing that two rarefaction fronts (or two contact discontinuities) of the same family never interact with each other, if any one particular family $i^* \in \{1, \ldots, n\}$ is singled out, together with (FT1) the following additional properties can be achieved.

(FT4) The speed of every genuinely non-linear shock satisfies (7.7) together with $\lambda_{k_\alpha}(u^+) < \dot{x}_\alpha < \lambda_{k_\alpha}(u^-)$. The speed of every genuinely non-linear rarefaction front satisfies (7.9) together with $\lambda_{k_\alpha}(u^-) < \dot{x}_\alpha < \lambda_{k_\alpha}(u^+)$. Each $i^*$-rarefaction front (or contact discontinuity) travels exactly with speed $\dot{x}_\alpha = \lambda_{i^*}(u^+)$.

## 7.4 Existence of solutions

To prove Theorem 7.1, fix a sequence $(\varepsilon_\nu)_{\nu\geq 1}$ decreasing to zero. For each $\nu \geq 1$, Theorem 7.2 yields the existence of an $\varepsilon_\nu$-approximate solution $u_\nu$ of the Cauchy problem (7.1)–(7.2). By the previous analysis, the functions $u_\nu(t, \cdot)$ have uniformly bounded total variation. Moreover, the maps $t \mapsto u_\nu(t, \cdot)$ are uniformly Lipschitz continuous with values in $\mathbf{L}^1(\mathbb{R}; \mathbb{R}^n)$. Indeed,

$$\begin{aligned} \|u_\nu(t) - u_\nu(s)\|_{\mathbf{L}^1} &= \mathcal{O}(1) \cdot (t-s) \cdot [\text{total strength of all wave-fronts}] \\ &\quad \cdot [\text{maximum speed}] \\ &\leq L\,(t-s) \end{aligned} \tag{7.79}$$

for some constant $L$ independent of $\nu$. We can thus apply Theorem 2.4 and extract a subsequence which converges to some limit function $u$ in $\mathbf{L}^1_{\text{loc}}$.

Since $\|u_\nu(0) - \bar{u}\|_{\mathbf{L}^1} \to 0$, by (7.79) the initial condition (7.2) clearly holds.

To prove that $u$ is a weak solution of the Cauchy problem, it remains to show that, for every $\phi \in \mathcal{C}^1_c$ with compact support contained in the open half plane where $t > 0$, one has

$$\int_0^\infty \int_{-\infty}^\infty \{\phi_t(t,x)u(t,x) + \phi_x(t,x)f(u(t,x))\}\,dx\,dt = 0. \tag{7.80}$$

Since the $u_\nu$ are uniformly bounded and $f$ is uniformly continuous on bounded sets, it suffices to prove that

$$\lim_{\nu\to 0}\left[\int_0^\infty\int_{-\infty}^\infty \{\phi_t(t,x)u_\nu(t,x)+\phi_x(t,x)f(u_\nu(t,x))\}\,dx\,dt\right]=0. \tag{7.81}$$

Choose $T>0$ such that $\phi(t,x)=0$ whenever $t\notin\,]0,T[$ . For a fixed $\nu$, at any time $t$ call $x_1(t)<\cdots<x_N(t)$ the points where $u_\nu(t,\cdot)$ has a jump, and set

$$\Delta u_\nu(t,x_\alpha)\doteq u_\nu(t,x_\alpha+)-u_\nu(t,x_\alpha-),$$
$$\Delta f(u_\nu(t,x_\alpha))\doteq f(u_\nu(t,x_\alpha+))-f(u_\nu(t,x_\alpha-)).$$

Observe that the polygonal lines $x=x_\alpha(t)$ subdivide the strip $[0,T]\times\mathbb{R}$ into finitely many regions $\Gamma_j$ where $u_\nu$ is constant. Introducing the vector

$$\Phi\doteq(\phi\cdot u_\nu,\ \phi\cdot f(u_\nu)),$$

by the divergence theorem the double integral in (7.81) can be written as

$$\sum_j\iint_{\Gamma_j}\operatorname{div}\Phi(t,x)\,dx\,dt=\sum_j\int_{\partial\Gamma_j}\Phi\cdot\mathbf{n}\,d\sigma. \tag{7.82}$$

Here $\partial\Gamma_j$ is the oriented boundary of $\Gamma_j$, while $\mathbf{n}$ denotes an outer normal (Fig. 7.14). Observe that $\mathbf{n}\,d\sigma=\pm(\dot x_\alpha,-1)\,dt$ along each polygonal line $x=x_\alpha(t)$, while $\phi(t,x)=0$ along the lines $t=0$, $t=T$. By (7.82) the expression within square brackets in (7.81) is computed by

$$\int_0^T\sum_\alpha[\dot x_\alpha(t)\cdot\Delta u_\nu(t,x_\alpha)-\Delta f(u_\nu(t,x_\alpha))]\phi(t,x_\alpha(t))\,dt. \tag{7.83}$$

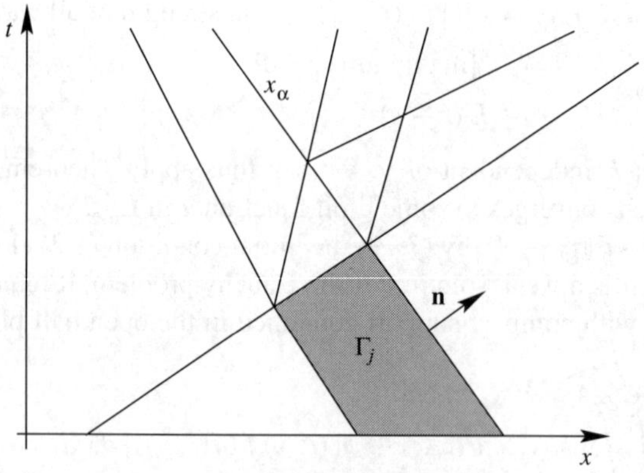


**Figure 7.14**

To estimate the above integral, let $|\sigma_\alpha|$ be the strength of the wave-front at $x_\alpha$. According to the estimates on $E_\alpha$ in Lemma 7.1, if this wave is a shock, a rarefaction or a contact discontinuity, one has

$$\dot{x}_\alpha(t) \cdot \Delta u_\nu(t, x_\alpha) - \Delta f(u_\nu(t, x_\alpha)) = \mathcal{O}(1) \cdot \varepsilon_\nu |\sigma_\alpha|.$$

On the other hand, if the wave at $x_\alpha$ is non-physical, then

$$\dot{x}_\alpha \cdot \Delta u_\nu(t, x_\alpha) - \Delta f(u_\nu(t, x_\alpha)) = \mathcal{O}(1) \cdot |\sigma_\alpha|.$$

Splitting the summation in (7.83) into physical (shocks, contacts or rarefactions) and non-physical wave-fronts, we conclude that

$$\begin{aligned}
&\limsup_{\nu\to\infty} \left| \sum_{\alpha\in\mathcal{S}\cup\mathcal{R}\cup\mathcal{NP}} [\dot{x}_\alpha(t) \cdot \Delta u_\nu(t, x_\alpha) - \Delta f(u_\nu(t, x_\alpha))]\phi(t, x_\alpha(t)) \right| \\
&\quad \le \left(\max_{t,x} |\phi(t,x)|\right) \cdot \limsup_{\nu\to\infty} \left\{ \sum_{\alpha\in\mathcal{S}\cup\mathcal{R}} \mathcal{O}(1) \cdot \varepsilon_\nu |\sigma_\alpha| + \sum_{\alpha\in\mathcal{NP}} \mathcal{O}(1) \cdot |\sigma_\alpha| \right\} \\
&\quad = 0. \qquad (7.84)
\end{aligned}$$

Indeed, for every $t > 0$ the total strength of waves in $u_\nu(t, \cdot)$ remains uniformly bounded, while the amount of non-physical waves is $< \varepsilon_\nu$; hence it approaches zero as $\nu \to \infty$. The limit (7.81) now follows from (7.84), showing that $u$ is a weak solution to the Cauchy problem.

Finally, let $\eta$ be a convex entropy for (7.1) with entropy flux $q$. In order to establish (7.4), it suffices to show that

$$\liminf_{\nu\to\infty} \int_0^\infty \int_{-\infty}^\infty \left\{ \eta(u_\nu)\varphi_t + q(u_\nu)\varphi_x \right\} dx\, dt \ge 0 \qquad (7.85)$$

for every non-negative $\varphi \in C_c^1$ with compact support contained in the half plane where $t > 0$. Choose $T > 0$ so that $\varphi$ vanishes outside the strip $[0, T] \times \mathbb{R}$. Using the divergence theorem again, for every $\nu$ the double integral in (7.85) can be computed as

$$\int_0^T \sum_\alpha [\dot{x}_\alpha(t) \cdot \Delta\eta(u_\nu(t, x_\alpha)) - \Delta q(u_\nu(t, x_\alpha))]\varphi(t, x_\alpha)\, dt, \qquad (7.86)$$

where the sum ranges over all jumps of $u_\nu(t, \cdot)$. We use here the notation

$$\Delta\eta \doteq \eta(u_\nu(t, x_\alpha+)) - \eta(u_\nu(t, x_\alpha-)), \qquad \Delta q \doteq q(u_\nu(t, x_\alpha+)) - q(u_\nu(t, x_\alpha-)).$$

Splitting the set of wave-fronts as in (7.84) and using the estimates for $E'_\alpha$ in Lemma 7.1, we now obtain

$$\begin{aligned}
&\liminf_{\nu\to\infty} \sum_{\alpha\in\mathcal{S}\cup\mathcal{R}\cup\mathcal{NP}} [\dot{x}_\alpha(t) \cdot \Delta\eta(u_\nu(t, x_\alpha)) - \Delta q(u_\nu(t, x_\alpha))]\varphi(t, x_\alpha)\, dt \\
&\quad \ge \left(\max_{t,x} \varphi(t,x)\right) \cdot \liminf_{\nu\to\infty} \left\{ -\sum_{\alpha\in\mathcal{R}\cup\mathcal{S}} \mathcal{O}(1) \cdot \varepsilon_\nu |\sigma_\alpha| - \sum_{\alpha\in\mathcal{NP}} \mathcal{O}(1) \cdot |\sigma_\alpha| \right\} \\
&\quad \ge 0. \qquad (7.87)
\end{aligned}$$

This completes the proof of Theorem 7.1. □

## 7.5 Further interaction estimates

In the proof of Theorem 7.1, a key role was played by the uniform estimates of the total amount of waves in a front tracking approximation. Indeed, we proved that the quantity $\Upsilon(t)$ in (7.58), determined by the wave-fronts in $u(t,\cdot)$, is non-increasing in time. In turn, by Helly's theorem this provides the compactness of the sequence of approximate solutions. The purpose of this section is to show that, more generally, these estimates apply to the wave-fronts of $u$ crossing arbitrary space-like curves. This result will be needed in Chapter 9, in the discussion of uniqueness for general entropy-admissible weak solutions.

**Definition 7.2.** Let $\hat{\lambda}$ be a fixed constant, strictly larger than the absolute value of all characteristic speeds. By a *space-like curve* we mean a curve of the form $\{t = \gamma(x);\ x \in\ ]a,b[\}$, with

$$|\gamma(x_2) - \gamma(x_1)| < \frac{x_2 - x_1}{\hat{\lambda}} \qquad \text{for all } a < x_1 < x_2 < b. \tag{7.88}$$

**Definition 7.3.** Given two space-like curves $\gamma : ]a,b[ \mapsto \mathbb{R}$ and $\gamma' : ]a',b'[ \mapsto \mathbb{R}$, we say that $\gamma$ *dominates* $\gamma'$ (and write $\gamma \prec \gamma'$) if $a \le a' < b' \le b$ and, moreover,

$$\gamma(x) \le \gamma'(x) \le \min\left\{\gamma(a) + \frac{x-a}{\hat{\lambda}},\ \gamma(b) + \frac{b-x}{\hat{\lambda}}\right\} \qquad \text{for all } x \in\ ]a',b'[\,. \tag{7.89}$$

Observe that, by (7.89), $\gamma'$ is entirely contained in a domain of determinacy for the curve $\gamma$ (Fig. 7.15).

Now consider a front tracking approximate solution $u$ and a space-like curve $\gamma$ not passing through any interaction point of $u$. We then define the functional $V(u;\gamma)$ as the sum of strengths of all fronts in $u$ crossing $\gamma$. Moreover, we define $Q(u;\gamma)$ as the wave interaction potential determined by the fronts in $u$ crossing $\gamma$. In analogy with (7.58), we also set

$$\Upsilon(u;\gamma) \doteq V(u;\gamma) + C_0 Q(u;\gamma).$$

Clearly, if $\gamma : \mathbb{R} \mapsto \mathbb{R}$ is the constant map $\gamma(x) \equiv t$, then $V(u;\gamma) = V(t)$ and $Q(u;\gamma) = Q(t)$, as defined in (7.53)–(7.54).

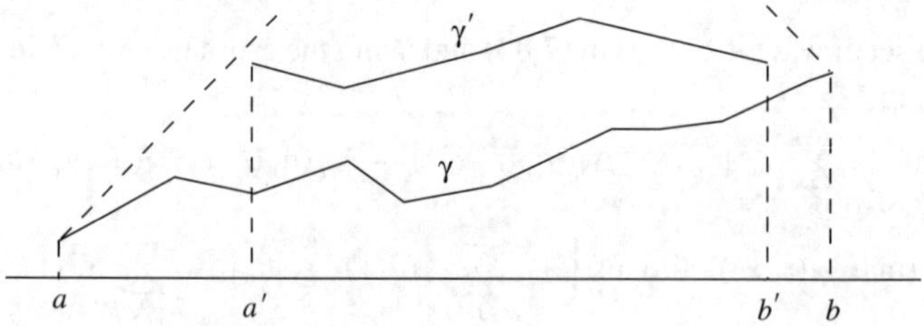


**Figure 7.15**

**Lemma 7.3.** *Let $u = u(t, x)$ be a piecewise constant approximate solution constructed by the front tracking algorithm, with suitably small total variation. Let $\gamma \prec \gamma'$ be any two space-like curves not passing through any interaction point of $u$. Then*

$$Q(u;\gamma') \leq Q(u;\gamma), \qquad \Upsilon(u;\gamma') \leq \Upsilon(u;\gamma). \tag{7.90}$$

*Moreover*

$$\text{Tot. Var.}\{u\,;\gamma'\} = \mathcal{O}(1)\cdot \text{Tot. Var.}\{u\,;\gamma\}. \tag{7.91}$$

*Proof.* For $a < x < b$, call

$$\phi(x) \doteq \gamma(a) + \frac{x-a}{\hat{\lambda}}, \qquad \psi(x) \doteq \gamma(b) + \frac{b-x}{\hat{\lambda}}$$

the boundaries of a domain of determinacy for the curve $\gamma$. Observe that, if $\tilde{\gamma}$ is the restriction of $\gamma$ to a subinterval $]\tilde{a}, \tilde{b}[\, \subset\, ]a, b[$ , then

$$V(u;\tilde{\gamma}) \leq V(u;\gamma), \qquad Q(u;\tilde{\gamma}) \leq Q(u;\gamma).$$

By possibly restricting $\gamma$ to a smaller interval, it is thus not restrictive to assume

$$\gamma'(a') = \phi(a'), \qquad \gamma'(b') = \psi(b').$$

Consider the domain

$$\Gamma \doteq \{(t,x);\ a < x < b,\ \gamma(x) \leq t \leq \min\{\phi(x),\ \gamma'(x),\ \psi(x)\}\}.$$

We can now construct a finite number of space-like curves $\gamma_j : ]a_j, b_j[ \mapsto \mathbb{R}$ such that (Fig. 7.16)

$$\gamma = \gamma_0 \prec \gamma_1 \prec \cdots \prec \gamma_m = \gamma',$$
$$\gamma_j(a_j) = \phi(a_j),\ \gamma_j(b_j) = \psi(b_j),$$

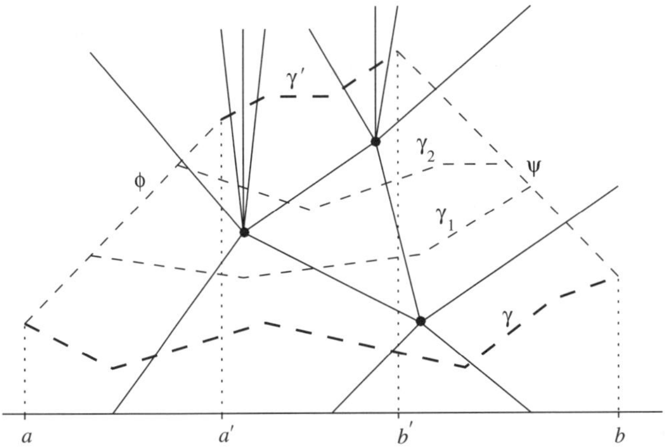


**Figure 7.16**

and such that each subregion $\Gamma_j \subseteq \Gamma$ bounded between two consecutive curves $\gamma_{j-1}$, $\gamma_j$ contains at most one interaction point of $u$. Since no wave-front can enter $\Gamma$ through the lateral boundaries, for each $j$ the estimates (7.55)–(7.57) yield

$$Q(u;\gamma_j) \leq Q(u;\gamma_{j-1}), \qquad \Upsilon(u;\gamma_j) \leq \Upsilon(u;\gamma_{j-1}),$$

provided that the total variation remains small. By induction on $j$ this yields (7.90).

The estimate (7.91) follows from (7.90) using (7.58)–(7.59) and recalling that $Q(u) \leq V^2(u)$. □

To prove the lower semicontinuity of the Glimm functionals $Q$ and $V + C_0 Q$ in Chapter 10 we shall need a more general interaction estimate. Observe that the bounds (7.31)–(7.32) refer to the interaction of two incoming wave-fronts. More generally, we now consider the case of three states $u_l, u_m, u_r$, and relate the waves generated by the Riemann problem $(u_l, u_r)$ with the waves generated by the two Riemann problems with data $(u_l, u_m)$ and $(u_m, u_r)$.

**Lemma 7.4.** *Consider any left, middle and right states $u_l, u_m, u_l$. Assume that the Riemann problem with data $(u_l, u_m)$ is solved by waves of sizes $\sigma_1', \ldots, \sigma_n'$, while the Riemann problem with data $(u_m, u_r)$ is solved by waves of sizes $\sigma_1'', \ldots, \sigma_n''$. Calling $\sigma_1, \ldots, \sigma_n$ the sizes of the waves in the Riemann problem $(u_l, u_r)$, one has the estimates*

$$\sum_i |\sigma_i - \sigma_i' - \sigma_i''| = \mathcal{O}(1) \cdot \sum_{\text{approaching}} |\sigma_j' \sigma_k''|. \tag{7.92}$$

The two waves $\sigma_j', \sigma_k''$ are here defined as *approaching* if either $j > k$ or else $j = k$ and at least one of the waves is a genuinely non-linear shock. The proof will be given in two steps.

1. We introduce the interaction potential

$$Q((u_l, u_m), (u_m, u_r)) \doteq \sum_{\text{approaching}} |\sigma_j' \sigma_k''|. \tag{7.93}$$

Assume first that the jump $(u_m, u_r)$ is solved in terms of a single $p$-wave, so that $\sigma_j'' = 0$ for $j \neq p$. Call $\sigma_+'$ the vector whose components are the waves approaching $\sigma_p''$. More precisely, we set

$$\sigma_-' \doteq (\sigma_1', \ldots, \sigma_{p-1}'), \qquad \sigma_+' \doteq (\sigma_p', \ldots, \sigma_n')$$

if one of the waves $\sigma_p', \sigma_p''$ is a genuinely non-linear shock, otherwise we set

$$\sigma_-' \doteq (\sigma_1', \ldots, \sigma_p'), \qquad \sigma_+' \doteq (\sigma_{p+1}', \ldots, \sigma_n').$$

Consider the functions

$$\Psi_i(\sigma_-', \sigma_+'; \sigma_p'') \doteq \sigma_i - \sigma_i' - \sigma_i''.$$

Observe that each $\Psi_i$ is a $\mathcal{C}^2$ map satisfying

$$\Psi_i(\sigma_-', 0; \sigma_p'') = \Psi_i(\sigma_-', \sigma_+'; 0) = 0.$$

By Lemma 2.5 it thus follows that

$$\Psi_i(\sigma'_-, \sigma'_+; \sigma''_p) = \mathcal{O}(1) \cdot |\sigma'_+|\,|\sigma''_p| = \mathcal{O}(1) \cdot Q, \tag{7.94}$$

proving the lemma in this special case.

2. Let $u_m = \omega_0, \omega_1, \ldots, \omega_n = u_r$ be the intermediate states in the solution of the Riemann problem $(u_m, u_r)$. For each $p = 0, \ldots, n$, call $\sigma_1^p, \ldots \sigma_n^p$ the waves in the Riemann problem $(u_l, \omega_p)$ and consider the interaction potential

$$Q_p \doteq Q((u_l, \omega_p),\ (\omega_p, u_r)),$$

defined as in (7.93). Observe that, for each $p$, the waves $\sigma_i^p$ are generated by the interaction of the waves $\sigma_j^{p-1}$ with the single $p$-wave $\sigma''_p$. Using the previous estimate (7.94) we thus obtain

$$\begin{aligned} |\sigma_p^p - \sigma_p^{p-1} - \sigma''_p| + \sum_{i\neq p} |\sigma_i^p - \sigma_i^{p-1}| &= \mathcal{O}(1) \cdot Q((u_l, \omega_{p-1}),\ (\omega_{p-1}, \omega_p)) \\ &\leq C_0\,(Q_{p-1} - Q_p) \end{aligned}$$

for some constant $C_0$, provided that the total strength of waves is sufficiently small. Indeed,

$$\begin{aligned} Q_p = Q_{p-1} &- Q((u_l, \omega_{p-1}), (\omega_{p-1}, \omega_p)) \\ &+ \mathcal{O}(1) \cdot \Big\{ |\sigma_p^p - \sigma_p^{p-1} - \sigma''_p| + \sum_{i\neq p} |\sigma_i^p - \sigma_i^{p-1}| \Big\} \cdot \sum_{i>p} |\sigma''_i|. \end{aligned}$$

Since $Q_0 = Q$, $Q_n = 0$, $\sigma_i^0 = \sigma'_i$, $\sigma_i^n = \sigma_i$ for every $i \in \{1, \ldots, n\}$, then summing the above estimates for $p = 1, \ldots, n$ we obtain (7.92). □

## 7.6 Approximate conservation of waves

According to the estimates (7.31)–(7.34) proved in Lemma 7.2, the magnitude of wave-fronts can change only as a result of interactions or cancellations. This property can be conveniently expressed as an approximate conservation principle for wave strengths. More precisely, let $u = u(t, x)$ be a front tracking approximate solution of (7.1). We then define a measure $\mu^I$ of *interaction* and a measure $\mu^{IC}$ of *interaction and cancellation* as follows.

Both measures are positive and purely atomic, concentrated on the set of points $P$ where two physical wave-fronts of $u$ interact. If the incoming fronts belong to the families $i, i' \in \{1, \ldots, n\}$ and have sizes $\sigma, \sigma'$, we define

$$\mu^I(\{P\}) \doteq |\sigma\sigma'|, \tag{7.95}$$

$$\mu^{IC}(\{P\}) \doteq |\sigma\sigma'| + \begin{cases} |\sigma| + |\sigma'| - |\sigma + \sigma'| & \text{if } i = i', \\ 0 & \text{otherwise.} \end{cases} \tag{7.96}$$

Now consider any open region $\Gamma \subset\, ]0, \infty[\, \times \mathbb{R}$ with polygonal boundary (Fig. 7.17). For $i = 1, \ldots, n$ call $W_{\text{in}}^{i\pm}$, $W_{\text{out}}^{i\pm}$ the total amount of positive and negative $i$-waves which

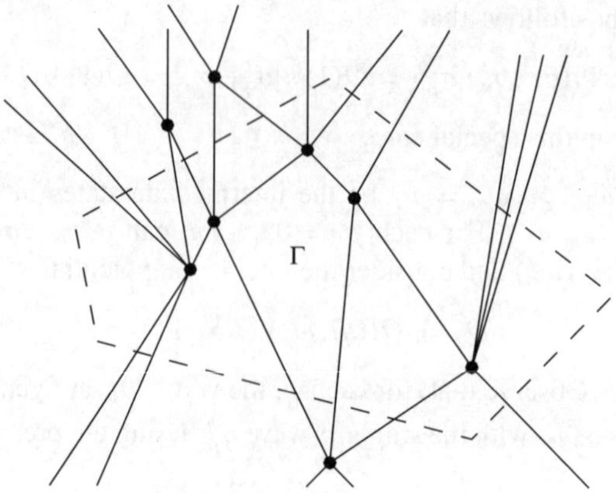


**Figure 7.17**

are entering or leaving $\Gamma$, respectively. Moreover, let us define

$$W^i_{\rm in} \doteq W^{i+}_{\rm in} - W^{i-}_{\rm in}, \qquad W^i_{\rm out} \doteq W^{i+}_{\rm out} - W^{i-}_{\rm out}.$$

From the estimates (7.31)–(7.32) it follows that the change in the strength of waves at each interaction point is controlled by the amount of interaction. Recalling that the size of physical fronts is unaffected when they cross a non-physical front, we thus have the estimates

$$|W^i_{\rm out} - W^i_{\rm in}| = \mathcal{O}(1) \cdot \mu^I(\Gamma), \tag{7.97}$$

$$|W^{i\pm}_{\rm out} - W^{i\pm}_{\rm in}| = \mathcal{O}(1) \cdot \mu^{IC}(\Gamma). \tag{7.98}$$

## 7.7 A positively invariant domain

The main existence theorem proved in this chapter applies to initial data with sufficiently small total variation. More precisely, we construct here a closed domain $\mathcal{D} \subset \mathbf{L}^1$ with the property that, for all initial data $\bar{u} \in \mathcal{D}$, the front tracking algorithm yields a solution of the Cauchy problem (7.1)–(7.2) with $u(t, \cdot) \in \mathcal{D}$ for all $t \geq 0$.

Let $u : \mathbb{R} \mapsto \mathbb{R}^n$ be a piecewise constant function with jumps at the points $x_1 < \cdots < x_N$. Call $\sigma_{\alpha,1}, \ldots, \sigma_{\alpha,n}$ the sizes of the waves determined by the Riemann problem with data $u(x_\alpha-)$, $u(x_\alpha+)$. The *total strength of waves* and the *interaction potential* of $u$ can now be defined as

$$\mathbf{V}(u) = \sum_{\alpha=1}^{N} \sum_{i=1}^{n} |\sigma_{\alpha,i}|, \qquad \mathbf{Q}(u) = \sum_{(\sigma_{\alpha,i},\sigma_{\beta,j})\in\mathcal{A}} |\sigma_{\alpha,i}|\,|\sigma_{\beta,j}|, \tag{7.99}$$

where the second summation ranges over all couples of approaching waves. Two waves $\sigma_{\alpha,i}$, $\sigma_{\beta,j}$ located at $x_\alpha < x_\beta$ are here defined as *approaching* if either $i > j$ or else if $i = j$ and at least one of the waves is a genuinely non-linear shock.

The reader should be aware of the relationships between the quantities defined in (7.99) and those introduced in (7.53)–(7.54):

- The quantities $V, Q$ in (7.53)–(7.54) refer only to a solution constructed by front tracking. They are defined only at times where no interaction takes place, so that at each $x_\alpha$ the jump occurs in one single characteristic family, or else it is a non-physical front. On the other hand, $\mathbf{V}, \mathbf{Q}$ in (7.99) refer to any piecewise constant function $u$.
- In the case of a front tracking $\varepsilon$-approximate solution, we have

$$|V(u(t)) - \mathbf{V}(u(t))| = \mathcal{O}(1) \cdot \varepsilon, \qquad |Q(u(t)) - \mathbf{Q}(u(t))| = \mathcal{O}(1) \cdot \varepsilon. \tag{7.100}$$

Indeed, let $u(t)$ have a non-physical front at $x_\alpha$. In the definition of $V, Q$ this is treated as a wave of an artificial $(n+1)$-th family, having strength $|\sigma_\alpha| \doteq |u(x_\alpha+) - u(x_\alpha-)|$. On the other hand, in the computation of $\mathbf{V}, \mathbf{Q}$ the non-physical front at $x_\alpha$ yields the $n$ waves $\sigma_{\alpha,1}, \ldots, \sigma_{\alpha,n}$ corresponding to the solution of the Riemann problem with data $u(x_\alpha\pm)$. Since the total amount of non-physical fronts is $\mathcal{O}(1) \cdot \varepsilon$, one obtains (7.100).

We can now define the domain

$$\mathcal{D} = cl\{u \in \mathbf{L}^1(\mathbb{R}; \mathbb{R}^n);\ u \text{ is piecewise constant, } \mathbf{V}(u) + C_0\mathbf{Q}(u) < \delta_0\}, \tag{7.101}$$

where $cl$ denotes closure in $\mathbf{L}^1$. From (7.100), recalling that the functional $t \mapsto V(u(t)) + C_0 Q(u(t))$ is non-increasing along $\varepsilon$-approximate front tracking solutions, we obtain the positive invariance of $\mathcal{D}$ w.r.t. the flow generated by the system of conservation laws.

**Proposition 7.1.** *For suitable constants $C_0, \delta_0 > 0$ in (7.101), the following holds. For all initial data $\bar{u} \in \mathcal{D}$, the Cauchy problem (7.1)–(7.2) has a global solution $u = u(t, x)$ obtained as the limit of front tracking approximations, with $u(t, \cdot) \in \mathcal{D}$ for all $t \geq 0$.*

Indeed, if $\bar{u} \in \mathcal{D}$, by definition there exists a sequence of piecewise constant functions $\bar{u}_\nu$ converging to $\bar{u}$, with

$$\mathbf{V}(\bar{u}_\nu) + C_0\mathbf{Q}(\bar{u}_\nu) = \delta_0 - \varepsilon_\nu^*$$

for some $\varepsilon_\nu^* > 0$. For each $\nu$ we can now construct a front tracking $\varepsilon_\nu$-approximate solution $u_\nu = u_\nu(t, x)$, with initial data $u_\nu(0, x) = \bar{u}_\nu(x)$. Choosing $\varepsilon_\nu > 0$ sufficiently small compared with $\varepsilon_\nu^*$, we obtain

$$\begin{aligned} \mathbf{V}(u_\nu(t)) + C_0\mathbf{Q}(u_\nu(t)) &= V(u_\nu(t)) + C_0 Q(u_\nu(t)) + \mathcal{O}(1) \cdot \varepsilon_\nu \\ &\leq V(u_\nu(0)) + C_0 Q(u_\nu(0)) + \mathcal{O}(1) \cdot \varepsilon_\nu \\ &= \delta_0 - \varepsilon_\nu^* + \mathcal{O}(1) \cdot \varepsilon_\nu \\ &< \delta_0 \end{aligned} \tag{7.102}$$

for every $t \geq 0$. By (7.102) and the definition (7.101), any solution $u$ obtained as the limit of the $u_n$ satisfies $u(t, \cdot) \in \mathcal{D}$ for all $t \geq 0$.

## Problems

(1) Let $u$ be an approximate solution of (7.1), constructed by the wave-front tracking algorithm. At any time $t \geq 0$, let the wave-fronts in $u(t, \cdot)$ be located at points $x_\alpha$ and have sizes $\sigma_\alpha$. Define the weights

$$W_\alpha \doteq 1 + \kappa_1 \sum_{\beta \in \mathcal{A}_\alpha} |\sigma_\beta| + \kappa_1 \kappa_2 \sum_{(\beta,\gamma)\in \mathcal{A}} |\sigma_\beta \sigma_\gamma|,$$

where $\mathcal{A}_\alpha$ is the set of all waves which are approaching the front at $x_\alpha$, and $\mathcal{A}$ is the set of all couples of approaching waves. Prove that, if the total variation of $u$ is sufficiently small, then one can choose constants $\kappa_1, \kappa_2$ large enough so that the quantity

$$\sum_\alpha W_\alpha |\sigma_\alpha|$$

decreases at every interaction.

(2) Let $K$ be a compact subset of $\Omega$. Under the standard assumptions (♣), show that, for some constant $\delta_0 > 0$ sufficiently small, the Cauchy problem (7.1)–(7.2) has a globally defined weak solution for all initial data $\bar{u}$ such that

$$\text{Tot. Var.}\{\bar{u}\} \leq \delta_0, \quad \bar{u}(0) \in K.$$

This result extends Theorem 7.1 to a class of bounded, not necessarily integrable, initial data.

Hint: in the definition of the $\varepsilon$-approximate front tracking solution, replace (7.12) with

$$\int_{-1/\varepsilon}^{1/\varepsilon} |u(0, x) - \bar{u}(x)|\, dx < \varepsilon.$$

Then retrace all the steps in the proofs of Theorems 7.1 and 7.2.

(3) Let $u$ be a $BV$ solution of the system (7.1), and assume that its point values are chosen so that $u(t, x) = u(t, x+)$. For every $a < b$ and $0 \leq \tau_0 < \tau$, show that

$$\int_a^b u(\tau, x)\, dx - \int_a^b u(\tau_0, x)\, dx = \int_{\tau_0}^{\tau} [f(u(t, a)) - f(u(t, b))]\, dt.$$

Hint: in analogy with (6.28), use (7.3) in connection with a family of smooth test functions $\phi$ which approximate the characteristic function of the rectangle $[\tau_0, \tau] \times [a, b]$.

(4) Consider the system of isentropic gas dynamics (5.50), with $p(v) = v^{-2}$. Prove that the Riemann problem generated by the collision of two shocks of the first family is solved in terms of a shock of the first family and a centred rarefaction wave of the second family. Similarly, the Riemann problem generated by the collision of two 2-shocks is solved by a 1-rarefaction and a 2-shock.

# 8

# Stability

The analysis in the previous chapter has shown the existence of a global entropy weak solution of the Cauchy problem for all initial data with sufficiently small total variation. More precisely, recalling the definitions of the functionals $\mathbf{V}$, $\mathbf{Q}$ at (7.101), consider a domain of the form

$$\mathcal{D} = cl\{u \in \mathbf{L}^1(\mathbb{R}; \mathbb{R}^n);\ u \text{ is piecewise constant, } \mathbf{V}(u) + C_0\,\mathbf{Q}(u) < \delta_0\}, \tag{8.1}$$

where $cl$ denotes closure in $\mathbf{L}^1$. With a suitable choice of the constants $C_0$ and $\delta_0 > 0$, the proofs of Theorems 7.1 and 7.2 show that, for every $\bar{u} \in \mathcal{D}$, one can construct a sequence of $\varepsilon$-approximate front tracking solutions converging to a weak solution $u$ taking values inside $\mathcal{D}$. However, since the proof of convergence relied on a compactness argument, no information was obtained on the uniqueness of the limit. The main goal of the present chapter is to show that this limit is unique and depends continuously on the initial data. The following theorem applies to solutions constructed by the front tracking algorithm described in Chapter 7. More generally, it refers to all $\varepsilon$-approximate solutions (in the sense of Definition 7.1) satisfying the additional conditions (FT1) and (FT3) stated in Remark 7.2. In Chapter 9 we shall see that these two conditions can be dropped altogether.

**Theorem 8.1 (Existence of a semigroup of solutions).** *For a suitably small $\delta_0 > 0$ in (8.1), the following holds. Given any $\bar{u} \in \mathcal{D}$, consider a sequence of $\varepsilon$-approximate front tracking solutions $u_\varepsilon : [0, \infty[\ \mapsto \mathcal{D}$ of the Cauchy problem (7.1)–(7.2), satisfying the conditions (FT1) and (FT3). Then, as $\varepsilon \to 0$, the sequence $u_\varepsilon$ converges to a unique limit solution $u : [0, \infty[\ \mapsto \mathcal{D}$. The map $(\bar{u}, t) \mapsto u(t, \cdot) \doteq S_t\bar{u}$ is a uniformly Lipschitz semigroup. Indeed, there exist constants $L$, $L'$ such that, for all $\bar{u}, \bar{v} \in \mathcal{D}$, $s, t \geq 0$, one has*

$$S_0\bar{u} = \bar{u}, \qquad S_s(S_t\bar{u}) = S_{s+t}\bar{u}, \tag{8.2}$$

$$\|S_t\bar{u} - S_s\bar{v}\|_{\mathbf{L}^1} \leq L \cdot \|\bar{u} - \bar{v}\|_{\mathbf{L}^1} + L' \cdot |t - s|. \tag{8.3}$$

## 8.1 Stability of front tracking approximations

To prove the uniqueness of the limit of front tracking approximations, we need to estimate the distance between any two $\varepsilon$-approximate solutions $u, v$ of (7.1). For this purpose, we introduce a functional $\Phi = \Phi(u, v)$, equivalent to the $\mathbf{L}^1$ distance, which is 'almost decreasing' along pairs of solutions. Given $u, v : \mathbb{R} \mapsto \mathbb{R}^n$, consider the scalar functions $q_i$ defined implicitly by

$$v(x) = S_n(q_n(x)) \circ \cdots \circ S_1(q_1(x))(u(x)). \tag{8.4}$$

As in (5.35), by $\sigma \mapsto S_i(\sigma)(u_0)$ we denote the parametrized $i$-shock surve through the point $u_0$. We adopt here the parametrization choice (♠) described in Chapter 5. That is, if the $i$-th characteristic field is linearly degenerate, the curve $S_i$ will be parametrized by arc-length. If the $i$-th field is genuinely non-linear, the parametrization is chosen according to (5.37).

Intuitively, $q_i(x)$ can be regarded as the strength of the $i$-th component in the jump $(u(x), v(x))$, where these components are measured along shock curves. On a compact neighbourhood of the origin, we clearly have

$$\frac{1}{C_1} \cdot |v(x) - u(x)| \leq \sum_{i=1}^{n} |q_i(x)| \leq C_1 \cdot |v(x) - u(x)| \tag{8.5}$$

for some constant $C_1$. We now consider the functional

$$\Phi(u, v) \doteq \sum_{i=1}^{n} \int_{-\infty}^{\infty} |q_i(x)| W_i(x)\, dx, \tag{8.6}$$

where the weights $W_i$ are defined by setting

$$\begin{aligned} W_i(x) &\doteq 1 + \kappa_1 \cdot [\text{total strength of waves in } u \text{ and in } v \text{ which approach the } i\text{-wave } q_i(x)] \\ &\qquad + \kappa_2 \cdot [\text{wave interaction potentials of } u \text{ and of } v] \\ &\doteq 1 + \kappa_1 A_i(x) + \kappa_1 \kappa_2 [Q(u) + Q(v)]. \end{aligned} \tag{8.7}$$

Here $Q$ is the interaction potential defined at (7.54), while $A_i(x)$ measures the total amount of (physical) waves in $u$ and in $v$ which approach the $i$-wave $q_i$ located at $x$. More precisely, if the $i$-th characteristic field is linearly degenerate, we simply define

$$A_i(x) \doteq \left[ \sum_{x_\alpha < x,\, i < k_\alpha \leq n} + \sum_{x_\alpha > x,\, 1 \leq k_\alpha < i} \right] |\sigma_\alpha|. \tag{8.8}$$

The summations here extend to waves both of $u$ and of $v$. On the other hand, if the $i$-th field is genuinely non-linear, our definition of $A_i$ will contain an additional term, accounting for waves in $u$ and in $v$ of the same $i$-th family:

$$\begin{aligned} A_i(x) \doteq & \left[ \sum_{\substack{\alpha \in \mathcal{J}(u) \cup \mathcal{J}(v) \\ x_\alpha < x,\, i < k_\alpha \leq n}} + \sum_{\substack{\alpha \in \mathcal{J}(u) \cup \mathcal{J}(v) \\ x_\alpha > x,\, 1 \leq k_\alpha < i}} \right] |\sigma_\alpha| \\ & + \begin{cases} \left[ \sum_{\substack{k_\alpha = i \\ \alpha \in \mathcal{J}(u),\ x_\alpha < x}} + \sum_{\substack{k_\alpha = i \\ \alpha \in \mathcal{J}(v),\ x_\alpha > x}} \right] |\sigma_\alpha| & \text{if } q_i(x) < 0, \\ \left[ \sum_{\substack{k_\alpha = i \\ \alpha \in \mathcal{J}(v),\ x_\alpha < x}} + \sum_{\substack{k_\alpha = i \\ \alpha \in \mathcal{J}(u),\ x_\alpha > x}} \right] |\sigma_\alpha| & \text{if } q_i(x) \geq 0. \end{cases} \end{aligned} \tag{8.9}$$

Here and in the sequel, $\mathcal{J}(u)$ and $\mathcal{J}(v)$ denote the sets of all jumps in $u$ and in $v$, while $\mathcal{J} \doteq \mathcal{J}(u) \cup \mathcal{J}(v)$. We recall that $k_\alpha \in \{1, \ldots, n+1\}$ is the family of the jump located at $x_\alpha$ with size $\sigma_\alpha$. By definition, $k_\alpha \doteq n+1$ in case of a non-physical front. Notice that the strengths of non-physical waves do enter in the definition of $Q$. Indeed, a non-physical front located at $x_\alpha$ approaches all shock and rarefaction fronts located at points $x_\beta > x_\alpha$. On the other hand, non-physical fronts play no role in the definition of $A_i$.

The values of the large constants $\kappa_1, \kappa_2$ in (8.7) will be specified later. Observe that, as soon as these constants have been assigned, we can then impose a suitably small bound on the total variation of $u, v$ so that

$$1 \le W_i(x) \le 2 \quad \text{for all } i, x. \tag{8.10}$$

From (8.5), (8.6) and (8.10) it thus follows that

$$\frac{1}{C_1} \cdot \|v - u\|_{\mathbf{L}^1} \le \Phi(u, v) \le 2C_1 \cdot \|v - u\|_{\mathbf{L}^1}. \tag{8.11}$$

Recalling the definition $\Upsilon(u) \doteq V(u) + C_0 Q(u)$, the basic $\mathbf{L}^1$ stability estimate for front tracking approximations can now be stated as follows.

**Theorem 8.2 (Stability of front tracking approximations).** *For suitable constants $C_2, \kappa_1, \kappa_2, \delta_0 > 0$ the following holds. Let $u, v$ be $\varepsilon$-approximate front tracking solutions of (7.1) satisfying the conditions (FT1) and (FT3), with*

$$\Upsilon(u(t)) < \delta_0, \quad \Upsilon(v(t)) < \delta_0 \qquad \textit{for all } t \ge 0. \tag{8.12}$$

*Then the functional $\Phi$ in (8.6)–(8.9) satisfies*

$$\Phi(u(t), v(t)) - \Phi(u(s), v(s)) \le C_2 \varepsilon (t - s) \quad \textit{for all } 0 \le s < t. \tag{8.13}$$

*Proof.* For a proof of Theorem 8.2, the key point is to understand how the functional $\Phi$ evolves in time. In connection with (8.4), at each $x$ we define the intermediate states $\omega_0(x) = u(x), \omega_1(x), \ldots, \omega_n(x) = v(x)$ by setting

$$\omega_i(x) \doteq S_i(q_i(x)) \circ S_{i-1}(q_{i-1}(x)) \circ \cdots \circ S_1(q_1(x))(u(x)). \tag{8.14}$$

Moreover, we call

$$\lambda_i(x) \doteq \lambda_i(\omega_{i-1}(x), \omega_i(x)) \tag{8.15}$$

the speed of the $i$-shock connecting $\omega_{i-1}(x)$ with $\omega_i(x)$, and let $x_1 < \cdots < x_N$ be all the points where $u(t)$ and $v(t)$ have a jump. A direct computation now yields

$$\frac{d}{dt}\Phi(u(t), v(t)) = \sum_{\alpha \in \mathcal{J}} \sum_{i=1}^{n} \{|q_i(x_\alpha -)| W_i(x_\alpha -) - |q_i(x_\alpha +)| W_i(x_\alpha +)\} \dot{x}_\alpha, \tag{8.16}$$

where the first sum extends over all jumps in $u$ and in $v$. For notational convenience, we shall write $q_i^{\alpha+} \doteq q_i(x_\alpha+)$, $q_i^{\alpha-} \doteq q_i(x_\alpha-)$ and similarly for $W_i^{\alpha\pm}, \lambda_i^{\alpha\pm}$. We regard the quantity $|q_i(x)|\lambda_i(x)$ as the flux of the $i$-th component of $|v-u|$ at $x$. Since $u, v$ are piecewise constant, for $x \in \,]x_{\alpha-1}, x_\alpha[$ one clearly has

$$|q_i(x)|\lambda_i(x) W_i(x) = |q_i^{(\alpha-1)+}| \lambda_i^{(\alpha-1)+} W_i^{(\alpha-1)+} = |q_i^{\alpha-}| \lambda_i^{\alpha-} W_i^{\alpha-}. \tag{8.17}$$

Moreover, the assumption $u(t), v(t) \in \mathbf{L}^1$ implies $q_i(t,x) \equiv 0$ for $x$ outside a bounded interval. We can thus add and subtract the terms (8.17) in (8.16), without changing the overall sum. This yields

$$\frac{d}{dt}\Phi(u(t),v(t)) = \sum_{\alpha\in\mathcal{J}}\sum_{i=1}^{n}\left\{W_i^{\alpha+}|q_i^{\alpha+}|(\lambda_i^{\alpha+}-\dot{x}_\alpha) - W_i^{\alpha-}|q_i^{\alpha-}|(\lambda_i^{\alpha-}-\dot{x}_\alpha)\right\}. \tag{8.18}$$

In connection with (8.18), for each jump point $\alpha \in \mathcal{J}$ and every $i = 1, \ldots, n$, we define

$$E_{\alpha,i} \doteq W_i^{\alpha+}|q_i^{\alpha+}|(\lambda_i^{\alpha+}-\dot{x}_\alpha) - W_i^{\alpha-}|q_i^{\alpha-}|(\lambda_i^{\alpha-}-\dot{x}_\alpha). \tag{8.19}$$

Our main goal will be to establish the bounds

$$\sum_{i=1}^{n} E_{\alpha,i} \le \mathcal{O}(1)\cdot|\sigma_\alpha| \quad \text{if } \alpha \in \mathcal{NP}, \tag{8.20}$$

$$\sum_{i=1}^{n} E_{\alpha,i} \le \mathcal{O}(1)\cdot\varepsilon|\sigma_\alpha| \quad \text{if } \alpha \in \mathcal{S}\cup\mathcal{R}. \tag{8.21}$$

Here $\mathcal{S}$ and $\mathcal{R}$ denote respectively the family of shocks (or contact discontinuities) and of rarefaction fronts in $u$ and in $v$, while $\mathcal{NP}$ refers to the set of non-physical fronts. As usual, by the Landau symbol $\mathcal{O}(1)$ we denote a quantity whose absolute value satisfies a uniform bound, depending only on the system (7.1). In particular, this bound will not depend on $\varepsilon$ or on the functions $u$, $v$. It is also independent of the choice of the constants $\kappa_1, \kappa_2$ in (8.7).

From (8.20)–(8.21), since the total strength of non-physical fronts in $u(t,\cdot)$ and in $v(t,\cdot)$ is always $\le \varepsilon$, one obtains the key estimate

$$\frac{d}{dt}\Phi(u(t),v(t)) \le \mathcal{O}(1)\cdot\varepsilon. \tag{8.22}$$

This describes the behaviour of the functional $\Phi$ outside interaction times. Next, consider a time $\tau$ where two fronts of $u$ interact, say of strengths $\sigma, \sigma'$. The maps $t \mapsto q_i(t,\cdot)$, regarded as functions from $[0,\infty[$ into $\mathbf{L}^1$, are continuous also at time $t = \tau$. Moreover, from (7.57) and the interaction estimates (7.31)–(7.34) we deduce that

$$Q(u(\tau+)) - Q(u(\tau-)) \le -\frac{|\sigma\sigma'|}{2},$$

$$A_i(\tau+,x) - A_i(\tau-,x) = \mathcal{O}(1)\cdot|\sigma\sigma'|.$$

Therefore, if $\kappa_2$ in (8.7) is large enough, all weight functions $W_i(x)$ will decrease across time $\tau$. The interaction of two fronts of $v$ is entirely similar. Integrating (8.22) over the interval $[s,t]$ we therefore obtain

$$\Phi(u(t),v(t)) \le \Phi(u(s),v(s)) + \mathcal{O}(1)\cdot\varepsilon(t-s), \tag{8.23}$$

proving the theorem.

It now remains to establish the estimates (8.20)–(8.21). A detailed proof will be given in the next section. To help the reader, we first give a rough sketch of the heart of the matter, with the aid of a few diagrams.

Given two piecewise constant functions $u, v$ with compact support, for $i = 1, \ldots, n$ one can define the scalar components $u_i, v_i$ as follows. Let $x_1 < \cdots < x_N$ be all the points where either $u$ or $v$ has a jump and let the functions $q_1, \ldots, q_n$ be as in (8.4). We start by setting $u_i(-\infty) = v_i(-\infty) = 0$, and then proceed by induction on $\alpha = 1, \ldots, N$. If $x_\alpha \in \mathcal{J}(u)$, then we let $v_i$ be constant across $x_\alpha$ and set

$$u_i(x_\alpha+) \doteq u_i(x_\alpha-) - [q_i(x_\alpha+) - q_i(x_\alpha-)].$$

On the other hand, if $x_\alpha \in \mathcal{J}(v)$, then we let $u_i$ be constant across $x_\alpha$ and set

$$v_i(x_\alpha+) \doteq v_i(x_\alpha-) + [q_i(x_\alpha+) - q_i(x_\alpha-)].$$

These definitions trivially imply

$$q_i(x) = v_i(x) - u_i(x) \quad \text{for all } x \in \mathbb{R},\ i = 1, \ldots, n. \tag{8.24}$$

If the $i$-th field is genuinely non-linear, according to the definition (8.9) the $i$-waves in $u$ and $v$ which approach $q_i(x)$ are those located within the thick portions of the graphs of $u_i$ and $v_i$ in Fig. 8.1. Conversely, for a given $i$-wave $\sigma_\alpha$ (say, in the function $v$) located at $x_\alpha$, the regions where the jumps $q_i(x)$ approach $\sigma_\alpha$ are represented by the shaded areas in Fig. 8.2.

Now let $v$ have a wave-front at $x_\alpha$ with strength $\sigma_\alpha$, say in the genuinely non-linear $k$-th family. To fix the ideas, assume that $u_k(x_\alpha) < v_k(x_\alpha+) < v_k(x_\alpha-)$. In connection

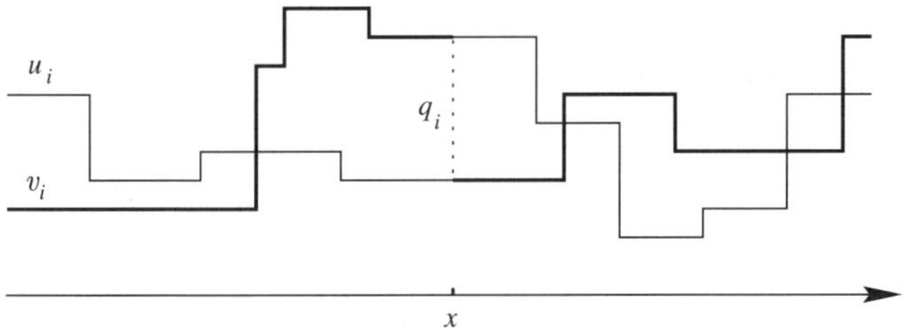


**Figure 8.1**

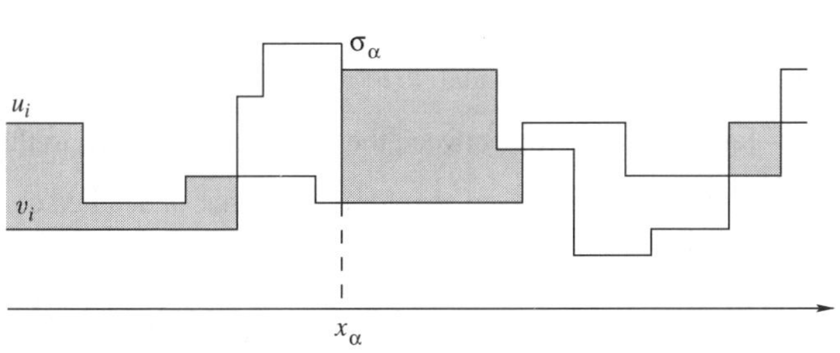


**Figure 8.2**

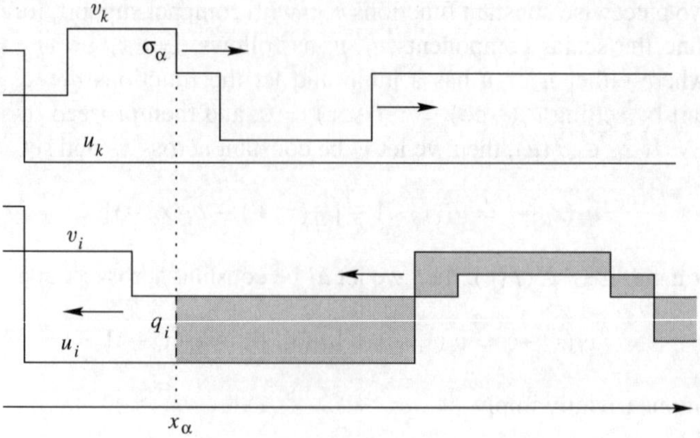


**Figure 8.3**

with this front (Fig. 8.3), for every $i < k$ the functional $\Phi(u, v)$ contains a term of the form

$\Phi_{\alpha,i} \doteq \kappa_1 \cdot |\sigma_\alpha| \cdot$ [area of the region between the graphs of $u_i$ and $v_i$, to the right of $x_\alpha$].

By strict hyperbolicity, the $i$-th and $k$-th characteristic speeds are strictly separated, say $\lambda_k - \lambda_i \geq c > 0$. If each component $u_i, v_i, i = 1, \ldots, n$, provided an exact solution to a scalar conservation law

$$(u_i)_t + F_i(u_i)_x = 0$$

(with $F_i' = \lambda_i$), uncoupled from all the other components, then we would have the estimate

$$\frac{d\Phi_{\alpha,i}}{dt} \leq -\kappa_1|\sigma_\alpha||q_i^{\alpha+}|(\dot{x}_\alpha - \lambda_i^{\alpha+}) \leq -c\kappa_1|\sigma_\alpha||q_i^{\alpha+}|. \tag{8.25}$$

Here $\lambda_i^{\alpha+} \doteq \lambda_i(u_i(x_\alpha+), v_i(x_\alpha+))$ is the speed of an $i$-shock of strength $q_i^{\alpha+}$. In general, the estimate (8.25) must be supplemented with coupling and error terms, whose size is estimated as

$$\mathcal{O}(1) \cdot \left(\varepsilon + |q_k^{\alpha+}|(|q_k^{\alpha+}| + |\sigma_\alpha|) + \sum_{j \neq k} |q_j^{\alpha+}|\right)|\sigma_\alpha|. \tag{8.26}$$

In addition, for every $i > k$ the functional $\Phi(u, v)$ contains a term of the form

$\Phi_{\alpha,i} \doteq \kappa_1 \cdot |\sigma_\alpha| \cdot$ [area of the region between the graphs of $u_i$ and $v_i$, to the left of $x_\alpha$].

Entirely similar estimates can be proved also for these terms. A detailed computation of the quantity in (8.19) thus yields

$$E_{\alpha,i} \leq -c\kappa_1|q_i^{\alpha+}||\sigma_\alpha| + \mathcal{O}(1) \cdot \left(\varepsilon + |q_k^{\alpha+}|(|q_k^{\alpha+}| + |\sigma_\alpha|) + \sum_{j \neq k} |q_j^{\alpha+}|\right)|\sigma_\alpha| \quad i \neq k. \tag{8.27}$$

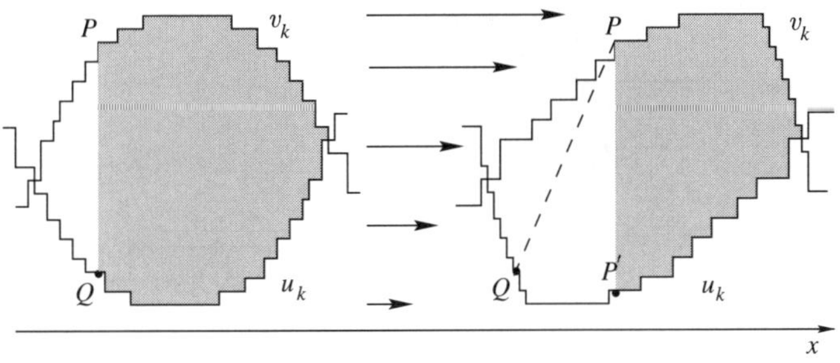


**Figure 8.4**

Next, according to (8.9) the functional $\Phi(u, v)$ also contains a term of the form

$$\Phi_{\alpha,k} \doteq \kappa_1 \cdot |\sigma_\alpha| \cdot [\text{area of the region between the graphs of } u_k \text{ and } v_k, \text{ to the right of } x_\alpha].$$

If the components $u_k$, $v_k$ were exact solutions of the scalar conservation law

$$(u_k)_t + F_k(u_k)_x = 0 \tag{8.28}$$

with $F_k' = \lambda_k$, $F_k'' = r_k \bullet \lambda_k \geq c' > 0$, then we would have the estimate

$$\frac{d\Phi_{\alpha,k}}{dt} \leq -\kappa_1|\sigma_\alpha||q_k^{\alpha+}|(\dot{x}_\alpha - \lambda_k^{\alpha+}) \leq -\kappa_1|\sigma_\alpha||q_k^{\alpha+}| \cdot \frac{c'}{2}(|q_k^{\alpha+}| + |\sigma_\alpha|). \tag{8.29}$$

Indeed, owing to genuine non-linearity, the points on the graphs of $u_k$ and $v_k$ move with different speeds (Fig. 8.4). As a consequence, the shape of these graphs changes in time. In particular, while the area enclosed by the two graphs may remain constant, the portion of this area located to the right of a given front $\sigma_\alpha$ will decrease. In Fig. 8.4, the two points $P$, $Q$ lie initially on the same vertical line. At a later time this is no longer true. The area of the shaded region, enclosed by the graphs of $u_k$ and $v_k$ and by a vertical line through $P$, has decreased by an amount roughly equal to the area of the triangular region with vertices $P$, $P'$, $Q$.

In general, the estimate (8.29) must be supplemented with coupling and error terms, whose size is again estimated by (8.26). A detailed computation thus yields

$$\begin{aligned} E_{\alpha,k} \leq &-\frac{c'\kappa_1}{2}|q_k^{\alpha+}||\sigma_\alpha|(|q_k^{\alpha+}| + |\sigma_\alpha|) \\ &+ \mathcal{O}(1) \cdot \left(\varepsilon + |q_k^{\alpha+}|(|q_k^{\alpha+}| + |\sigma_\alpha|) + \sum_{j \neq k} |q_j^{\alpha+}|\right)|\sigma_\alpha|. \end{aligned} \tag{8.30}$$

Choosing $\kappa_1$ sufficiently large, we find that (8.27) and (8.30) together yield (8.21).

A different kind of estimate is needed in the case where the jump in $v_k$ crosses the graph of $u_k$, say $v_k(x_\alpha+) < u_k(x_\alpha) < v_k(x_\alpha-)$. To fix the ideas, assume

$$|q_k^{\alpha+}| = |v_k(x_\alpha+) - u_k(x_\alpha)| \leq |v_k(x_\alpha-) - u_k(x_\alpha)| = |q_k^{\alpha-}|. \tag{8.31}$$

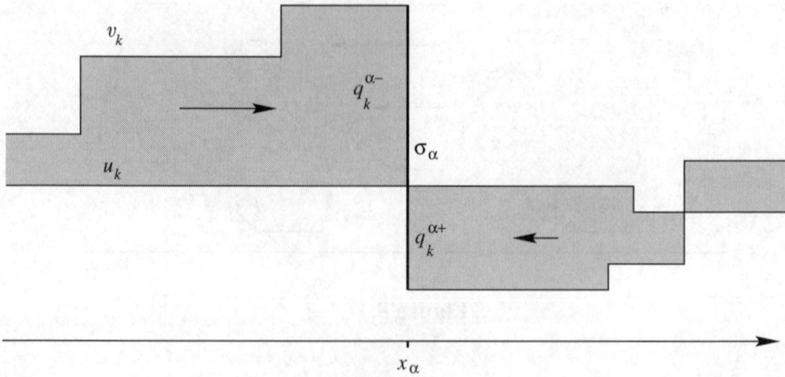


**Figure 8.5**

In this case, the estimates (8.27) remain valid. In connection with the $k$-th field, the functional $\Phi$ contains a term of the form (Fig. 8.5)

$$\Phi_k^* \doteq 1 \cdot [\text{area of the region between the graphs of } u_k \text{ and } v_k], \tag{8.32}$$

where the above area includes points both on the right and on the left of $x_\alpha$.

If $q_k^{\alpha+} = q_k^{\alpha-} + \sigma_\alpha$ and if the components $u_k$, $v_k$ provided an exact solution to the genuinely non-linear scalar conservation law (8.32), the assumption of genuine non-linearity would imply

$$\frac{d\Phi_k^*}{dt} \le -\frac{c'}{2}|q_k^{\alpha-}| \cdot |q_k^{\alpha+}| \le -\frac{c'}{4}|\sigma_\alpha||q_k^{\alpha+}|. \tag{8.33}$$

Indeed, by (8.31), $|\sigma_\alpha| = |q_k^{\alpha-}| + |q_k^{\alpha+}| \le 2|q_k^{\alpha-}|$. In general, the estimate (8.33) must be supplemented with coupling and error terms, whose size is again estimated by (8.26). A detailed computation thus yields

$$E_{\alpha,k} \le -\frac{c'}{4}|q_k^{\alpha+}||\sigma_\alpha| + \mathcal{O}(1) \cdot \left( \varepsilon + |q_k^{\alpha+}|(|q_k^{\alpha+}| + |\sigma_\alpha|) + \sum_{j \neq k} |q_j^{\alpha+}| \right) |\sigma_\alpha|. \tag{8.34}$$

Assuming that the total strength of waves remains sufficiently small, we have

$$|q_k^{\alpha+}| + |\sigma_\alpha| \ll \frac{c'}{4}.$$

Therefore, choosing $\kappa_1$ sufficiently large, we find that (8.27) and (8.34) together yield (8.21).

## 8.2 Proof of the main estimates

In this section we establish the key estimates (8.20)–(8.21), thus completing the proof of Theorem 8.2. We shall always assume that $x_\alpha$ is a point where $v$ has a jump in the $k_\alpha$-th family, of size $\sigma_\alpha$. The case of a jump in $u$ is entirely similar. In the following, since all computations refer to a fixed jump $\alpha \in \mathcal{J}(v)$, we shall often drop the superscript $\alpha$ and simply write $W_i^+ \doteq W_i^{\alpha+}$, $q_{k_\alpha}^- \doteq q_{k_\alpha}^{\alpha-}$, $v^- \doteq v(x_\alpha-)$, $v^+ \doteq v(x_\alpha+)$, etc. The proof will be given in several steps.

**1. Non-physical fronts.** If $\alpha \in \mathcal{NP}$, the jump travels with the fixed speed $\dot{x}_\alpha = \hat{\lambda}$ and $\sigma_\alpha \doteq |v^+ - v^-|$. Hence, for $i = 1, \ldots, n$ one has

$$q_i^+ - q_i^- = \mathcal{O}(1) \cdot \sigma_\alpha, \qquad \lambda_i^+ - \lambda_i^- = \mathcal{O}(1) \cdot \sigma_\alpha. \tag{8.35}$$

Moreover, from the definitions (8.7)–(8.9) and the bound (8.10) it follows that

$$W_i^+ - W_i^- = 0 \quad \text{if } q_i^+ q_i^- > 0, \tag{8.36}$$

$$|W_i^+ - W_i^-| \le 1 \quad \text{for all } i. \tag{8.37}$$

If $q_i^-$ and $q_i^+$ have opposite signs, then (8.35) implies

$$|q_i^-| + |q_i^+| = \mathcal{O}(1) \cdot |\sigma_\alpha|.$$

If instead $q_i^-$ and $q_i^+$ have the same sign, then (8.36) applies. Writing

$$E_{\alpha,i} = W_i^+(|q_i^+| - |q_i^-|)(\lambda_i^+ - \dot{x}_\alpha) + (W_i^+ - W_i^-)|q_i^-|(\lambda_i^+ - \dot{x}_\alpha) + W_i^-|q_i^-|(\lambda_i^+ - \lambda_i^-),$$

the estimate (8.20) is now clear.

**2. Reduction to the shock case.** In the case of a physical front $\alpha \in \mathcal{R} \cup \mathcal{S}$, consider the auxiliary state and speed

$$v^\diamond \doteq S_{k_\alpha}(\sigma_\alpha)(v^-), \qquad \dot{x}_\alpha^\diamond \doteq \lambda_{k_\alpha}(v^-, v^\diamond). \tag{8.38}$$

Define the components $q_i^\diamond$ in terms of the implicit relation

$$v^\diamond = S_n(q_n^\diamond) \circ \cdots \circ S_1(q_1^\diamond)(u(x_\alpha)). \tag{8.39}$$

Similarly, define the intermediate states $\omega_0^\diamond = u(x_\alpha)$, $\omega_1^\diamond, \ldots, \omega_n^\diamond = v^\diamond$ and the shock speeds $\lambda_i^\diamond$ by setting

$$\begin{aligned} \omega_i^\diamond(x) &\doteq S_i(q_i^\diamond) \circ S_{i-1}(q_{i-1}^\diamond) \circ \cdots \circ S_1(q_1^\diamond)(u(x_\alpha)), \\ \lambda_i^\diamond &\doteq \lambda_i(\omega_{i-1}^\diamond, \omega_i^\diamond). \end{aligned} \tag{8.40}$$

We now consider two cases.

CASE 1: The jump at $x_\alpha$ is a shock or a contact discontinuity. In this case $v^\diamond = v^+$ and hence

$$q_i^\diamond = q_i^+, \qquad \omega_i^\diamond = \omega_i^+, \qquad \lambda_i^\diamond = \lambda_i^+ \tag{8.41}$$

for all $i$. Since $\dot{x}_\alpha^\diamond \doteq \lambda_{k_\alpha}(v^-, v^+)$, the assumption (7.7) implies

$$|\dot{x}_\alpha^\diamond - \dot{x}_\alpha| < \varepsilon. \tag{8.42}$$

CASE 2: The jump at $x_\alpha$ is a rarefaction and hence $0 < \sigma_\alpha \le \varepsilon$. In this case, since shock and rarefaction curves have a tangency of second order, we have $v^\diamond - v^+ = \mathcal{O}(1) \cdot |\sigma_\alpha|^3$

and hence

$$q_i^\diamond - q_i^+ = \mathcal{O}(1)\cdot|\sigma_\alpha|^3, \qquad \omega_i^\diamond - \omega_i^+ = \mathcal{O}(1)\cdot|\sigma_\alpha|^3,$$
$$\lambda_i^\diamond - \lambda_i^+ = \mathcal{O}(1)\cdot|\sigma_\alpha|^3, \tag{8.43}$$

for all $i$. Moreover, by (7.9), the bound (8.42) remains valid.

We now compute (8.19) as

$$\begin{aligned}
E_{\alpha,i} &= W_i^+|q_i^+|(\lambda_i^+ - \dot{x}_\alpha) - W_i^-|q_i^-|(\lambda_i^- - \dot{x}_\alpha)\\
&= W_i^+|q_i^+|(\lambda_i^+ - \dot{x}_\alpha^\diamond) - W_i^-|q_i^-|(\lambda_i^- - \dot{x}_\alpha^\diamond)\\
&\quad + (\dot{x}_\alpha^\diamond - \dot{x}_\alpha)\{W_i^+|q_i^+| - W_i^-|q_i^-|\}\\
&= \{W_i^+|q_i^\diamond|(\lambda_i^\diamond - \dot{x}_\alpha^\diamond) - W_i^-|q_i^-|(\lambda_i^- - \dot{x}_\alpha^\diamond)\}\\
&\quad + \{W_i^+|q_i^\diamond|(\lambda_i^+ - \lambda_i^\diamond) + W_i^+(|q_i^+| - |q_i^\diamond|)(\lambda_i^+ - \dot{x}_\alpha^d)\}\\
&\quad + (\dot{x}_\alpha^\diamond - \dot{x}_\alpha)\{W_i^+(|q_i^+| - |q_i^-|) + (W_i^+ - W_i^-)|q_i^-|\}\\
&\doteq E'_{\alpha,i} + E''_{\alpha,i} + E'''_{\alpha,i}.
\end{aligned} \tag{8.44}$$

In view of the estimates (8.43), for each $i = 1, \dots, n$ we have

$$E''_{\alpha,i} = \begin{cases} 0 & \text{if } \sigma_\alpha < 0,\\ \mathcal{O}(1)\cdot|\sigma_\alpha|^3 & \text{if } \sigma_\alpha \in [0,\varepsilon]. \end{cases} \tag{8.45}$$

For a bound for $E'''_{\alpha,i}$, observe that the definitions (8.7)–(8.9) of the weights imply

$$W_i^+ - W_i^- = \mathcal{O}(1)\cdot\kappa_1\sigma_\alpha \quad \text{if } q_i^+ q_i^- > 0.$$

Moreover, we have the estimates

$$q_i^+ - q_i^- = \mathcal{O}(1)\cdot|\sigma_\alpha|,$$
$$q_i^- = \mathcal{O}(1)\cdot|\sigma_\alpha| \quad \text{if } q_i^+ q_i^- \le 0.$$

By (8.10), the bound (8.37) always holds. Using (8.42) together with the above estimates we deduce that

$$E'''_{\alpha,i} = \mathcal{O}(1)\cdot\varepsilon\,|\sigma_\alpha|. \tag{8.46}$$

From (8.45)–(8.46), the proof of (8.21) is reduced to showing that

$$\sum_{i=1}^{n} E'_{\alpha,i} \le \mathcal{O}(1)\cdot\varepsilon|\sigma_\alpha|, \tag{8.47}$$

with $E'_{\alpha,i}$ defined at (8.44). In our future estimates, we can thus replace $v^+$ with $v^\diamond$ and $\dot{x}_\alpha$ with $\dot{x}_\alpha^\diamond$. In essence, this reduces the problem to the case (8.40) where the right state $v^\diamond$ is connected to the left state $v^-$ by a $k_\alpha$-shock, travelling with the exact Rankine–Hugoniot speed. From now on, we will work towards a proof of (8.47) in the case $\alpha \in \mathcal{R}\cup\mathcal{S}$.

**3. Elementary estimates.** We describe here a simple technique for deriving a number of a priori bounds. First of all, observe that, as soon as the state $u^- = u(x_\alpha-)$ is assigned, all quantities $v^-, v^+, \lambda_i^-, q_i^\diamond, \lambda_i^\diamond, \omega_i^-, \omega_i^\diamond \dots$ can then be recovered as functions of $q_1^-, \ldots, q_n^-$ and $\sigma_\alpha$. Indeed, $v^-$ is obtained from (8.4), while $v^\diamond$ and $\dot{x}_\alpha^\diamond$ are obtained by (8.38). In turn, the components $q_i^\diamond$ are implicitly determined by (8.39), while (8.14) and (8.40) define the intermediate states $\omega_i^-$ and $\omega_i^\diamond$, respectively. Introducing the variables

$$\tilde{q} \doteq (q_1^-, \ldots, q_{k_\alpha-1}^-, q_{k_\alpha+1}^-, \ldots, q_n^-) \in \mathbb{R}^{n-1}, \qquad q^* \doteq q_{k_\alpha}^-, \qquad \sigma \doteq \sigma_\alpha,$$

we can thus apply Lemma 2.6 or Lemma 2.7 to functions of the form

$$\Psi = \Psi(q_1^-, \ldots, q_n^-, \sigma_\alpha) = \Psi(\tilde{q}, q^*, \sigma)$$

and derive useful a priori estimates.

For example, if the $k_\alpha$-th field is linearly degenerate, then each of the functions

$$\Psi_i \doteq q_i^\diamond - q_i^- \quad i \neq k_\alpha,$$
$$\Psi_{k_\alpha} \doteq q_{k_\alpha}^\diamond - q_{k_\alpha}^- - \sigma_\alpha$$

satisfies the assumptions (2.71) and (2.73) of Lemma 2.6. Therefore, (2.74) yields the estimate

$$|q_{k_\alpha}^\diamond - q_{k_\alpha}^- - \sigma_\alpha| + \sum_{i \neq k_\alpha} |q_i^\diamond - q_i^-| = \mathcal{O}(1) \cdot \Bigg( \sum_{i \neq k_\alpha} |q_i^-| \Bigg) |\sigma_\alpha|. \tag{8.48}$$

On the other hand, if the $k_\alpha$-th field is genuinely non-linear, then the above functions $\Psi_i$, $\Psi_{k_\alpha}$ satisfy only the assumptions (2.71) of Lemma 2.6. In this case, (2.72) yields the weaker estimate

$$|q_{k_\alpha}^\diamond - q_{k_\alpha}^- - \sigma_\alpha| + \sum_{i \neq k_\alpha} |q_i^\diamond - q_i^-| = \mathcal{O}(1) \cdot \Bigg( |q_{k_\alpha}^-||q_{k_\alpha}^- + \sigma_\alpha| + \sum_{i \neq k_\alpha} |q_i^-| \Bigg) |\sigma_\alpha|. \tag{8.49}$$

As a further example, observe that the two functions

$$\Psi_\alpha'(q_1^-, \ldots, q_n^-, \sigma_\alpha) \doteq \dot{x}_\alpha^\diamond - \lambda_{k_\alpha}^-, \qquad \Psi_\alpha''(q_1^-, \ldots, q_n^-, \sigma_\alpha) \doteq \dot{x}_\alpha^\diamond - \lambda_{k_\alpha}^\diamond$$

satisfy the assumptions (2.78) and (2.80), respectively. Using Lemma 2.7 we thus obtain the estimates

$$\dot{x}_\alpha^\diamond - \lambda_{k_\alpha}^- = \frac{q_{k_\alpha}^- + \sigma_\alpha}{2} + \mathcal{O}(1) \cdot \Bigg( |q_{k_\alpha}^- + \sigma_\alpha|(|q_{k_\alpha}^-| + |\sigma_\alpha|) + \sum_{i \neq k_\alpha} |q_i^-| \Bigg) \tag{8.50}$$

$$\dot{x}_\alpha^\diamond - \lambda_{k_\alpha}^\diamond = \frac{q_{k_\alpha}^-}{2} + \mathcal{O}(1) \cdot \Bigg( |q_{k_\alpha}^-|(|q_{k_\alpha}^-| + |\sigma_\alpha|) + \sum_{i \neq k_\alpha} |q_i^-| \Bigg). \tag{8.51}$$

**4. Linearly degenerate fields.** Here and in the next step, our goal is to estimate the terms $E'_{\alpha,i}$ introduced at (8.44). We start with the case where the $k_\alpha$-th field is linearly degenerate, so that (8.41)–(8.42) hold. For notational convenience, consider the two sets of indices

$$\begin{aligned} \mathcal{I} &\doteq \{i \in \{1, \ldots, n\},\ i \neq k_\alpha,\ q_i^+ q_i^- > 0\}, \\ \mathcal{I}' &\doteq \{i \in \{1, \ldots, n\},\ i \neq k_\alpha,\ q_i^+ q_i^- \leq 0\}. \end{aligned} \tag{8.52}$$

Our definition of weights implies $W^+_{k_\alpha} = W^-_{k_\alpha}$ while

$$W^+_i = W^-_i - \kappa_1|\sigma_\alpha|\,\mathrm{sgn}(i - k_\alpha) \quad \text{if } i \in \mathcal{I}. \tag{8.53}$$

On the other hand, if $i \in \mathcal{I}'$, by (8.48) there exists a constant $C$ such that

$$|q_i^\diamond| + |q_i^-| \le C \cdot \left(\sum_{i \neq k_\alpha} |q_i^-|\right)|\sigma_\alpha|. \tag{8.54}$$

Hence

$$E'_{\alpha,i} = \mathcal{O}(1) \cdot \left(\sum_{i \neq k_\alpha} |q_i^-|\right)|\sigma_\alpha|. \tag{8.55}$$

A bound on $E'_{\alpha,k_\alpha}$ will be obtained using the estimates

$$\lambda^-_{k_\alpha} - \dot{x}^\diamond_\alpha = \lambda_{k_\alpha}(\omega^-_{k_\alpha}) - \lambda_{k_\alpha}(\omega^-_n) = \mathcal{O}(1) \cdot \sum_{i > k_\alpha} |q_i^-|,$$
$$q^\diamond_{k_\alpha} - q^-_{k_\alpha} = \mathcal{O}(1) \cdot |\sigma_\alpha|.$$

Observing that the quantity

$$\Psi_{k_\alpha}(q_1^-, \ldots, q_n^-, \sigma_\alpha) \doteq \lambda^-_{k_\alpha} - \lambda^\diamond_{k_\alpha} = \lambda_{k_\alpha}(\omega^-_{k_\alpha}) - \lambda_{k_\alpha}(\omega^\diamond_{k_\alpha})$$

satisfies both assumptions (2.71) and (2.73), using Lemma 2.6 and the above estimates we obtain

$$\begin{aligned} E'_{\alpha,k_\alpha} &= W^+_{k_\alpha} \cdot \left\{|q^\diamond_{k_\alpha}|(\lambda^\diamond_{k_\alpha} - \dot{x}^\diamond_\alpha) - |q^-_{k_\alpha}|(\lambda^-_{k_\alpha} - \dot{x}^\diamond_\alpha)\right\} \\ &\le W^+_{k_\alpha} \cdot \left\{|q^\diamond_{k_\alpha}||\lambda^-_{k_\alpha} - \lambda^\diamond_{k_\alpha}| + ||q^\diamond_{k_\alpha}| - |q^-_{k_\alpha}||\,|\lambda^-_{k_\alpha} - \dot{x}^\diamond_\alpha|\right\} \\ &= \mathcal{O}(1) \cdot |\sigma_\alpha| \sum_{i \neq k_\alpha} |q_i^-|. \end{aligned} \tag{8.56}$$

For $i \in \mathcal{I}$, observing that the quantity

$$\Psi_i(q_1^-, \ldots, q_n^-, \sigma_\alpha) \doteq q_i^\diamond(\lambda_i^\diamond - \dot{x}^\diamond_\alpha) - q_i^-(\lambda_i^- - \dot{x}^\diamond_\alpha) \tag{8.57}$$

satisfies both assumptions (2.71) and (2.73), using Lemma 2.6 we obtain

$$\begin{aligned} E'_{\alpha,i} &= -\kappa_1|\sigma_\alpha||q_i^-||\lambda_i^- - \dot{x}^\diamond_\alpha| + W_i^+\left\{|q_i^\diamond|(\lambda_i^\diamond - \dot{x}^\diamond_\alpha) - |q_i^-|(\lambda_i^- - \dot{x}^\diamond_\alpha)\right\} \\ &\le -c\kappa_1|\sigma_\alpha||q_i^-| + \mathcal{O}(1) \cdot \left(\sum_{i \neq k_\alpha} |q_i^-|\right)|\sigma_\alpha|. \end{aligned} \tag{8.58}$$

Since the total variation of $u, v$ is small, in (8.52) it is not restrictive to assume that $nC|\sigma_\alpha| < 1/2$. Therefore,

$$\sum_{i \in \mathcal{I}'} |q_i^-| \le nC\left(\sum_{i \neq k_\alpha} |q_i^-|\right)|\sigma_\alpha| < \frac{1}{2} \sum_{i \in \mathcal{I} \cup \mathcal{I}'} |q_i^-|, \qquad \sum_{i \in \mathcal{I}} |q_i^-| > \frac{1}{2} \cdot \sum_{i \neq k_\alpha} |q_i^-|. \tag{8.59}$$

Summing together the inequalities (8.55), (8.56), (8.58) and using (8.59) we obtain

$$\begin{aligned}\sum_{i=1}^{n} E'_{\alpha,i} &\le -c\kappa_1|\sigma_\alpha| \cdot \sum_{i\in\mathcal{I}} |q_i^-| + \mathcal{O}(1)\cdot\Big(\sum_{i\neq k_\alpha}|q_i^-|\Big)|\sigma_\alpha| \\ &= -c\kappa_1|\sigma_\alpha| \cdot \sum_{i\in\mathcal{I}} |q_i^-| + \mathcal{O}(1)\cdot\Big(\sum_{i\in\mathcal{I}}|q_i^-|\Big)|\sigma_\alpha| \\ &\le 0, \end{aligned} \tag{8.60}$$

provided that the constant $\kappa_1$ is chosen large enough.

**5. Genuinely non-linear fields.** We now assume that the $k_\alpha$-th field is genuinely non-linear. We define the two sets of indices

$$\begin{aligned} \mathcal{I} &\doteq \{i \in \{1,\dots,n\},\ i\neq k_\alpha,\ q_i^-, q_i^+, q_i^\diamond \text{ all have the same sign}\}, \\ \mathcal{I}' &\doteq \{i \in \{1,\dots,n\},\ i\neq k_\alpha,\ i\notin\mathcal{I}\}. \end{aligned} \tag{8.61}$$

For $i \in \mathcal{I}$, the weights $W_i^\pm$ still satisfy (8.53). Repeating the argument at (8.58) we deduce that

$$\begin{aligned} E'_{\alpha,i} &= -\kappa_1|\sigma_\alpha||q_i^-||\lambda_i^- - \dot{x}_\alpha^\diamond| + W_i^+\big\{|q_i^\diamond|(\lambda_i^\diamond - \dot{x}_\alpha^\diamond) - |q_i^-|(\lambda_i^- - \dot{x}_\alpha^\diamond)\big\} \\ &\le -c\kappa_1|\sigma_\alpha||q_i^-| + \mathcal{O}(1)\cdot\Big(|q_{k_\alpha}^-||q_{k_\alpha}^- + \sigma_\alpha| + \sum_{i\neq k_\alpha}|q_i^-|\Big)|\sigma_\alpha|. \end{aligned} \tag{8.62}$$

Indeed, the quantity $\Psi_i$ in (8.57) now satisfies only the assumptions (2.71) of Lemma 2.6; hence it can be estimated in terms of (2.72).

On the other hand, if $i \in \mathcal{I}'$, by (8.41), (8.43) and (8.49) we deduce that

$$\begin{aligned} |q_i^\diamond| + |q_i^-| &\le 2(|q_i^- - q_i^\diamond| + |q_i^+ - q_i^\diamond|) \\ &\le C\cdot\Big(\varepsilon|\sigma_\alpha| + |q_{k_\alpha}^-||q_{k_\alpha}^- + \sigma_\alpha| + \sum_{i\neq k_\alpha}|q_i^-|\Big)|\sigma_\alpha| \end{aligned} \tag{8.63}$$

for some constant $C$. Therefore,

$$E'_{\alpha,i} = \mathcal{O}(1)\cdot\Big(\varepsilon + |q_{k_\alpha}^-||q_{k_\alpha}^- + \sigma_\alpha| + \sum_{i\neq k_\alpha}|q_i^-|\Big)|\sigma_\alpha|. \tag{8.64}$$

The term $E'_{\alpha,k_\alpha}$ will be estimated separately in three different cases.

CASE 1: $|\sigma_\alpha| \le \varepsilon$, $|q_{k_\alpha}^-| \le 2|\sigma_\alpha|$. In this case, using the bounds

$$q_{k_\alpha}^\diamond - q_{k_\alpha}^- = \mathcal{O}(1)\cdot|\sigma_\alpha|, \qquad \lambda_{k_\alpha}^\diamond - \lambda_{k_\alpha}^- = \mathcal{O}(1)\cdot|\sigma_\alpha|,$$

$$\lambda_{k_\alpha}^\diamond - \dot{x}_\alpha^\diamond = \mathcal{O}(1)\cdot\Big(|\sigma_\alpha| + \sum_{i=1}^{n}|q_i^-|\Big) = \mathcal{O}(1)\cdot\Big(\varepsilon + \sum_{i\neq k_\alpha}|q_i^-|\Big),$$

we can write

$$\begin{aligned}E'_{\alpha,k_\alpha} &\le W^+_{k_\alpha}|q^\diamond_{k_\alpha} - q^-_{k_\alpha}||\lambda^\diamond_{k_\alpha} - \dot{x}^\diamond_\alpha| + |W^+_{k_\alpha} - W^-_{k_\alpha}||q^-_{k_\alpha}||\lambda^\diamond_{k_\alpha} - \dot{x}^\diamond_\alpha| \\ &\quad + W^-_{k_\alpha}|q^-_{k_\alpha}||\lambda^\diamond_{k_\alpha} - \lambda^-_{k_\alpha}| \\ &= \mathcal{O}(1)\cdot\Big(\varepsilon + \sum_{i\neq k_\alpha}|q_i^-|\Big)|\sigma_\alpha|. \end{aligned} \tag{8.65}$$

By (8.63), when $i \in \mathcal{I}'$ it follows that

$$|q_i^-| \le C\cdot\Big(\varepsilon|\sigma_\alpha| + \sum_{i\neq k_\alpha}|q_i^-|\Big)|\sigma_\alpha|$$

for some constant $C$. Since the total variation is small, we can again assume that $nC|\sigma_\alpha| < 1/2$. This yields

$$\sum_{i\in\mathcal{I}'}|q_i^-| \le \frac{1}{2}\Big(\varepsilon|\sigma_\alpha| + \sum_{i\in\mathcal{I}\cup\mathcal{I}'}|q_i^-|\Big), \qquad \sum_{i\neq k_\alpha}|q_i^-| \le \varepsilon|\sigma_\alpha| + 2\sum_{i\in\mathcal{I}}|q_i^-|. \tag{8.66}$$

From (8.62), (8.64) and (8.65), using (8.66) we conclude that

$$\begin{aligned}\sum_{i=1}^n E'_{\alpha,i} &\le -c\kappa_1|\sigma_\alpha|\sum_{i\in\mathcal{I}}|q_i^-| + \mathcal{O}(1)\cdot|\sigma_\alpha|\Big(\varepsilon + \sum_{i\in\mathcal{I}}|q_i^-|\Big) \\ &\le \mathcal{O}(1)\cdot\varepsilon|\sigma_\alpha|, \end{aligned} \tag{8.67}$$

provided that the constant $\kappa_1$ is large enough.

CASE 2: $q^-_{k_\alpha}, q^+_{k_\alpha}, q^\diamond_{k_\alpha}$ all have the same sign. In this case we have

$$W^+_{k_\alpha} = W^-_{k_\alpha} + \kappa_1|\sigma_\alpha|\,\mathrm{sgn}(q^-_{k_\alpha}).$$

Since the quantity

$$\Psi_{k_\alpha}(q_1^-,\ldots,q_n^-,\sigma_\alpha) \doteq q^\diamond_{k_\alpha}(\lambda^\diamond_{k_\alpha} - \dot{x}^\diamond_\alpha) - q^-_{k_\alpha}(\lambda^-_{k_\alpha} - \dot{x}^\diamond_\alpha)$$

satisfies the assumptions (2.71), by Lemma 2.6 it can be estimated as

$$q^\diamond_{k_\alpha}(\lambda^\diamond_{k_\alpha} - \dot{x}^\diamond_\alpha) - q^-_{k_\alpha}(\lambda^-_{k_\alpha} - \dot{x}^\diamond_\alpha) = \mathcal{O}(1)\cdot\Big(|q^-_{k_\alpha}||q^-_{k_\alpha} + \sigma_\alpha| + \sum_{i\neq k_\alpha}|q_i^-|\Big)|\sigma_\alpha|. \tag{8.68}$$

From the estimate (8.50) we deduce that

$$\begin{aligned}&\kappa_1|\sigma_\alpha|\,\mathrm{sgn}(q^-_{k_\alpha})\,|q^-_{k_\alpha}|(\lambda^-_{k_\alpha} - \dot{x}^\diamond_\alpha) \\ &\le -\frac{1}{2}\kappa_1|\sigma_\alpha|\,|q^-_{k_\alpha}||q^-_{k_\alpha} + \sigma_\alpha| + \mathcal{O}(1)\cdot\Big(|q^-_{k_\alpha}||q^-_{k_\alpha} + \sigma_\alpha| + \sum_{i\neq k_\alpha}|q_i^-|\Big)|\sigma_\alpha|. \end{aligned} \tag{8.69}$$

Using (8.68) and (8.69) we obtain

$$\begin{aligned}E'_{\alpha,k_\alpha} &= (W^+_{k_\alpha} - W^-_{k_\alpha})|q^-_{k_\alpha}|(\lambda^-_{k_\alpha} - \dot{x}^\diamond_\alpha) + W^+_{k_\alpha}\cdot\big\{|q^\diamond_{k_\alpha}|(\lambda^\diamond_{k_\alpha} - \dot{x}^\diamond_\alpha) - |q^-_{k_\alpha}|(\lambda^-_{k_\alpha} - \dot{x}^\diamond_\alpha)\big\} \\ &\le -\frac{\kappa_1}{2}|\sigma_\alpha|\,|q^-_{k_\alpha}|\,|q^-_{k_\alpha} + \sigma_\alpha| + \mathcal{O}(1)\cdot\Big(|q^-_{k_\alpha}||q^-_{k_\alpha} + \sigma_\alpha| + \sum_{i\neq k_\alpha}|q_i^-|\Big)|\sigma_\alpha|. \end{aligned} \tag{8.70}$$

Since the total variation is small, in (8.63) we can assume that $nC|\sigma_\alpha| < 1/2$. Hence

$$\sum_{i\in\mathcal{I}'} |q_i^-| \le \frac{1}{2}\cdot\left(\varepsilon|\sigma_\alpha| + |q_{k_\alpha}^-||q_{k_\alpha}^- + \sigma_\alpha| + \sum_{i\in\mathcal{I}\cup\mathcal{I}'} |q_i^-|\right),$$
$$\sum_{i\neq k_\alpha} |q_i^-| \le \varepsilon|\sigma_\alpha| + |q_{k_\alpha}^-||q_{k_\alpha}^- + \sigma_\alpha| + 2\sum_{i\in\mathcal{I}} |q_i^-|. \tag{8.71}$$

From (8.62), (8.64) and (8.70), using (8.71) we conclude that

$$\begin{aligned}\sum_{i=1}^{n} E'_{\alpha,i} &\le -c\kappa_1|\sigma_\alpha|\sum_{i\in\mathcal{I}}|q_i^-| - \frac{\kappa_1}{2}|\sigma_\alpha|\,|q_{k_\alpha}^-|\,|q_{k_\alpha}^- + \sigma_\alpha|\\ &\quad + \mathcal{O}(1)\cdot|\sigma_\alpha|\left(\varepsilon + |q_{k_\alpha}^-||q_{k_\alpha}^- + \sigma_\alpha| + \sum_{i\in\mathcal{I}}|q_i^-|\right)\\ &\le \mathcal{O}(1)\cdot\varepsilon|\sigma_\alpha|,\end{aligned} \tag{8.72}$$

provided that the constant $\kappa_1$ is large enough.

CASE 3: $q_{k_\alpha}^+ < 0 < q_{k_\alpha}^-$. Notice that in this case $\sigma_\alpha < 0$; hence the front is a shock or contact discontinuity and the identities (8.41) hold. By (8.49) we can also assume that

$$\tfrac{1}{2}|\sigma_\alpha| < |q_{k_\alpha}^-| + |q_{k_\alpha}^+| < 2|\sigma_\alpha|. \tag{8.73}$$

Recalling that $W_{k_\alpha}^\pm \in [1,2]$, using (8.50)–(8.51), (8.73) and the fact that $q_{k_\alpha}^+ = q_{k_\alpha}^\diamond = -|q_{k_\alpha}^\diamond|$, we obtain

$$\begin{aligned}E'_{\alpha,k_\alpha} &= W_{k_\alpha}^+|q_{k_\alpha}^\diamond|(\lambda_{k_\alpha}^\diamond - \dot{x}_\alpha^\diamond) - W_{k_\alpha}^-|q_{k_\alpha}^-|(\lambda_{k_\alpha}^- - \dot{x}_\alpha^\diamond)\\ &= W_{k_\alpha}^+|q_{k_\alpha}^\diamond|\left(-\frac{q_{k_\alpha}^-}{2} + \mathcal{O}(1)\cdot|q_{k_\alpha}^-|\big(|q_{k_\alpha}^-| + |\sigma_\alpha|\big) + \mathcal{O}(1)\cdot\sum_{i\neq k_\alpha}|q_i^-|\right)\\ &\quad - W_{k_\alpha}^-|q_{k_\alpha}^-|\left(-\frac{q_{k_\alpha}^- + \sigma_\alpha}{2} + \mathcal{O}(1)\cdot|q_{k_\alpha}^- + \sigma_\alpha|\big(|q_{k_\alpha}^-| + |\sigma_\alpha|\big) + \mathcal{O}(1)\cdot\sum_{i\neq k_\alpha}|q_i^-|\right)\\ &\le -|q_{k_\alpha}^- + \sigma_\alpha||q_{k_\alpha}^-| + \mathcal{O}(1)\cdot|q_{k_\alpha}^-||q_{k_\alpha}^\diamond - q_{k_\alpha}^- - \sigma_\alpha|\\ &\quad + \mathcal{O}(1)\cdot|q_{k_\alpha}^- + \sigma_\alpha||q_{k_\alpha}^-|\big(|q_{k_\alpha}^-| + |\sigma_\alpha|\big) + \mathcal{O}(1)\cdot\big(|q_{k_\alpha}^\diamond| + |q_{k_\alpha}^-|\big)\sum_{i\neq k_\alpha}|q_i^-|\\ &\le -|q_{k_\alpha}^- + \sigma_\alpha|\,|q_{k_\alpha}^-| + \mathcal{O}(1)\cdot|\sigma_\alpha|\left(|q_{k_\alpha}^-||q_{k_\alpha}^- + \sigma_\alpha| + \sum_{i\neq k_\alpha}|q_i^-|\right).\end{aligned} \tag{8.74}$$

When $i \in \mathcal{I}$, the estimates (8.62) remain valid. When $i \in \mathcal{I}'$, by (8.41) and (8.49) we deduce that

$$|q_i^\diamond| + |q_i^-| = |q_i^\diamond - q_i^-| \le C\cdot\left(|q_{k_\alpha}^-||q_{k_\alpha}^- + \sigma_\alpha| + \sum_{i\neq k_\alpha}|q_i^-|\right)|\sigma_\alpha| \tag{8.75}$$

for some constant $C$. Therefore,

$$E'_{\alpha,i} = \mathcal{O}(1) \cdot \left( |q^-_{k_\alpha}||q^-_{k_\alpha} + \sigma_\alpha| + \sum_{i \neq k_\alpha} |q^-_i| \right) |\sigma_\alpha|. \tag{8.76}$$

Since the total variation is small, in (8.75) we can assume that $nC|\sigma_\alpha| < 1/2$. Hence

$$\begin{aligned} \sum_{i \in \mathcal{I}'} |q^-_i| &\leq \frac{1}{2} \cdot \left( |q^-_{k_\alpha}| \, |q^-_{k_\alpha} + \sigma_\alpha| + \sum_{i \in \mathcal{I} \cup \mathcal{I}'} |q^-_i| \right), \\ \sum_{i \neq k_\alpha} |q^-_i| &\leq |q^-_{k_\alpha}| \, |q^-_{k_\alpha} + \sigma_\alpha| + 2 \sum_{i \in \mathcal{I}} |q^-_i|. \end{aligned} \tag{8.77}$$

By the smallness of the total variation we can also assume that

$$\mathcal{O}(1) \cdot |\sigma_\alpha| \ll 1.$$

Summing together (8.62), (8.74) and (8.76) and using (8.77) we obtain

$$\begin{aligned} \sum_{i=1}^{n} E'_{\alpha,i} &\leq -c\kappa_1 |\sigma_\alpha| \sum_{i \in \mathcal{I}} |q^-_i| - |q^-_{k_\alpha}| \, |q^-_{k_\alpha} + \sigma_\alpha| \\ &\quad + \mathcal{O}(1) \cdot |\sigma_\alpha| \left( |q^-_{k_\alpha}| \, |q^-_{k_\alpha} + \sigma_\alpha| + \sum_{i \in \mathcal{I}} |q^-_i| \right) \\ &\leq 0, \end{aligned} \tag{8.78}$$

provided that the constant $\kappa_1$ is chosen suitably large.

The above three cases cover all possible situations. Indeed, if $q^-_{k_\alpha} \leq 0 \leq q^+_{k_\alpha}$, then the front is a rarefaction and has strength $\sigma_\alpha \in \,]0, \varepsilon]$. Hence CASE 1 applies.

In all cases, the above analysis shows that the bound (8.47) holds, provided that we first choose the constants $\kappa_1, \kappa_2$ suitably large, and then let the total variation be sufficiently small. This completes the proof of Theorem 8.2. □

## 8.3 Proof of $\mathbf{L}^1$ stability

Relying on the estimate (8.13), we now prove Theorem 8.1. Let $\bar{u} \in \mathcal{D}$ be given. Consider any sequence $(u_\nu)_{\nu \geq 1}$, such that each $u_\nu$ is a front tracking $\varepsilon_\nu$-approximate solution of the Cauchy problem (7.1)–(7.2), with

$$\lim_{\nu \to \infty} \varepsilon_\nu = 0, \qquad \Upsilon\big(u_\nu(t)\big) < \delta_0 \qquad \text{for all } t \geq 0, \ \nu \geq 1.$$

For every $\mu, \nu \geq 1$ and $t \geq 0$, by (8.11) and (8.13) it now follows that

$$\begin{aligned} \|u_\mu(t) - u_\nu(t)\|_{\mathbf{L}^1} &\leq C_1 \cdot \Phi(u_\mu(t), u_\nu(t)) \\ &\leq C_1 \cdot [\Phi(u_\mu(0), u_\nu(0)) + C_2 t \cdot \max\{\varepsilon_\mu, \varepsilon_\nu\}] \\ &\leq 2C_1^2 \|u_\mu(0) - u_\nu(0)\|_{\mathbf{L}^1} + C_1 C_2 t \cdot \max\{\varepsilon_\mu, \varepsilon_\nu\}. \end{aligned} \tag{8.79}$$

Since the right hand side of (8.79) approaches zero as $\mu, \nu \to \infty$, the sequence is Cauchy and converges to a unique limit. The semigroup property (8.2) is an immediate consequence of uniqueness. Finally, let $\bar{u}, \bar{v} \in \mathcal{D}$ be given. For each $\nu \geq 1$, let $u_\nu, v_\nu$ be front tracking $\varepsilon_\nu$-approximate solutions of (7.1) with

$$\|u_\nu(0) - \bar u\|_{\mathbf{L}^1} < \varepsilon_\nu, \qquad \|v_\nu(0) - \bar v\|_{\mathbf{L}^1} < \varepsilon_\nu, \qquad \lim_{\nu\to\infty} \varepsilon_\nu = 0. \tag{8.80}$$

Using (8.11) and (8.13) again we deduce that

$$\begin{aligned} \|u_\nu(t) - v_\nu(t)\|_{\mathbf{L}^1} &\le C_1 \cdot \Phi(u_\nu(t), v_\nu(t)) \\ &\le C_1 \cdot [\Phi(u_\nu(0), v_\nu(0)) + C_2 t\varepsilon_\nu] \\ &\le 2C_1^2 \|u_\nu(0) - v_\nu(0)\|_{\mathbf{L}^1} + C_1C_2 t\varepsilon_\nu. \end{aligned} \tag{8.81}$$

Letting $\nu \to \infty$, by (8.80) it follows that

$$\|u(t) - v(t)\|_{\mathbf{L}^1} \le 2C_1^2 \cdot \|\bar u - \bar v\|_{\mathbf{L}^1}. \tag{8.82}$$

This establishes the uniform Lipschitz continuity of the semigroup with respect to the initial data. By (7.79) all $\varepsilon$-approximate solutions are uniformly Lipschitz continuous as functions from $[0, \infty[$ into $\mathbf{L}^1$. Hence the Lipschitz continuity of the semigroup $S$ with respect to time is clear. This completes the proof of Theorem 8.1. □

## Problem

(1) On the space $\mathcal{C}_c^1$ of continuously differentiable functions with compact support, consider the functional

$$\Psi(w) \doteq \iint_{x<y} |w_x(x)| w(y)\, dx\, dy.$$

Let $u = u(t, x)$ be a non-negative, smooth solution of the scalar Cauchy problem

$$u_t + F(u)_x = 0, \qquad u(0, x) = \bar u(x) \tag{8.83}$$

defined on the strip $[0, T] \times \mathbb{R}$. Assume that $\bar u \in \mathcal{C}_c^1$ and $\bar u(x) \ge 0$ for every $x \in \mathbb{R}$. Prove the identity

$$\frac{d}{dt}\Psi(u(t)) = \int_{-\infty}^{\infty} [F(u) - F(0) - F'(u)\, u] \cdot |u_x|\, dx. \tag{8.84}$$

Show that the time derivative in (8.84) is constant on $[0, T]$. If the flux function $F$ is convex, deduce that the functional $\Psi$ is non-increasing in time, along every positive solution of (8.83).

Hint: observe that the solution $u$ is constant along each characteristic line of the form $x = x_0 + tF'(\bar u(x_0))$. For each $x_0 \in \mathbb{R}$, compute

$$\frac{d}{dt} \int_{x_0 + tF'(\bar u(x_0))}^{\infty} u(t, y)\, dy.$$

# 9
# Uniqueness

According to the analysis in the previous chapter, the solution of the Cauchy problem (7.1)–(7.2) obtained as a limit of front tracking approximations is unique and depends Lipschitz-continuously on the initial data. This basic result, however, leaves open the question of whether other entropy weak solutions may exist, possibly constructed by different approximation algorithms. We address this problem in the present chapter, proving the uniqueness of a solution $u = u(t, x)$ in the following two general situations:

(I) A constructive algorithm is given, which produces a whole semigroup of trajectories $(t, u_0) \mapsto \tilde{S}_t u_0$ depending Lipschitz-continuously on the initial data $u_0$ (w.r.t. the $\mathbf{L}^1$ distance). Moreover, the algorithm yields the correct entropy-admissible self-similar solution of (7.1) when applied to a single Riemann problem. The solution $t \mapsto u(t, \cdot)$ that we are considering coincides with one particular trajectory of this semigroup $\tilde{S}$.

(II) The given function $u = u(t, x)$ provides a weak solution to the Cauchy problem (7.1)–(7.2) and satisfies the Lax entropy conditions together with a suitable regularity assumption.

In both cases, we will show that $u$ coincides with the particular solution of the Cauchy problem constructed as a limit of front tracking approximations; hence it is unique.

The first case is considered in Theorem 9.1, stating the uniqueness of Riemann semigroups. Remarkably, in this case there is no need to check that each trajectory of the semigroup $\tilde{S}$ is actually a weak solution, or satisfies the usual entropy conditions. All these properties follow from the correct behaviour of the semigroup on Riemann data.

The second case is covered by Theorem 9.4. To obtain uniqueness, two additional regularity assumptions are introduced here. The *tame oscillation condition* restricts the oscillation of the solution $u = u(t, x)$ on a domain of determinacy of a horizontal segment in the $t$–$x$ plane. The *bounded variation condition* simply requires $u$ to have bounded variation on all space-like curves with suitably small slope. It is not difficult to show that both of these conditions are satisfied by solutions obtained via front tracking approximations. According to Theorem 9.4, either one of these assumptions suffices to achieve the uniqueness of solutions.

Throughout the following, we continue to use the basic assumptions (♣) stated at the beginning of Chapter 7. Moreover, we shall always work with solutions $u = u(t, x)$ which are continuous when regarded as maps $t \mapsto u(t, \cdot)$ with values in $\mathbf{L}^1_{loc}$, and such that Tot. Var. $\{u(t, \cdot)\} < \infty$ at each time $t$. By possibly changing the point values of $u$ on a set of measure zero, we will thus assume that every such solution is right continuous, i.e.

$$u(t,x) = u(t,x+) \doteq \lim_{y\to x+} u(t,y) \quad \text{for all } t, x.$$

## 9.1 Uniqueness of the semigroup

Consider piecewise constant initial data $u(0,x) = \bar{u}(x)$ with jumps at points $x_1 < \cdots < x_N$. An exact solution of the Cauchy problem (7.1)–(7.2), valid on a small time interval $t \in [0, \delta]$, can then be constructed simply by piecing together the solutions to the various Riemann problems generated by the jumps of $\bar{u}$. More precisely (Fig. 9.1), for each $\alpha = 1, \ldots, N$, call $\omega_\alpha = \omega_\alpha(t,x)$ the self-similar solution of the Riemann problem

$$\omega_t + f(\omega)_x = 0, \qquad \omega(0,x) = \begin{cases} u^- & \text{if } x < 0, \\ u^+ & \text{if } x > 0, \end{cases} \tag{9.1}$$

with

$$u^- = \bar{u}(x_\alpha -), \qquad u^+ = \bar{u}(x_\alpha +). \tag{9.2}$$

Choose points $y_\alpha$, with $x_\alpha < y_\alpha < x_{\alpha+1}$, $y_0 \doteq -\infty$, $y_N \doteq \infty$, and define

$$u(t,x) \doteq \omega_\alpha(t,\ x - x_\alpha) \quad \text{if } y_{\alpha-1} < x \leq y_\alpha. \tag{9.3}$$

For $t > 0$ small, as long as the waves from the various Riemann problems do not cross any one of the lines $x = y_\alpha$ and interact with each other, the function in (9.3) clearly provides an entropy solution to our Cauchy problem. It is straightforward to show that this solution can be obtained as a limit of a sequence of front tracking approximations.

In the following we shall consider an arbitrary Lipschitz semigroup defined on a domain of $BV$ functions. We will show that, by assigning the local flow in connection with piecewise constant initial data, the whole semigroup can be uniquely determined.

**Definition 9.1.** Let $\mathcal{D} \subset \mathbf{L}^1(\mathbb{R}; \mathbb{R}^n)$ be a closed domain. A map $S : \mathcal{D} \times [0, \infty[ \mapsto \mathcal{D}$ is a *standard Riemann semigroup* (SRS) generated by the system of conservation

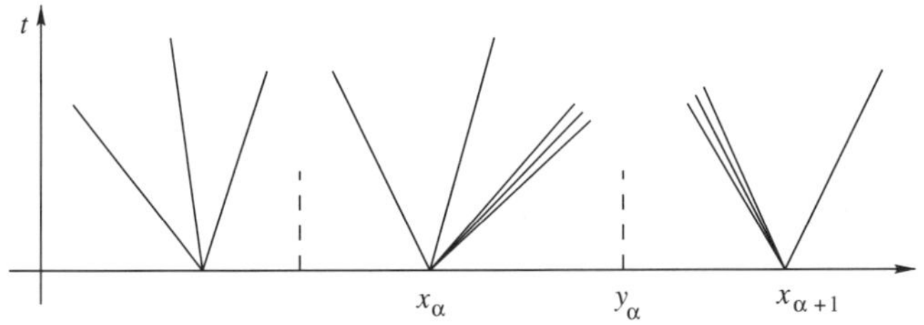


**Figure 9.1**

laws (7.1) if the following three conditions hold:

**(1) Semigroup property:** For every $\bar{u} \in \mathcal{D}$, $t, s \geq 0$ one has

$$S_0\bar{u} = \bar{u}, \qquad S_t S_s \bar{u} = S_{s+t}\bar{u}. \tag{9.4}$$

**(2) Lipschitz continuity:** There exist constants $L, L'$ such that, for all $\bar{u}, \bar{v} \in \mathcal{D}$, $s, t \geq 0$ one has

$$\|S_t\bar{u} - S_s\bar{v}\|_{\mathbf{L}^1} \leq L\|\bar{u} - \bar{v}\|_{\mathbf{L}^1} + L'|t - s|. \tag{9.5}$$

**(3) Consistency with Riemann solver:** For all piecewise constant initial data $\bar{u} \in \mathcal{D}$, there exists a small $\delta > 0$ such that, for all $t \in [0, \delta]$, the trajectory $u(t, \cdot) = S_t\bar{u}$ coincides with the solution of the Cauchy problem (7-1)–(7.2) obtained by piecing together the standard entropy solutions of the Riemann problems determined by the jumps of $\bar{u}$.

Observe that, in the definition of an SRS, we do not require that each trajectory $t \mapsto S_t\bar{u}$ be a solution of the corresponding Cauchy problem (7.1)–(7.2). This property, however, turns out to be a consequence of the above conditions. Before stating a precise result in this direction, we prove a lemma concerning approximate solutions to the Riemann problem. As in the previous chapters, we assume that $f$ is a smooth map defined on an open domain $\Omega \subseteq \mathbb{R}^n$, and $\hat{\lambda}$ is a fixed speed, strictly larger than the absolute values of all characteristic speeds.

**Lemma 9.1.** *For $u^+, u^- \in \Omega$, $|\lambda| \leq \hat{\lambda}$ and $t > 0$, call $\omega = \omega(t, x)$ the self-similar solution of the Riemann problem (9.1) and consider the function*

$$v(t, x) \doteq \begin{cases} u^- & \text{if } x < \lambda t, \\ u^+ & \text{if } x \geq \lambda t. \end{cases} \tag{9.6}$$

*(i) In the general case, one has*

$$\frac{1}{t} \cdot \int_{-\infty}^{\infty} |v(t, x) - \omega(t, x)|\, dx = O(1) \cdot |u^+ - u^-|. \tag{9.7}$$

*(ii) Assuming the additional relations $u^+ = R_i(\sigma)(u^-)$ and $\lambda = \lambda_i(u^+)$, for some $\sigma > 0$, $i \in \{1, \ldots, n\}$, one has the sharper estimate*

$$\frac{1}{t} \cdot \int_{-\infty}^{\infty} |v(t, x) - \omega(t, x)|\, dx = O(1) \cdot \sigma^2. \tag{9.8}$$

(iii) *Let $u^* \in \Omega$ and call $\lambda_1^* < \cdots < \lambda_n^*$ the eigenvalues of the matrix $A^* \doteq Df(u^*)$. If, for some $i$, $A^*(u^+ - u^-) = \lambda_i^*(u^+ - u^-)$ holds and $\lambda = \lambda_i^*$ in (9.6), then one has*

$$\frac{1}{t} \cdot \int_{-\infty}^{\infty} |v(t,x) - \omega(t,x)|\, dx = O(1) \cdot |u^+ - u^-|(|u^+ - u^*| + |u^- - u^*|). \tag{9.9}$$

*Proof.* The bound (9.7) follows from

$$\frac{1}{t} \int_{-\infty}^{\infty} |v(t,x) - \omega(t,x)|\, dx = \frac{1}{t} \int_{-t\hat{\lambda}}^{t\hat{\lambda}} |v(t,x) - \omega(t,x)|\, dx = O(1) \cdot |u^+ - u^-|.$$

Similarly, (9.8) follows from

$$\begin{aligned} \frac{1}{t} \int_{-\infty}^{\infty} |v(t,x) - \omega(t,x)|\, dx &= \frac{1}{t} \int_{t\lambda_i(u^-)}^{t\lambda_i(u^+)} |u^- - \omega(t,x)|\, dx \\ &= O(1) \cdot (\lambda_i(u^+) - \lambda_i(u^-)) \cdot |u^+ - u^-| \\ &= O(1) \cdot \sigma^2. \end{aligned}$$

For a proof of (9.9), let $u^+ = u^+(\theta) = u^- + \theta r_i(u^*)$ and call $\sigma_1(\theta), \ldots, \sigma_n(\theta)$ the strengths of the waves generated by the corresponding Riemann problem (9.1). In the special case where $u^* = u^-$, computing the derivatives at $\theta = 0$ we find that

$$\frac{\partial \sigma_j}{\partial \theta}(0) = \delta_{ij} \doteq \begin{cases} 1 & \text{if } i = j, \\ 0 & \text{if } i \neq j. \end{cases}$$

By the smoothness of the coefficients, in the general case $u^* \neq u^-$ we thus have

$$\begin{aligned} \left| \frac{\partial \sigma_j}{\partial \theta}(0) - \delta_{ij} \right| &= \mathcal{O}(1) \cdot |u^* - u^-|, \\ \left| \frac{\partial \sigma_j}{\partial \theta}(\theta) - \delta_{ij} \right| &= \mathcal{O}(1) \cdot (|u^* - u^-| + |\theta|) \\ &= \mathcal{O}(1) \cdot (|u^+ - u^*| + |u^- - u^*|). \end{aligned}$$

Call $\omega_0 = u^-, \omega_1, \ldots, \omega_n = u^+$ the intermediate states occurring in the solution of the Riemann problem (9.1). Observing that $|\theta| = \mathcal{O}(1) \cdot |u^+ - u^-|$, the previous estimates imply

$$\begin{aligned} |\omega_j - \omega_{j-1}| = \mathcal{O}(1) \cdot |\sigma_j| &= \mathcal{O}(1) \cdot |u^+ - u^-|(|u^+ - u^*| + |u^- - u^*|) \quad (j \neq i), \\ |\omega_i - u^+| + |\omega_{i-1} - u^-| &= \mathcal{O}(1) \cdot |u^+ - u^-|(|u^+ - u^*| + |u^- - u^*|), \\ |\lambda_i(\omega_i) - \lambda_i^*| + |\lambda_i(\omega_{i-1}) - \lambda_i^*| &= \mathcal{O}(1) \cdot (|u^+ - u^*| + |u^- - u^*|). \end{aligned} \tag{9.10}$$

Define the speeds

$$\lambda_i^- \doteq \min\{\lambda_i^*,\ \lambda_i(\omega_{i-1}),\ \lambda_i(\omega_i)\}, \qquad \lambda_i^+ \doteq \max\{\lambda_i^*, \lambda_i(\omega_{i-1}), \lambda_i(\omega_i)\}.$$

The $i$-wave in the solution of the Riemann problem (9.1) may be a rarefaction, a shock or a contact discontinuity. In all cases, the corresponding speed ranges within the interval $[\lambda_i^-,\ \lambda_i^+]$. The bound (9.9) now follows from

$$\frac{1}{t}\int_{-\infty}^{\infty} |v(t,x) - \omega(t,x)|\, dx = \frac{1}{t}\left(\int_{-t\hat{\lambda}}^{t\lambda_i^-} + \int_{t\lambda_i^+}^{t\hat{\lambda}} + \int_{t\lambda_i^-}^{t\lambda_i^+}\right)|v(t,x) - \omega(t,x)|\, dx$$

$$= O(1)\cdot \sum_{j\neq i} |\omega_j - \omega_{j-1}| + O(1)\cdot |u^+ - u^-|\cdot (|\lambda_i(\omega_i) - \lambda_i^*| + |\lambda_i(\omega_{i-1}) - \lambda_i^*|),$$

using the estimates (9.10). □

We can now state the main result concerning the uniqueness of the semigroup generated by a system of conservation laws. Throughout this chapter, we assume that the constants $C_0, \delta_0 > 0$ in (8.1) are suitably chosen so that the domain $\mathcal{D}$ is positively invariant for the flow generated by the system of conservation laws.

**Theorem 9.1 (Uniqueness of the Riemann semigroup).** *Let $\mathcal{D} \subset \mathbf{L}^1(\mathbb{R};\mathbb{R}^n)$ be a domain of the form (8.1) and let $S : \mathcal{D}\times[0,\infty[\mapsto \mathcal{D}$ be the semigroup of entropy weak solutions of (7.1) constructed as limits of front tracking approximations. In addition, let $\tilde{S} : \tilde{\mathcal{D}}\times[0,\infty[\mapsto \tilde{\mathcal{D}}$ be any SRS, defined on a domain $\tilde{\mathcal{D}} \supseteq \mathcal{D}$. Then*

$$\tilde{S}_t\bar{u} = S_t\bar{u} \quad \textit{for all } \bar{u}\in\mathcal{D},\ t\geq 0. \tag{9.11}$$

*Proof.* By Theorem 7.2, for each $\bar{u} \in \mathcal{D}$ and every $\varepsilon > 0$ the Cauchy problem (7.1)–(7.2) admits an $\varepsilon$-approximate front tracking solution $u$. Using the algorithm described in Chapter 7, one can indeed construct an $\varepsilon$-approximate solution satisfying the additional property $u(t,\cdot)\in\mathcal{D}$ for all $t\geq 0$.

We now provide a bound on the distance between this approximate solution and the corresponding semigroup trajectory $t\mapsto \tilde{S}_t\bar{u}$. Call $L$ the Lipschitz constant of the semigroup $\tilde{S}$. By Theorem 2.9, for every $t\geq 0$ one has

$$\|u(t) - \tilde{S}_t u(0)\|_{\mathbf{L}^1} \leq L\cdot \int_0^t \left\{\liminf_{h\to 0+} \frac{\|u(\tau+h) - \tilde{S}_h u(\tau)\|_{\mathbf{L}^1}}{h}\right\} d\tau. \tag{9.12}$$

To estimate the integrand on the right hand side of (9.12), consider any time $\tau\in[0,t]$ where no wave-front interaction takes place. Let $u(\tau,\cdot)$ have jumps at points $x_1 < \cdots < x_N$. Call $\mathcal{S}$ the set of indices $\alpha\in\{1,\ldots,N\}$ such that $u(\tau, x_\alpha-)$ and $u(\tau, x_\alpha+)$ are connected by a shock or by a contact discontinuity, so that (7.6)–(7.7) hold. Moreover, call $\mathcal{R}$ the set of indices $\alpha$ corresponding to a rarefaction wave of a genuinely non-linear

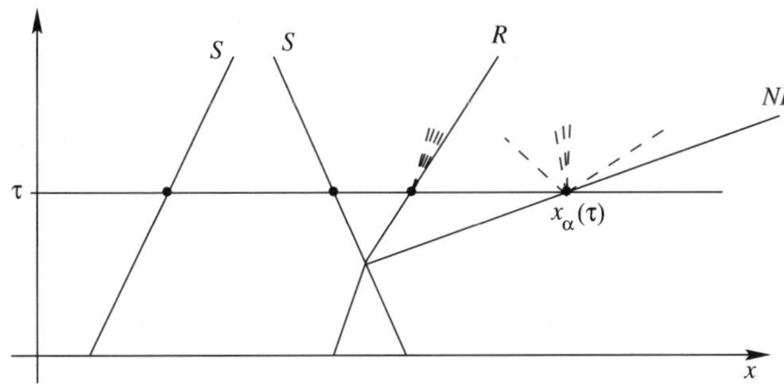


**Figure 9.2**

family, so that (7.8)–(7.9) hold. Finally, call $\mathcal{NP}$ the set of indices $\alpha$ corresponding to non-physical fronts.

For each $\alpha$, call $\omega_\alpha$ the self-similar solution of the Riemann problem (9.1) with data $u^\pm = u(\tau,\ x_\alpha\pm)$. By assumption, for $h > 0$ small enough the semigroup trajectory $h \mapsto \tilde{S}_h u(\tau)$ is obtained by piecing together the solutions of these Riemann problems (Fig. 9.2). Recalling the properties (7.6)–(7.11), an application of parts (i)–(ii) of Lemma 9.1 now yields

$$\begin{aligned}
&\lim_{h\to 0+} \frac{\|u(\tau+h) - \tilde{S}_h u(\tau)\|_{\mathbf{L}^1}}{h} \\
&\quad = \sum_{\alpha\in\mathcal{R}\cup\mathcal{S}\cup\mathcal{NP}} \left( \lim_{h\to 0+} \frac{1}{h} \int_{x_\alpha-\rho}^{x_\alpha+\rho} |u(\tau+h, x) - \omega_\alpha(h,\ x - x_\alpha)|\, dx \right) \\
&\quad = \sum_{\alpha\in\mathcal{S}} \mathcal{O}(1)\cdot\varepsilon|\sigma_\alpha| + \sum_{\alpha\in\mathcal{R}} \mathcal{O}(1)\cdot|\sigma_\alpha|(|\sigma_\alpha|+\varepsilon) + \sum_{\alpha\in\mathcal{NP}} \mathcal{O}(1)\cdot|\sigma_\alpha| \\
&\quad \le C\varepsilon
\end{aligned} \tag{9.13}$$

for some constant $C$. As $\rho$ we can choose here any suitably small positive number.

We now consider a sequence $\varepsilon_\nu \to 0$ and, for each $\nu$, we let $u_\nu$ be a front tracking $\varepsilon_\nu$-approximate solution of the Cauchy problem (7.1)–(7.2). The analysis in Chapter 8 showed that this sequence converges to the semigroup trajectory $u(t,\cdot) = S_t\bar{u}$. On the other hand, by (7.12) and (9.12)–(9.13) it follows that

$$\begin{aligned}
\limsup_{\nu\to\infty} \|u_\nu(t) - \tilde{S}_t\bar{u}\|_{\mathbf{L}^1} &\le \limsup_{\nu\to\infty} \left( \|u_\nu(t) - \tilde{S}_t u_\nu(0)\|_{\mathbf{L}^1} + \|\tilde{S}_t u_\nu(0) - \tilde{S}_t\bar{u}\|_{\mathbf{L}^1} \right) \\
&\le \limsup_{\nu\to\infty} \left( L\cdot\int_0^t C\varepsilon_\nu\, d\tau + L\cdot\|u_\nu(0) - \bar{u}\|_{\mathbf{L}^1} \right) \\
&= 0.
\end{aligned}$$

Hence also $u(t,\cdot) = \tilde{S}_t\bar{u}$ for all $t \ge 0$. This achieves a proof of Theorem 9.1. □

With Theorem 8.1 we proved the convergence to a unique limit for all sequences of piecewise constant approximations constructed by our front tracking algorithm. More generally, the result applied to $\varepsilon$-approximate solutions satisfying Definition 7.1 together with the additional conditions (FT1) and (FT3) stated in Remark 7.2. The above analysis further improves this result. Indeed, in the proof of Theorem 9.1, none of the conditions (FT1)–(FT4) is ever used. We thus have:

**Corollary 9.1.** *For every positively invariant domain $\mathcal{D}$ of the form (8.1), with $\delta_0 > 0$ sufficiently small, there exists a unique SRS $S : \mathcal{D} \times [0, \infty[ \mapsto \mathcal{D}$ generated by the hyperbolic system (7.1). Every trajectory $t \mapsto S_t\bar{u}$ is a weak, entropy-admissible solution of (7.1).*

*Moreover, for each $\nu \geq 1$, let $u_\nu$ be an $\varepsilon_\nu$-approximate front tracking solution of the Cauchy problem (7.1)–(7.2). If $\varepsilon_\nu \to 0$ and $u_\nu(t, \cdot) \in \mathcal{D}$ for all $t, \nu$, then the $\mathbf{L}^1$ convergence $u_\nu(t, \cdot) \to S_t\bar{u}$ holds for every $t \geq 0$.*

As a key step in order to apply Theorem 9.1, one needs to check that the semigroup $\tilde{S}$ acts correctly on piecewise constant initial data. We now show that the assumption (3) of consistency with the Riemann solver can be considerably weakened. Indeed, it suffices to assume that the semigroup behaves correctly only on a much smaller class of initial data. In several applications, this can considerably reduce the burden of the proof.

**Definition 9.2.** In connection with the hyperbolic system (7.1), we say that a piecewise constant function $\bar{u} : \mathbb{R} \mapsto \mathbb{R}^n$ has *simple jumps* if, at every point $x_\alpha$ where $\bar{u}$ is discontinuous, the corresponding Riemann problem (9.1)–(9.2) is solved in terms of one single wave.

In other words, we require that, for some $k_\alpha \in \{1, \ldots, n\}$, either the states $u(x_\alpha-)$, $u(x_\alpha+)$ are connected by an admissible $k_\alpha$-shock (or $k_\alpha$-contact discontinuity), or else the Riemann problem is solved by a single centred rarefaction wave of the $k_\alpha$-th family.

In the definition of SRS, we can now replace (3) with the weaker condition

(3′) For all piecewise constant initial data $\bar{u} \in \mathcal{D}$ with simple jumps, there exists a small $\delta > 0$ such that, for all $t \in [0, \delta]$, the trajectory $u(t, \cdot) = S_t\bar{u}$ coincides with the solution of the Cauchy problem (7-1)–(7.2) obtained by piecing together the standard entropy solutions of the Riemann problems determined by the jumps of $\bar{u}$.

**Corollary 9.2.** *Let $\mathcal{D} \subset \mathbf{L}^1(\mathbb{R}; \mathbb{R}^n)$ be a domain of the form (8.1) and let $S : \mathcal{D} \times [0, \infty[ \mapsto \mathcal{D}$ be the semigroup of entropy weak solutions of (7.1) constructed as limits of front tracking approximations. In addition, consider a map $\tilde{S} : \tilde{\mathcal{D}} \times [0, \infty[ \mapsto \tilde{\mathcal{D}}$, defined on a domain $\tilde{\mathcal{D}} \supseteq \mathcal{D}$, which satisfies (9.4)–(9.5) together with the assumption (3′) of consistency with the Riemann solver. Then the identity (9.11) still holds.*

*Proof.* Let $u = u(t, x)$ be a front tracking $\varepsilon$-approximate solution of the Cauchy problem (7.1)–(7.2). Let $t$ be a time where no interaction takes place. At each point $x_\alpha$ where $u(t, \cdot)$ has a shock, a rarefaction front or a contact discontinuity, the corresponding Riemann problem is obviously solved by a single wave. On the other hand, at a point

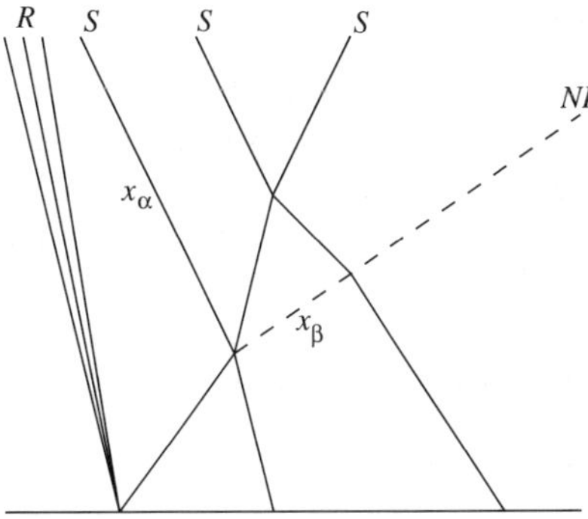


**Figure 9.3a**

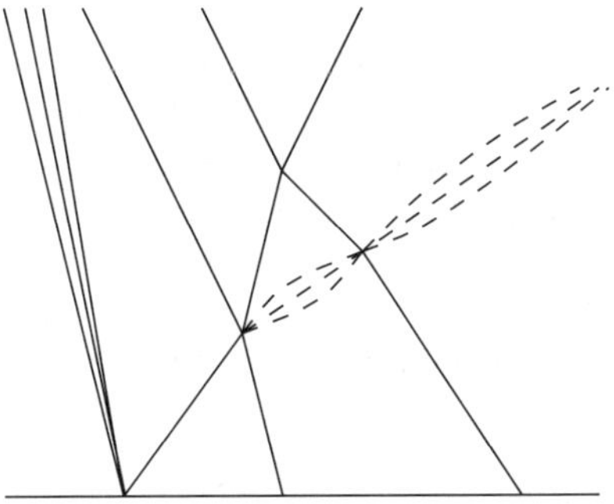

**Figure 9.3b**

$x_\beta$ where $u(t, \cdot)$ has a non-physical front, the solution to the corresponding Riemann problem may well contain waves of $n$ distinct families.

To cope with this situation, we shall slightly modify the function $u$, constructing a new approximate solution $\tilde{u}$ as follows. Let $u$ have non-physical front along the line $x = x_\beta(t)$, connecting the left and right states $u^-, u^+$ (Fig. 9.3a). Call $u^- = \omega_0, \omega_1, \ldots, \omega_n = u^+$ the constant states present in the solution of the Riemann problem (9.1). We now replace the line $x_\beta$ with $n$ lines of jump $x_{\beta,1}, \ldots, x_{\beta,n}$ (Fig. 9.3b), whose speeds satisfy $|\dot{x}_{\beta,i}(t)| \leq 2\hat{\lambda}$ for all $i, t$. Moreover, we define

$$\tilde{u}(t, x) = \omega_i \quad \text{if } x \in \,]x_{\beta,i}(t), x_{\beta,i+1}(t)[\;\; i = 1, \ldots, n-1.$$

We can clearly apply this same procedure to all non-physical fronts of $u$, and arrange things so that

$$\|\tilde{u}(t, \cdot) - u(t, \cdot)\|_{\mathbf{L}^1} < \varepsilon$$

for every $t \geq 0$. By construction, for almost every time $t$ the modified function $\tilde{u}$ is piecewise constant with simple jumps. By the assumption (3′), the estimates (9.12)–(9.13) now hold provided that $u$ is replaced by $\tilde{u}$. In particular, they yield

$$\|\tilde{u}(t) - \tilde{S}_t \tilde{u}(0)\|_{\mathbf{L}^1} = \mathcal{O}(1) \cdot t\varepsilon.$$

We now consider a sequence $\varepsilon_\nu \to 0$ and, for each $\nu$, we let $u_\nu$ be a front tracking $\varepsilon_\nu$-approximate solution of the Cauchy problem (7.1)–(7.2). By Theorem 8.1, this sequence converges to the semigroup trajectory $u(t, \cdot) = S_t \bar{u}$. In addition, for each $\nu$ we can now consider a piecewise constant approximate solution $\tilde{u}_\nu$ with simple jumps, obtained from $u_\nu$ by the above procedure. We then have

$$S_t \bar{u} = \lim_{\nu \to \infty} u_\nu(t, \cdot) = \lim_{\nu \to \infty} \tilde{u}_\nu(t, \cdot) = \tilde{S}_t \bar{u}.$$

Since this construction can be carried out for all initial data $\bar{u} \in \mathcal{D}$, the corollary is proved. □

## 9.2 A characterization of semigroup trajectories

Given the system of conservation laws (7.1), let $\mathcal{D} \subset \mathbf{L}^1$ be a positively invariant domain of the form (8.1). Let $S : \mathcal{D} \times [0, \infty[ \mapsto \mathcal{D}$ be the SRS generated by (7.1). By Theorem 9.1 this semigroup is unique. In this section we derive some necessary and sufficient conditions in order that a function $t \mapsto u(t) \in \mathcal{D}$ can coincide with a semigroup trajectory.

We begin by introducing some notation. Given a function $u = u(t, x)$ and a point $(\tau, \xi)$, we denote by $U^\sharp_{(u;\tau,\xi)}$ the solution of the Riemann problem (9.1) with data

$$u^- = \lim_{x\to\xi-} u(\tau, x), \qquad u^+ = \lim_{x\to\xi+} u(\tau, x). \tag{9.14}$$

Moreover, we define $U^\flat_{(u;\tau,\xi)}$ as the solution of the linear Cauchy problem with constant coefficients where the initial data are assigned at time $t = \tau$

$$w_t + \widetilde{A} w_x = 0, \qquad w(\tau, x) = u(\tau, x), \tag{9.15}$$

with $\widetilde{A} \doteq Df(u(\tau, \xi))$. Observe that (9.15) is obtained from the quasilinear system

$$u_t + A(u)u_x = 0$$

by 'freezing' the coefficients of the matrix $A(u) = Df(u)$ at the point $(\tau, \xi)$. The total variation of the function $u(\tau, \cdot)$ over a set $I$ will be denoted as Tot. Var. $\{u(\tau); I\}$.

**Theorem 9.2 (Characterization of semigroup trajectories).** *Let $S : \mathcal{D} \times [0, \infty[ \mapsto \mathcal{D}$ be the semigroup generated by (7.1) and let $\hat\lambda$ be an upper bound for all wave speeds. Then every semigroup trajectory $u(t, \cdot) = S_t\bar u$ satisfies the following conditions at every $\tau \geq 0$:*

*(i) For every $\xi$, one has*

$$\lim_{h\to 0+} \frac{1}{h} \int_{\xi-h\hat\lambda}^{\xi+h\hat\lambda} \left|u(\tau + h, x) - U^\sharp_{(u;\tau,\xi)}(h,\ x - \xi)\right| dx = 0. \tag{9.16}$$

*(ii) There exists a constant $C$ such that, for every $a < \xi < b$ and $0 < h < (b-a)/2\hat\lambda$, one has*

$$\frac{1}{h} \int_{a+h\hat\lambda}^{b-h\hat\lambda} \left|u(\tau + h,\ x) - U^\flat_{(u;\tau,\xi)}(\tau + h,\ x)\right| dx \leq C \cdot (\text{Tot. Var.}\, \{u(\tau); \,]a, b[\,\})^2. \tag{9.17}$$

*Conversely, let $u : [0, T] \mapsto \mathcal{D}$ be Lipschitz continuous as a map with values in $\mathbf{L}^1$, and assume that the conditions (i)–(ii) hold at almost every time $\tau$. Then $u$ coincides with a semigroup trajectory:*

$$u(t, \cdot) = S_t u(0) \quad \textit{for all } t \in [0, T]. \tag{9.18}$$

Roughly speaking, (9.16) says that, in a forward neighbourhood of each point $(\tau, \xi)$, a solution $u$ is well approximated by the solution of the corresponding Riemann problem. On the other hand, in a region where the total variation is small, (9.17) shows that $u$ can be accurately approximated by the solution of a linear hyperbolic system with constant coefficients.

Before proving the theorem, we establish a couple of preliminary results. They show that, owing to the uniform boundedness of all wave speeds, the Lipschitz-continuous dependence of solutions on their initial data can be localized to bounded intervals.

**Lemma 9.2.** *Let* $]a, b[$ *be a (possibly unbounded) open interval, and let* $\hat{\lambda}$ *be an upper bound for all wave speeds. If* $\bar{u}, \bar{v} \in \mathcal{D}$, *then for all* $t \geq 0$ *one has*

$$\int_{a+\hat{\lambda}t}^{b-\hat{\lambda}t} |S_t\bar{u}(x) - S_t\bar{v}(x)|\, dx \leq L \cdot \int_a^b |\bar{u}(x) - \bar{v}(x)|\, dx. \tag{9.19}$$

*Proof.* If two initial conditions $\bar{u}, \bar{w} \in \mathcal{D}$ coincide on $]a, b[$, then one can construct sequences of approximate solutions $u_\nu, w_\nu$, with $u_\nu(0, \cdot) \to \bar{u}$, $w_\nu(0, \cdot) \to \bar{w}$, and such that $u_\nu(t, x) = w_\nu(t, x)$ for $x \in\, ]a + \hat{\lambda}t,\ b - \hat{\lambda}t[$ . Letting $\nu \to \infty$ we conclude that $S_t\bar{u}(x) = S_t\bar{w}(x)$ for all $x \in\, ]a + \hat{\lambda}t,\ b - \hat{\lambda}t[$ . Next, if $\bar{u}, \bar{v} \in \mathcal{D}$, for every $m \geq 1$ the function

$$\bar{u}_m(x) = \begin{cases} \bar{u}(x) & \text{if } a \leq x \leq b, \\ \bar{u}(a - m(a - x)) & \text{if } x < a, \\ \bar{u}(b + m(x - b)) & \text{if } x > b \end{cases}$$

is also contained in $\mathcal{D}$. Letting $m \to \infty$ and recalling that $\bar{u} \in \mathbf{L}^1 \cap BV$, we obtain

$$\hat{u} \doteq \lim_{m\to\infty} \bar{u}_m = \bar{u} \cdot \chi_{[a,b]} \in \mathcal{D}.$$

Similarly, $\hat{v} \doteq \bar{v} \cdot \chi_{[a,b]} \in \mathcal{D}$. The uniform Lipschitz continuity of the semigroup $S$ now implies

$$\int_{a+\hat{\lambda}t}^{b-\hat{\lambda}t} |S_t\bar{u}(x) - S_t\bar{v}(x)|\, dx = \int_{a+\hat{\lambda}t}^{b-\hat{\lambda}t} |S_t\hat{u}(x) - S_t\hat{v}(x)|\, dx \leq \|S_t\hat{u} - S_t\hat{v}\|_{\mathbf{L}^1}$$

$$\leq L \cdot \|\hat{u} - \hat{v}\|_{\mathbf{L}^1} = L \cdot \int_a^b |\bar{u}(x) - \bar{v}(x)|\, dx.$$

□

Using (9.19), one can 'localize' the error estimate (2.57), owing to the uniform boundedness of all wave speeds.

**Corollary 9.3.** *Let* $S : \mathcal{D} \times [0, \infty[\, \mapsto \mathcal{D}$ *be the semigroup generated by (7.1) and let* $\hat{\lambda}$ *be an upper bound for all wave speeds. Given any interval* $I_0 \doteq [a, b]$, *define the* *intervals of determinacy*

$$I_t \doteq [a + \hat{\lambda}t, b - \hat{\lambda}t] \quad t < \frac{b-a}{2\hat{\lambda}}. \tag{9.20}$$

*For every Lipschitz-continuous map $w : [0, T] \mapsto \mathcal{D}$, calling $L$ the Lipschitz constant of the semigroup, the following holds:*

$$\|w(t) - S_t w(0)\|_{\mathbf{L}^1(I_t)} \leq L \cdot \int_0^t \left\{ \liminf_{h \to 0+} \frac{\|w(\tau + h) - S_h w(\tau)\|_{\mathbf{L}^1(I_{\tau+h})}}{h} \right\} d\tau. \tag{9.21}$$

The proof is the same as for Theorem 2.9, using (9.19).

We can now give a proof of Theorem 9.2.

**Part 1: necessity.** Given a semigroup trajectory $u(t, \cdot) = S_t \bar{u}$, we now show that the conditions (9.16)–(9.17) hold for every $\tau \geq 0$, $a < \xi < b$.

To prove (9.16), let $h, \varepsilon > 0$ be given and define the open intervals (Fig. 9.4)

$$J_t \doteq ]\xi - (2h - t + \tau)\hat{\lambda}, \xi + (2h - t + \tau)\hat{\lambda}[ \quad t \in [\tau, \tau + h]. \tag{9.22}$$

Choose a piecewise constant function $\bar{v} \in \mathcal{D}$ such that

$$\bar{v}(\xi) = u(\tau, \xi), \qquad \bar{v}(\xi\pm) = u(\tau, \ \xi\pm), \tag{9.23}$$

$$\int_{J_\tau} |\bar{v}(x) - u(\tau, x)| \, dx \leq \varepsilon, \tag{9.24}$$

$$\text{Tot. Var.} \{\bar{v}; J_\tau\} \leq \text{Tot. Var.} \{u(\tau); J_\tau\}. \tag{9.25}$$

For $t \geq 0$ define the function

$$v(t, x) = \begin{cases} U^\sharp_{(u;\tau,\xi)}(t - \tau, \ x - \xi) & \text{if } |x - \xi| \leq \hat{\lambda}(t - \tau), \\ \bar{v}(x - (t - \tau)\hat{\lambda}) & \text{if } x > \xi + \hat{\lambda}(t - \tau), \\ \bar{v}(x + (t - \tau)\hat{\lambda}) & \text{if } x < \xi - \hat{\lambda}(t - \tau). \end{cases}$$

Observe that the function $v$ provides an exact solution of (7.1) in the region where $|x - \xi| \leq \hat{\lambda}(t - \tau)$. Moreover, for every $t$, the fronts of $v(t, \cdot)$ contained in the interval

$$]\xi + (t - \tau)\hat{\lambda}, \xi + (2h - t + \tau)\hat{\lambda}[$$

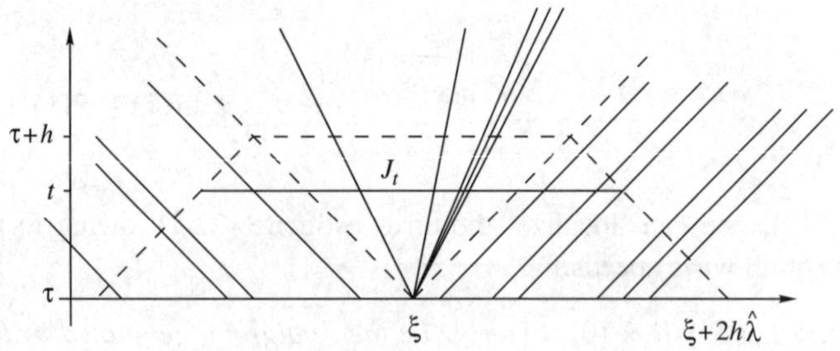


**Figure 9.4**

coincide with the fronts of $\bar{v}$ located inside $]\xi,\ \xi+2(h-t+\tau)\hat{\lambda}[$. Applying the estimate (9.21) to the function $v$ and then using (9.7), we obtain

$$\int_{J_{\tau+h}} |v(\tau+h,x) - (S_h\bar{v})(x)|\,dx$$

$$\leq L \cdot \int_{\tau}^{\tau+h} \left\{ \liminf_{\epsilon\to 0+} \frac{\|v(t+\epsilon) - S_\epsilon v(t)\|_{\mathbf{L}^1(J_{t+\epsilon})}}{\epsilon} \right\} dt$$

$$= L \cdot \int_{\tau}^{\tau+h} \mathcal{O}(1) \cdot \text{Tot. Var.}\,\{\bar{v};\ ]\xi - 2(h-t+\tau)\hat{\lambda}, \xi[\ \cup\ ]\xi, \xi + 2(h-t+\tau)\hat{\lambda}[\}\,dt.$$

$$= Lh \cdot \mathcal{O}(1) \cdot \text{Tot. Var.}\{\bar{v};\ ]\xi - 2h\hat{\lambda}, \xi[\ \cup\ ]\xi, \xi + 2h\hat{\lambda}[\}. \tag{9.26}$$

Assuming that $u(\tau+h) = S_h u(\tau)$, since $v(\tau+h,\ x) = U^\sharp_{(u;\tau,\xi)}(h,\ x-\xi)$ for all $x \in J_{\tau+h} = ]\xi - h\hat{\lambda},\ \xi + h\hat{\lambda}[$, we have

$$\int_{\xi-h\hat{\lambda}}^{\xi+h\hat{\lambda}} |u(\tau+h,\ x) - U^\sharp_{(u;\tau,\xi)}(h,\ x-\xi)|\,dx$$

$$\leq \|S_h u(\tau) - S_h\bar{v}\|_{\mathbf{L}^1(J_{\tau+h})} + \|S_h\bar{v} - v(\tau+h)\|_{\mathbf{L}^1(J_{\tau+h})}\,dx$$

$$\leq L \cdot \int_{\xi-2h\hat{\lambda}}^{\xi+2h\hat{\lambda}} |u(\tau,x) - \bar{v}(x)|\,dx + \mathcal{O}(1) \cdot h \cdot \text{Tot. Var.}\,\{\bar{v};\ ]\xi - 2h\hat{\lambda}, \xi[\cup]\xi, \xi + 2h\hat{\lambda}[\}.$$

Since $\varepsilon$ can be chosen arbitrarily small, by (9.24) and (9.25) it follows that

$$\frac{1}{h} \cdot \int_{\xi-h\hat{\lambda}}^{\xi+h\hat{\lambda}} |u(\tau+h,\ x) - U^\sharp_{(u;\tau,\xi)}(h,\ x-\xi)|\,dx$$

$$= \mathcal{O}(1) \cdot \text{Tot. Var.}\,\{u(\tau);\ ]\xi - 2h\hat{\lambda},\ \xi[\cup]\xi,\ \xi + 2h\hat{\lambda}[\,\}.$$

Letting $h \to 0+$ we obtain (9.16).

To prove (9.17), let $h > 0$ and a point $(\tau, \xi)$ be given, together with an open interval $]a, b[$ containing $\xi$. Fix $\varepsilon > 0$ and choose a piecewise constant function $\bar{v} \in \mathcal{D}$ satisfying (9.23) together with

$$\int_a^b |\bar{v}(x) - u(\tau,x)|\,dx \leq \varepsilon, \tag{9.27}$$

$$\text{Tot. Var.}\,\{\bar{v};\ ]a,\ b[\,\} \leq \text{Tot. Var.}\,\{u(\tau);\ ]a,\ b[\,\}. \tag{9.28}$$

Let $v = v(t,x)$ be the solution of the linear hyperbolic Cauchy problem

$$v_t + \widetilde{A} v_x = 0, \qquad v(\tau,x) = \bar{v}(x), \tag{9.29}$$

with $\widetilde{A} \doteq Df\big(u(\tau,\xi)\big)$. Call $\tilde{\lambda}_1 < \cdots < \tilde{\lambda}_n$ the eigenvalues of $\widetilde{A}$ and let $\tilde{l}_i, \tilde{r}_i$, $i = 1,\dots,n$, be left and right eigenvectors, normalized as in (5.4). By (3.8), we have the explicit representation

$$v(t,x) = \sum_{i=1}^{n} (\tilde{l}_i \cdot \bar{v}(x - (t-\tau)\tilde{\lambda}_i))\tilde{r}_i. \tag{9.30}$$

From (9.30) and the analogous formula for the solution of (9.15), by (9.27) we obtain

$$\int_{a+h\hat{\lambda}}^{b-h\hat{\lambda}} |v(\tau+h,x) - U^{\flat}_{(u;\tau,\xi)}(\tau+h,x)|\,dx \le C' \cdot \int_a^b |\bar{v}(x) - u(\tau,x)|\,dx \le C'\cdot\varepsilon \tag{9.31}$$

for some constant $C'$. By (9.19) it follows that

$$\int_{a+h\hat{\lambda}}^{b-h\hat{\lambda}} |S_h\bar{v}(x) - (S_h u(\tau))(x)|\,dx \le L\cdot\int_a^b |\bar{v}(x) - u(\tau,x)|\,dx \le L\varepsilon. \tag{9.32}$$

Moreover, using (9.21) and then the estimate (9.9), we deduce that

$$\begin{aligned}\frac{1}{h}\int_{a+h\hat{\lambda}}^{b-h\hat{\lambda}} &|v(\tau+h,x) - (S_h\bar{v})(x)|\,dx\\ &\le \frac{L}{h}\cdot\int_{\tau}^{\tau+h}\left(\liminf_{\epsilon\to 0+}\frac{1}{\epsilon}\int_{a+(t-\tau+\epsilon)\hat{\lambda}}^{b-(t-\tau+\epsilon)\hat{\lambda}} |v(t+\epsilon,x) - (S_\epsilon v(t))(x)|dx\right)dt\\ &= O(1)\cdot(\text{Tot. Var.}\{\bar{v};\,]a,b[\,\})^2.\end{aligned} \tag{9.33}$$

Indeed, by (9.30) the function $v$ is piecewise constant with finitely many lines of discontinuity in the $t$–$x$ plane (Fig. 9.5). If $x = x_\alpha(t)$ describes a line of jump, then for almost every $t$ one has

$$(\widetilde{A} - \tilde{\lambda}_i)(v(t,x_\alpha+) - v(t,x_\alpha-)) = 0, \qquad \dot{x}_\alpha(t) = \tilde{\lambda}_i$$

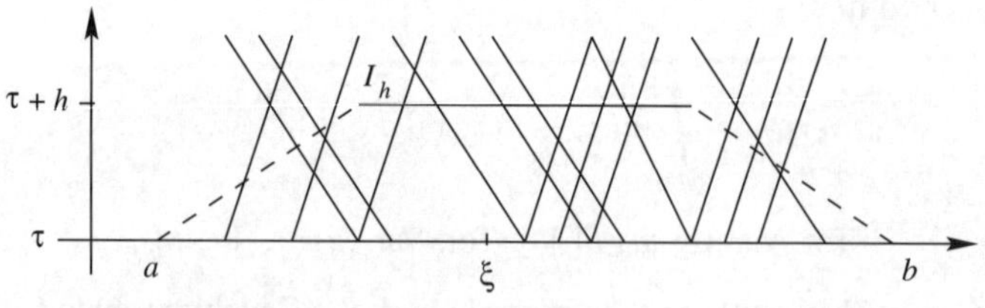


**Figure 9.5**

for some $i \in \{1, \ldots, n\}$. Observing that

$$|v(t, x_\alpha\pm) - u(\tau, \xi)| = O(1) \cdot \text{Tot. Var.}\{\bar{v}; ]a, b[\},$$
$$\sum_\alpha |v(t, x_\alpha+) - v(t, x_\alpha-)| = O(1) \cdot \text{Tot. Var.}\{\bar{v}; ]a, b[\}$$

and using (9.9) we obtain the last estimate in (9.33). For notational convenience, we define the interval

$$I_h \doteq [a + h\hat{\lambda}, \; b - h\hat{\lambda}].$$

Assuming that $u(\tau + h, \cdot) = S_h u(\tau)$, by (9.31)–(9.33) we now have

$$\frac{1}{h}\int_{a+h\hat{\lambda}}^{b-h\hat{\lambda}} |u(\tau+h, \, x) - U^\flat_{(u;\tau,\xi)}(\tau+h, \, x)|\, dx \le \frac{1}{h}\Big\{\|S_h u(\tau) - S_h\bar{v}\|_{\mathbf{L}^1(I_h)}$$
$$+ \|S_h\bar{v} - v(\tau+h, \cdot)\|_{\mathbf{L}^1(I_h)} + \|v(\tau+h, \cdot) - U^\flat_{(u;\tau,\xi)}(\tau+h, \cdot)\|_{\mathbf{L}^1(I_h)}\Big\}$$
$$\le \frac{L\varepsilon}{h} + \frac{C'\varepsilon}{h} + \mathcal{O}(1) \cdot (\text{Tot. Var.}\{u(\tau); ]a, b[\})^2.$$

Since $\varepsilon$ can be taken arbitrarily small, this establishes (9.17).

**Part 2: sufficiency.** We now consider a function $u = u(t, x)$ which satisfies the conditions (9.16)–(9.17) at almost every $\tau \in [0, T]$ and prove that $u(T, \cdot) = S_T u(0)$. It suffices to show that, for every $a < b$, one has

$$\int_{a+T\hat{\lambda}}^{b-T\hat{\lambda}} |u(T, x) - (S_T u(0))(x)|\, dx$$
$$\le L \cdot \int_0^T \left\{ \liminf_{h\to 0+} \frac{1}{h} \int_{a+(\tau+h)\hat{\lambda}}^{b-(\tau+h)\hat{\lambda}} |u(\tau+h, x) - (S_h u(\tau))(x)|\, dx \right\} d\tau = 0. \tag{9.34}$$

The first inequality follows from Corollary 9.3. To prove the last equality in (9.34) we need to show that the integrand is zero for almost every $\tau$.

Fix a time $\tau \in [0, T]$ at which the estimates (9.16)–(9.17) hold and let $\varepsilon > 0$ be given. Since the total variation of $u(\tau, \cdot)$ is finite, we can choose finitely many points (Fig. 9.6)

$$a + \tau\hat{\lambda} = x_0 < x_1 < \cdots < x_N = b - \tau\hat{\lambda}$$

such that

$$\text{Tot. Var.}\{u(\tau, \cdot); ]x_{i-1}, x_i[\} < \varepsilon.$$

Observe that, by the necessity part of the theorem, which has been already proved, the function $w(t, \cdot) \doteq S_{t-\tau} u(\tau)$ also satisfies the estimates (9.16)–(9.17). We now choose the midpoints $y_i \doteq (x_{i-1} + x_i)/2$. Using the estimate (9.16) at each of the points $\xi = x_i$

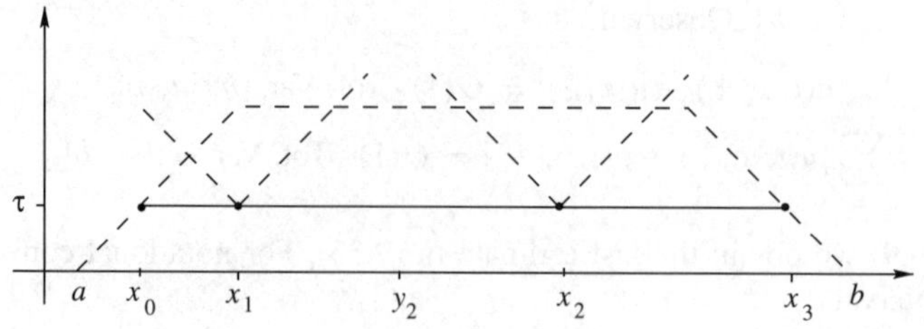


**Figure 9.6**

and the estimate (9.17) with $\xi \doteq y_i$ on each of the intervals $]x_{i-1},\ x_i[$, we compute

$$
\begin{aligned}
&\limsup_{h\to 0+} \frac{1}{h} \int_{a+(\tau+h)\hat\lambda}^{b-(\tau+h)\hat\lambda} |u(\tau+h,x) - (S_h u(\tau))(x)|\,dx \\
&\le \sum_{i=1}^{N-1} \limsup_{h\to 0+} \frac{1}{h} \int_{x_i-h\hat\lambda}^{x_i+h\hat\lambda} \big(|u(\tau+h,x) - U^\sharp_{(u;\tau,x_i)}(\tau+h,x)| \\
&\quad + |U^\sharp_{(u;\tau,x_i)}(\tau+h,\ x) - (S_h u(\tau))(x)|\big)\,dx \\
&\quad + \sum_{i=1}^{N} \limsup_{h\to 0+} \frac{1}{h} \int_{x_{i-1}+h\hat\lambda}^{x_i-h\hat\lambda} \big(|u(\tau+h,x) - U^\flat_{(u;\tau,y_i)}(h,\ x-x_i)| \\
&\quad + |U^\flat_{(u;\tau,y_i)}(h,\ x-x_i) - (S_h u(\tau))(x)|\big)\,dx \\
&\le 0 + \sum_{i=1}^{N} \mathcal{O}(1)\cdot(\text{Tot. Var.}\{u(\tau);\ ]x_{i-1},\ x_i[\})^2 \\
&= \mathcal{O}(1)\cdot\varepsilon\cdot\text{Tot. Var.}\{u(\tau);\ ]x_0,\ x_N[\} \\
&= \mathcal{O}(1)\cdot\varepsilon.
\end{aligned}
$$

Since $\varepsilon > 0$ was arbitrary, this shows that the integrand in (9.34) vanishes at time $\tau$. Since this is true for almost every $\tau \in [0,T]$, the claim (9.34) is proved.

## 9.3 Uniqueness of entropy weak solutions

We now consider an arbitrary weak solution of the hyperbolic Cauchy problem (7.1)–(7.2), satisfying the Lax entropy conditions (4.38). Our goal is to prove that any such solution coincides with the corresponding semigroup trajectory $t \mapsto S_t\bar u$. For the sake of clarity, a complete set of assumptions is listed below.

**(A1)** **(Conservation equations)** The function $u = u(t,x)$ is a weak solution of the Cauchy problem (7.1)–(7.2), taking values within the domain $\mathcal{D}$ of a semigroup

$S$. More precisely, $u : [0, T] \mapsto \mathcal{D}$ is continuous w.r.t. the $\mathbf{L}^1$ distance. The initial condition (7.2) holds, together with

$$\iint \{u\phi_t + f(u)\phi_x\}\, dx\, dt = 0 \tag{9.35}$$

for every $\mathcal{C}^1$ function $\phi$ with compact support contained inside the open strip $]0, T[\times\mathbb{R}$.

**(A2)** **(Lax entropy condition)** Let $u$ have an approximate jump discontinuity at some point $(\tau, \xi) \in\, ]0, T[\times\mathbb{R}$. More precisely, let there exist states $u^-, u^+ \in \Omega$ and a speed $\lambda \in \mathbb{R}$ such that, calling

$$U(t, x) \doteq \begin{cases} u^- & \text{if } x < \lambda t, \\ u^+ & \text{if } x > \lambda t, \end{cases} \tag{9.36}$$

the following holds:

$$\lim_{\rho\to 0+} \frac{1}{\rho^2} \int_{\tau-\rho}^{\tau+\rho} \int_{\xi-\rho}^{\xi+\rho} |u(t, x) - U(t - \tau, x - \xi)|\, dx\, dt = 0. \tag{9.37}$$

Then, for some $i \in \{1, \dots, n\}$, one has the entropy inequality:

$$\lambda_i(u^-) \geq \lambda \geq \lambda_i(u^+). \tag{9.38}$$

**(A3)** **(Bounded variation condition)** There exists $\delta > 0$ such that, for every bounded space-like curve $\{t = \gamma(x);\ x \in [a, b]\}$ with

$$|\gamma(x_1) - \gamma(x_2)| \leq \delta |x_1 - x_2| \quad \text{for all } x_1, x_2 \in [a, b],$$

the function $x \mapsto u(\gamma(x), x)$ has bounded variation.

**(A4)** **(Tame oscillation condition)** There exist constants $C, \delta > 0$ such that the following holds. For every $t > 0$ and every $a < b$, the oscillation of $u$ over the triangle

$$\Delta \doteq \{(s, y) : s \geq t,\ a + (s - t)/\delta < y < b - (s - t)/\delta\}$$

satisfies

$$\text{Osc.}\,\{u;\ \Delta\} \leq C \cdot \text{Tot. Var.}\,\{u(t, \cdot);\ ]a, b[\,\}. \tag{9.39}$$

We recall that the oscillation of a function $u$ on a set $\Delta$ is defined as

$$\text{Osc.}\,\{u;\ \Delta\} \doteq \sup_{(s,y),(s',y')\in\Delta} |u(s, y) - u(s', y')|.$$

We first show that all properties (A1)–(A4) are satisfied by solutions obtained as limits of front tracking approximations (i.e. by trajectories of the SRS), provided that we take their right continuous representatives in $\mathbf{L}^1_{\text{loc}}$.

**Theorem 9.3 (Properties of semigroup trajectories).** *Let $u = u(t, x)$ be a weak solution of the hyperbolic system (7.1) obtained as the limit of a sequence of front tracking* *approximations. Assume that the point values of u are chosen so that $u(t,x) = u(t,x+)$ for all $t, x$. Then u satisfies all conditions (A1)–(A4).*

*Proof.* Consider a sequence $\varepsilon_\nu \to 0+$ and, for each $\nu \geq 1$, let $u_\nu$ be an $\varepsilon_\nu$-approximate front tracking solution of (7.1), with $u_\nu \to u$ in $\mathbf{L}^1_{\mathrm{loc}}$. The proof of Theorem 7.1 shows that $u$ is then a weak solution of (7.1), satisfying (A1). Moreover, the function $t \mapsto u(t,\cdot)$ is Lipschitz continuous with values in $\mathbf{L}^1$.

Let $(\tau, \xi)$ be a point where $u$ has an approximate jump. By Theorem 2.6, (2.36) holds with $U$ given by (2.34)–(2.35). Call $U^\sharp \doteq U^\sharp_{(u;\tau,\xi)}$ the solution of the Riemann problem with data $u^\pm = u(\tau, \xi\pm)$. Observe that both $U$ and $U^\sharp$ are self-similar, i.e. they are constant along each ray through the origin. Using (9.16) and (2.36), for every $\lambda^* > 0$ arbitrarily large we thus obtain

$$0 = \limsup_{h\to 0+} \frac{1}{h^2} \int_\tau^{\tau+h} \int_{\xi-h\lambda^*}^{\xi+h\lambda^*} \{|U^\sharp(t-\tau, x-\xi) - u(t,x)|$$

$$+ |u(t,x) - U(t-\tau, x-\xi)|\}\, dx\, dt$$

$$\geq \limsup_{h\to 0+} \frac{1}{h^2} \int_\tau^{\tau+h} \int_{\xi-(t-\tau)\lambda^*}^{\xi+(t-\tau)\lambda^*} |U^\sharp(t-\tau, x-\xi) - U(t-\tau, x-\xi)|\, dx\, dt$$

$$= \frac{1}{2} \int_{-\lambda^*}^{\lambda^*} |U^\sharp(1,y) - U(1,y)|\, dy.$$

Therefore $U = U^\sharp$. By construction, $U^\sharp$ satisfies the Lax entropy conditions, and hence the same is true for $U$. This proves (9.38), showing that (A2) holds.

For a proof of (A3) and (A4), we first show that every limit of front tracking approximations satisfies the following stronger condition:

**(A5) (Tame variation condition)** For some constant $C$ the following holds. Fix $\tau \geq 0$ and a (possibly unbounded) open interval $]a_0, b_0[$. Let $\gamma : ]a, b[ \mapsto \mathbb{R}$ be any space-like curve (Fig. 9.7) such that $a_0 \leq a < b \leq b_0$ and

$$\tau \leq \gamma(x) < \min\left\{\tau + \frac{x - a_0}{\hat\lambda},\ \tau + \frac{b_0 - x}{\hat\lambda}\right\} \quad \text{for all } x \in ]a, b[.$$

Then

$$\text{Tot. Var.}\{u;\gamma\} \leq C \cdot \text{Tot. Var.}\{u(\tau);\ ]a_0, b_0[\,\}. \tag{9.40}$$

Indeed, consider a solution $u$ obtained as the limit of a sequence of front tracking approximations $u_\nu$. To prove (9.40), let any points $P_i = (\gamma(x_i), x_i)$ be given, with $x_0 < x_1 < \cdots < x_m$. At each time $t_i = \gamma(x_i)$, by possibly taking a subsequence we can assume that the functions $u_\nu(t_i, \cdot)$ converge to $u(t_i, \cdot)$ a.e. Since each $u(t_i, \cdot)$ is right continuous with bounded variation, given any $\varepsilon > 0$ we can find $x_i' \in [x_i, x_i + \varepsilon]$

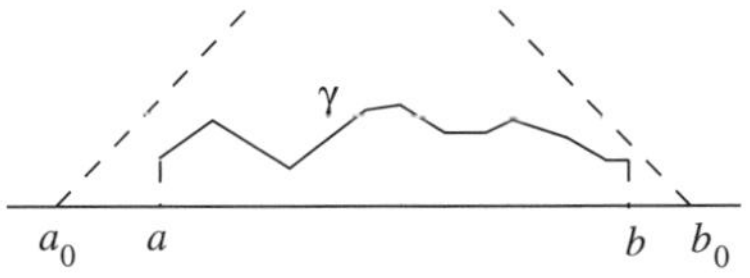


**Figure 9.7**

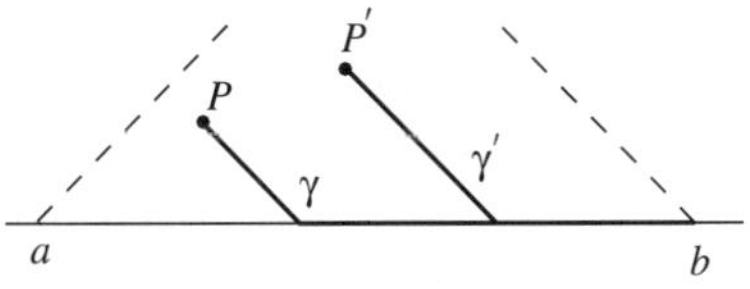


**Figure 9.8**

such that

$$|u(t_i, x'_i) - u(t_i, x_i)| < \varepsilon, \qquad \lim_{\nu\to\infty} u_\nu(t_i, x'_i) = u(t_i, x'_i),$$

$$\hat{\lambda}|t_i - t_{i-1}| < x'_i - x'_{i-1}, \qquad \hat{\lambda}|t_m - \tau| < b - x_m.$$

Calling $\gamma'$ the polygonal line with vertices at the points $P'_i \doteq (t_i, x'_i)$ and $\bar{\gamma}$ the constant map $\bar{\gamma}(x) \equiv \tau$ for all $x \in ]a, b[$, the above properties imply that $\bar{\gamma} \prec \gamma'$. Using Lemma 7.3 we thus have

$$\begin{aligned}
\sum_{i=1}^{m} |u(t_i, x_i) - u(t_{i-1}, x_{i-1})| &\le 2m\varepsilon + \sum_{i=1}^{m} |u(t_i, x'_i) - u(t_{i-1}, x'_{i-1})| \\
&\le 2m\varepsilon + \limsup_{\nu\to\infty}(\text{Tot. Var. }\{u_\nu; \gamma'\}) \\
&\le 2m\varepsilon + \mathcal{O}(1) \cdot \limsup_{\nu\to\infty}(\text{Tot. Var. }\{u_\nu(\tau); ]a, b[\,\}) \\
&= 2m\varepsilon + \mathcal{O}(1) \cdot \text{Tot. Var. }\{u(\tau); ]a, b[\,\}.
\end{aligned}$$

Since $\varepsilon > 0$ and the points $P_i$ were arbitrary, this proves (9.40).

From the property (A5), the conditions (A3)–(A4) follow easily. Indeed, taking $a_0 = -\infty$, $b_0 = \infty$, $\tau = 0$ we obtain

$$\text{Tot. Var. }\{u; \gamma\} \le C \cdot \text{Tot. Var. } u(0, \cdot) < \infty$$

for every space-like curve $\gamma$. This proves (A3), choosing any $\delta < 1/\hat{\lambda}$. Concerning (A4), let $\tau \ge 0$ and $a < b$ be given. Define the triangular domain

$$\Gamma \doteq \{(t, x);\, t \ge \tau, a + 2\hat{\lambda}(t - \tau) < x < b - 2\hat{\lambda}(t - \tau)\}.$$

For every couple of points $P = (s, y)$, $P' = (s', y')$ in $\Gamma$, we construct the space-like curves $\gamma : ]y, b[\, \mapsto \mathbb{R}$, $\gamma' : ]y', b[\, \mapsto \mathbb{R}$, by setting (Fig. 9.8)

$$\gamma(x) \doteq \max\left\{s - \frac{x - y}{2\hat{\lambda}}, \tau\right\}, \qquad \gamma'(x) \doteq \max\left\{s' - \frac{x - y'}{2\hat{\lambda}}, \tau\right\}.$$

Since $\gamma(x) = \gamma'(x) = \tau$ for all $x$ sufficiently close to $b$, we have

$$\begin{aligned}
|u(s, y) - u(s', y')| &\le \text{Tot. Var. }\{u; \gamma\} + \text{Tot. Var. }\{u; \gamma'\} \\
&= \mathcal{O}(1) \cdot \text{Tot. Var. }\{u(\tau); ]a, b[\,\}.
\end{aligned}$$

This establishes the tame oscillation condition (A4), with $\delta = 1/2\hat{\lambda}$.

We can now state the main result of this chapter, showing that the entropy weak solution of the Cauchy problem (7.1)–(7.2) is unique within the class of functions that satisfy either the additional regularity condition (A3), or (A4).

**Theorem 9.4 (Uniqueness of entropy solutions).** *Let $S : \mathcal{D} \times [0, \infty[ \mapsto \mathcal{D}$ be the semigroup generated by (7.1). Let $u : [0, T] \mapsto \mathcal{D}$ be continuous as a map with values into $\mathbf{L}^1$. If (A1), (A2) and (A3) hold, then*

$$u(t, \cdot) = S_t \bar{u} \quad \textit{for all } t \in [0, T]. \tag{9.41}$$

*In particular, the weak solution of the Cauchy problem (7.1)–(7.2) that satisfies these conditions is unique. The same conclusion holds if the assumption (A3) is replaced by (A4).*

The following technical result will be used in the proof.

**Lemma 9.3.** *Let $w : ]a, b[ \mapsto \mathbb{R}^n$ be an integrable function such that, for some measure $\mu$, one has*

$$\left| \int_{\zeta_1}^{\zeta_2} w(x)\, dx \right| \leq \mu([\zeta_1, \zeta_2]), \quad \textit{whenever } a < \zeta_1 < \zeta_2 < b. \tag{9.42}$$

*Then*

$$\int_a^b |w(x)|\, dx \leq \mu(]a, b[).$$

*Proof.* Observe that, in (9.42), one can replace closed intervals $[\zeta_1, \zeta_2]$ with open ones. Indeed,

$$\left| \int_{\zeta_1}^{\zeta_2} w(x)\, dx \right| = \lim_{\varepsilon \to 0} \left| \int_{\zeta_1+\varepsilon}^{\zeta_2-\varepsilon} w(x)\, dx \right| \leq \mu(]\zeta_1, \zeta_2[).$$

Next, fix any $\varepsilon > 0$. Then there exists a piecewise constant function $v$ such that

$$\int_a^b |w(x) - v(x)|\, dx \leq \varepsilon.$$

Calling $a = x_0 < \cdots < x_N = b$ the points of discontinuity of $v$, we compute

$$\begin{aligned} \int_a^b |w(x)|\, dx &\leq \int_a^b |w(x) - v(x)|\, dx + \int_a^b |v(x)|\, dx \\ &\leq \varepsilon + \sum_j \int_{x_{j-1}}^{x_j} |v(x)|\, dx \end{aligned}$$

$$
\begin{aligned}
&= \varepsilon + \sum_j \left| \int_{x_{j-1}}^{x_j} v(x)\,dx \right| \\
&\le \varepsilon + \sum_j \left| \int_{x_{j-1}}^{x_j} w(x)\,dx \right| + \sum_j \left| \int_{x_{j-1}}^{x_j} (v(x) - w(x))\,dx \right| \\
&\le \varepsilon + \sum_j \mu(]x_{j-1}, x_j[) + \int_a^b |v(x) - w(x)|\,dx \\
&\le 2\varepsilon + \mu(]a, b[).
\end{aligned}
$$

Since $\varepsilon$ was arbitrary, this proves the lemma. □

The proof of Theorem 9.4 will be given in several steps.

1. We claim that the map $u : [0, T] \mapsto \mathcal{D}$ is actually Lipschitz continuous w.r.t. the $\mathbf{L}^1$ distance, i.e.

$$
\|u(\tau, \cdot) - u(\tau', \cdot)\|_{\mathbf{L}^1} \le L^* |\tau - \tau'| \tag{9.43}
$$

for some constant $L^*$ and all $\tau, \tau' \in [0, T]$.

To prove (9.43), let $0 < \tau < \tau'$ be given and construct a smooth approximation of the characteristic function of the interval $[\tau, \tau']$. For this purpose, take a smooth non-decreasing function $\alpha : \mathbb{R} \to [0, 1]$ such that

$$
\alpha(x) = \begin{cases} 0 & \text{if } x \le 0, \\ 1 & \text{if } x \ge 1, \end{cases}
$$

and define $\alpha_\epsilon(x) \doteq \alpha(x/\epsilon)$. As $\epsilon \to 0+$, $\alpha_\epsilon$ thus approaches the Heaviside function. Consider any smooth function $\psi = \psi(x)$ with compact support, and define

$$
\varphi_\epsilon(t, x) = [\alpha_\epsilon(t - \tau) - \alpha_\epsilon(t - \tau')]\psi(x).
$$

Since $u$ is a weak solution of (7.1), using $\varphi_\epsilon$ in (9.35) and letting $\epsilon \to 0$, then by the $\mathbf{L}^1$ continuity of the function $t \mapsto u(t, \cdot)$ we obtain

$$
\int \psi(x)[u(\tau, x) - u(\tau', x)]\,dx + \int_\tau^{\tau'} \int \psi_x(x) f(u)\,dx\,dt = 0. \tag{9.44}
$$

By assumption, $u(t, \cdot) \in \mathcal{D}$ for all $t$. As a consequence, one has the uniform bounds

$$
\text{Tot. Var.}\{u(t, \cdot)\} \le M, \qquad |u(t, x)| \le M, \tag{9.45}
$$

for some constant $M$. Using (9.44) and (9.45) we thus obtain

$$\begin{aligned}\|u(\tau',\cdot)-u(\tau,\cdot)\|_{\mathbf{L}^1} &= \sup_{\psi\in\mathcal{C}_c^1,|\psi|\le 1} \int \psi(x)[u(\tau',x)-u(\tau,x)]\,dx \\ &\le \int_\tau^{\tau'} \text{Tot. Var.}\{f(u(t,\cdot))\}\,dt \\ &\le M\cdot Lip(f)\cdot|\tau'-\tau|,\end{aligned}$$

where by $Lip(f)$ we denote the Lipschitz constant of the function $f$ on the region where $|u|\le M$. This establishes (9.43).

2. By (9.43) and (9.45) we can now apply Theorem 2.6, proving that $u=u(t,x)$ is a *BV* function of the two variables $t,x$. Using Theorem 2.5 we obtain the existence of a set $\mathcal{N}\subset[0,T]$ of measure zero, containing the endpoints 0 and $T$, such that, at every point $(\tau,\xi)\in[0,T]\times\mathbb{R}$ with $\tau\notin\mathcal{N}$, setting $u^-\doteq u(\tau,\xi-)$, $u^+\doteq u(\tau,\xi+)$ the following property holds:

(P) Either $u^+=u^-$, in which case (9.36)–(9.37) hold with $\lambda$ arbitrary, or else $u^+\ne u^-$, in which case (9.36)–(9.37) hold for some particular $\lambda\in\mathbb{R}$. In this second case, $\lambda$ is an eigenvalue of the averaged matrix $A(u^-,u^+)$, i.e. $\lambda=\lambda_i(u^-,u^+)$ for some $i\in\{1,\dots,n\}$. The Rankine–Hugoniot equations and the Lax entropy condition hold:

$$\lambda\cdot(u^+-u^-)=f(u^+)-f(u^-),\qquad \lambda_i(u^-)\ge\lambda\ge\lambda_i(u^+). \tag{9.46}$$

Observe that the first part of (9.46) is a consequence of Theorem 4.1. The second part follows from the assumption (A2). For every $\hat\lambda>0$, (9.37) implies

$$\lim_{\rho\to 0+}\frac{1}{\rho^2}\int_{\tau-\rho}^{\tau+\rho}\int_{\xi-\rho\hat\lambda}^{\xi+\rho\hat\lambda}|u(t,x)-U(t-\tau,x-\xi)|\,dx\,dt=0. \tag{9.47}$$

3. According to Corollary 9.3, for every $R,t>0$ we have the error estimate

$$\int_{-R+t\hat\lambda}^{R-t\hat\lambda}|u(t,x)-(S_t\bar u)(x)|\,dx\le L\cdot\int_0^t\left\{\liminf_{h\to 0+}\frac{\|u(\tau+h)-S_h u(\tau)\|_{\mathbf{L}^1(J_{\tau+h})}}{h}\right\}d\tau, \tag{9.48}$$

where $L$ is the Lipschitz constant of the semigroup and

$$J_\tau\doteq[-R+\tau\hat\lambda,\,R-\tau\hat\lambda].$$

To prove the theorem it thus suffices to show that, for every $\tau \notin \mathcal{N}$ and every bounded interval $[-r, r]$, the following holds:

$$\liminf_{h\to 0+} \frac{1}{h} \int_{-r+h\hat{\lambda}}^{r-h\hat{\lambda}} |u(\tau + h,\ x) - (S_h u(\tau))(x)|\, dx = 0. \tag{9.49}$$

4. For every $\tau \notin \mathcal{N}$ and $\xi \in \mathbb{R}$, by (9.46) the function $U$ defined at (9.36) coincides with the solution $U^{\sharp}_{(u;\tau,\xi)}$ of the Riemann problem (9.1) with data (9.14). Therefore, by (2.37) the function $u$ satisfies

$$\lim_{h\to 0+} \frac{1}{h} \int_{\xi-h\hat{\lambda}}^{\xi+h\hat{\lambda}} |u(\tau + h, x) - U^{\sharp}_{(u;\tau,\xi)}(h, x - \xi)|\, dx = 0. \tag{9.50}$$

5. For a given point $(\tau, \xi)$, call $U^{\flat} = U^{\flat}_{(u;\tau,\xi)}$ the solution of the linear Cauchy problem (9.15). Consider any open interval $]a, b[$ containing the point $\xi$ and fix a speed $\hat{\lambda}$ strictly larger than the absolute values of all characteristic speeds. For $t \geq \tau$ define the open intervals

$$J(t) \doteq ]a + (t - \tau)\hat{\lambda}, b - (t - \tau)\hat{\lambda}[,$$

and consider the region (Fig. 9.9)

$$\Gamma \doteq \{(t, x);\ t \in [\tau, \tau'], x \in J(t)\}. \tag{9.51}$$

With the above notation we claim that, for every $\tau' \geq \tau$ with $\tau' - \tau < (b - a)/2\hat{\lambda}$, one has

$$\int_{J(\tau')} |u(\tau', x) - U^{\flat}(\tau', x)|\, dx = \mathcal{O}(1) \cdot \Big( \sup_{(t,x)\in\Gamma} \big|u(t, x) - u(\tau, \xi)\big|\Big) \cdot \int_{\tau}^{\tau'} \text{Tot. Var.}\,\{u(t, \cdot);\ J(t)\}\, dt. \tag{9.52}$$

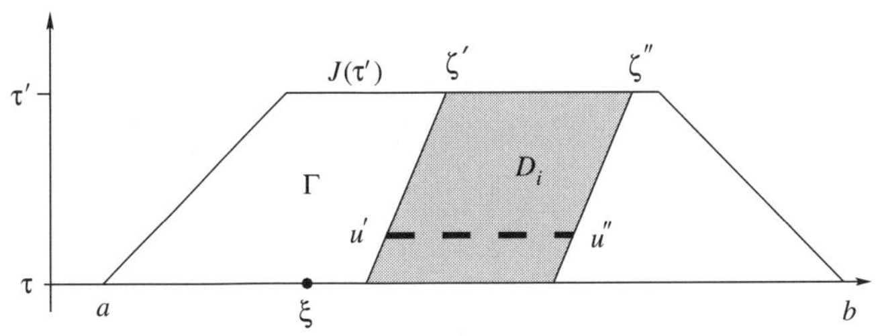


**Figure 9.9**

To prove (9.52), call $\tilde{\lambda}_i, \tilde{l}_i, \tilde{r}_i$ respectively the $i$-th eigenvalues and left and right eigenvectors of the matrix $\tilde{A} \doteq Df(u(\tau, \xi))$. By (9.15) it follows that

$$\tilde{l}_i \cdot U^\flat(\tau', x) = \tilde{l}_i \cdot U^\flat(\tau, x - (\tau' - \tau)\tilde{\lambda}_i) = \tilde{l}_i \cdot u(\tau, x - (\tau' - \tau)\tilde{\lambda}_i).$$

Now fix any $\zeta', \zeta'' \in J(\tau')$ and consider the quantity

$$\begin{aligned} E_i(\zeta', \zeta'') &\doteq \tilde{l}_i \cdot \int_{\zeta'}^{\zeta''} (u(\tau', x) - U^\flat(\tau', x))\, dx \\ &= \tilde{l}_i \cdot \int_{\zeta'}^{\zeta''} (u(\tau', x) - u(\tau, x - (\tau' - \tau)\tilde{\lambda}_i))\, dx. \end{aligned} \tag{9.53}$$

According to Remark 4.2, we can now apply the divergence theorem to the vector $(u,\ f(u))$ on the domain

$$D_i \doteq \{(t, x);\, t \in [\tau, \tau'],\, \zeta' + (t - \tau')\tilde{\lambda}_i \le x \le \zeta'' + (t - \tau')\tilde{\lambda}_i\}.$$

Since $u$ satisfies the conservation equation (7.1), the difference between the integral of $u$ at the top and at the bottom of the domain $D_i$ is thus measured by the inflow from the left side minus the outflow from the right side of $D_i$ (the shaded region in Fig. 9.9). From (9.53) it thus follows that

$$\begin{aligned} E_i(\zeta', \zeta'') &= \int_{\tau}^{\tau'} \tilde{l}_i \cdot ((f(u) - \tilde{\lambda}_i u)(t, \zeta' + (t - \tau')\tilde{\lambda}_i))\, dt \\ &\quad - \int_{\tau}^{\tau'} \tilde{l}_i \cdot ((f(u) - \tilde{\lambda}_i u)(t, \zeta'' + (t - \tau')\tilde{\lambda}_i))\, dt. \end{aligned} \tag{9.54}$$

To estimate the quantity in (9.54), consider the states

$$u'(t) \doteq u(t, \zeta' + (t - \tau')\tilde{\lambda}_i), \qquad u''(t) \doteq u(t, \zeta'' + (t - \tau')\tilde{\lambda}_i), \qquad \tilde{u} \doteq u(\tau, \xi).$$

We then have

$$\begin{aligned} \tilde{l}_i \cdot [f(u'') - f(u') - \tilde{\lambda}_i(u'' - u')] &= \tilde{l}_i \cdot [Df(\tilde{u}) \cdot (u'' - u') - \tilde{\lambda}_i(u'' - u')] \\ &\quad + \tilde{l}_i \cdot A^* \cdot (u'' - u'), \end{aligned} \tag{9.55}$$

where $A^*$ is the averaged matrix

$$A^* \doteq \int_0^1 [Df(su'' + (1 - s)u') - Df(\tilde{u})]\, ds.$$

Since the first term on the right hand side of (9.55) vanishes, we thus obtain

$$\begin{aligned}
&|\tilde{l}_i \cdot (f(u'') - f(u') - \tilde{\lambda}_i(u'' - u'))| \\
&\quad = \mathcal{O}(1) \cdot |u'' - u'| \cdot (|u'' - \tilde{u}| + |u' - \tilde{u}|) \\
&\quad = \mathcal{O}(1) \cdot \text{Tot. Var.}\,\{u(t); [\zeta' + (t - \tau')\tilde{\lambda}_i, \zeta'' + (t - \tau')\tilde{\lambda}_i]\} \\
&\qquad \cdot \sup_{(t,x)\in\Gamma} |u(t,x) - u(\tau,\xi)|.
\end{aligned}$$

In turn, this yields

$$|E_i(\zeta', \zeta'')| \le C_i \cdot \sup_{(t,x)\in\Gamma} |u(t,x) - u(\tau,\xi)| \cdot \int_\tau^{\tau'} \text{Tot. Var.}\,\{u(t,\cdot); [\zeta' + (t - \tau')\tilde{\lambda}_i, \zeta'' + (t - \tau')\tilde{\lambda}_i]\}\, dt \tag{9.56}$$

for some constants $C_i$. Define the measures $\mu_i$ on the interval $J(\tau')$ by setting

$$\mu_i(]a, b[) \doteq C_i' \cdot \int_\tau^{\tau'} \text{Tot. Var.}\,\{u(t,\cdot); [a + (t - \tau')\tilde{\lambda}_i, b + (t - \tau')\tilde{\lambda}_i]\}\, dt,$$

where

$$C_i' \doteq C_i \cdot \sup_{(t,x)\in\Gamma} |u(t,x) - u(\tau,\xi)|.$$

By Lemma 9.3 we thus have

$$\begin{aligned}
&\int_{J(\tau')} |\tilde{l}_i \cdot (u(\tau', x) - U^\flat(\tau', x))|\, dx \\
&\quad \le \mu_i(J(\tau')) \\
&\quad \le C_i \cdot \Big( \sup_{(t,x)\in\Gamma} \big|u(t,x) - u(\tau,\xi)\big|\Big) \cdot \int_\tau^{\tau'} \text{Tot. Var.}\,\{u(t,\cdot); J(t)\}\, dt.
\end{aligned} \tag{9.57}$$

Since this is true for all $i = 1, \ldots, n$ and the $\tilde{l}_i$ form a basis of left eigenvectors, (9.57) implies (9.52).

6. Let now $\tau \notin \mathcal{N}$ and $r, \varepsilon > 0$ be given. From now on, we also fix a constant $\hat{\lambda}$ strictly larger than all wave speeds, satisfying the additional relation $\hat{\lambda} \ge 1/\delta$.

Assuming either one of the conditions (A3) or (A4), we claim that one can find points

$$-r = x_0 < x_1 < \cdots < x_m = r$$

and $h^* > 0$ such that the following holds (Fig. 9.10). Defining the intervals $I_j \doteq ]x_{j-1}, x_j[$ and the sets

$$\Gamma_j \doteq \{(t,x);\ t \in [\tau, \tau + h^*], x_{j-1} + (t - \tau)\hat{\lambda} < x < x_j - (t - \tau)\hat{\lambda}\},$$

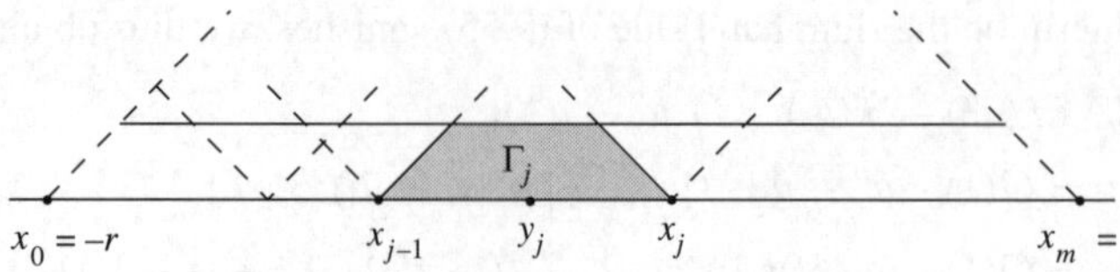


**Figure 9.10**

for every $j = 1, \dots, m$ we have

$$\text{Tot. Var.}\,\{u(\tau); I_j\} \le \varepsilon, \tag{9.58}$$

$$\text{Osc.}\,\{u; \Gamma_j\} \le \varepsilon. \tag{9.59}$$

If the tame oscillation assumption (A4) holds, then it suffices to choose points $x_j$ such that

$$\text{Tot. Var.}\,\{u(\tau); \,]x_{j-1}, x_j[\,\} \le \frac{\varepsilon}{1+C}.$$

Indeed, by (9.39) the condition (9.59) is clearly satisfied.

In the case where the assumption (A3) holds, the proof of our claim requires more work. First, we show that for every $\xi \in \mathbb{R}$ one can find $\rho > 0$ such that

$$|u(t,x) - u(\tau, \xi+)| \le \frac{\varepsilon}{2} \quad \text{for all } x \in \,]\xi, \xi+\rho], \quad t \in [\tau, \tau + \delta(x-\xi)[. \tag{9.60}$$

Otherwise, one could construct by induction two sequences of points $P_\nu = (t_\nu, x_\nu)$ and $Q_\nu = (\tau, y_\nu)$ with the following properties (Fig. 9.11):

$$x_\nu > y_\nu > x_{\nu+1} > \cdots > \xi,$$

$$t_\nu - \tau \le \delta(x_\nu - y_\nu), \qquad t_\nu - \tau \le \delta(y_\nu - x_{\nu+1}),$$

$$|u(t_\nu, x_\nu) - u(\tau, \xi+)| \ge \frac{\varepsilon}{2}$$

for all $\nu \ge 1$. Call $x \mapsto \gamma(x)$ the Lipschitz function defined on $[\xi, x_1]$, whose graph consists of all segments joing the points $P_1, Q_1, P_2, Q_2, \dots$. By construction, $\gamma$ is continuous with Lipschitz constant $\delta$. However, the total variation of $u$ along the curve $x \mapsto (\gamma(x), x)$ is infinite. Indeed,

$$\liminf_{\nu\to\infty} |u(P_\nu) - u(\tau, \xi+)| \ge \frac{\varepsilon}{2}, \qquad \lim_{\nu\to\infty} u(Q_\nu) = u(\tau, \xi+).$$

We thus reach a contradiction with the assumption (A3), proving our claim. Similarly, for every $\xi$ there exists $\rho > 0$ such that

$$|u(t,x) - u(\tau, \xi-)| \le \frac{\varepsilon}{2} \quad \text{for all} \quad x \in \,]\xi-\rho, \xi], \quad t \in \big[\tau, \tau + \delta(\xi - x)\big[. \tag{9.61}$$

By compactness, we can now cover the interval $[-r, r]$ with finitely many open intervals $]x_i - \rho_i, x_i + \rho_i[$ so that (9.60)–(9.61) hold for each $i$, and, moreover,

$$\text{Tot. Var.}\,\{u(\tau,\cdot); \,]x_i, x_i+\rho_i[\,\} < \frac{\varepsilon}{2}, \qquad \text{Tot. Var.}\,\{u(\tau,\cdot); \,]x_i-\rho_i, x_i[\,\} < \frac{\varepsilon}{2}. \tag{9.62}$$

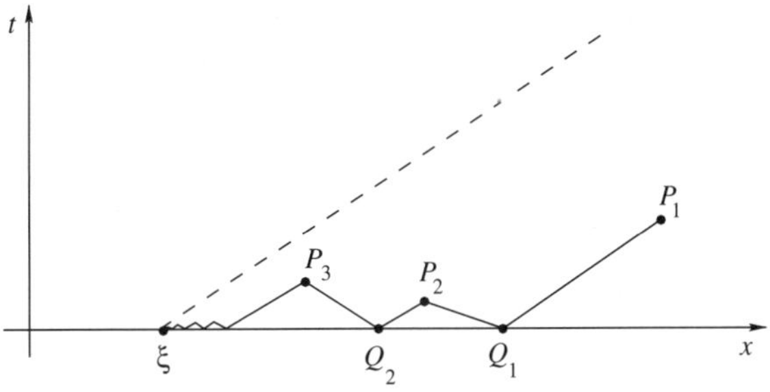


**Figure 9.11**

By possibly adding the points $x_0 \doteq -r$ and $x_m = r$ and relabelling the $x_i$ in increasing order, it is easily checked that the properties (9.58)–(9.59) hold, with $h^* \doteq \delta \cdot \min_i \rho_i$.

7. Introduce the midpoints $y_j \doteq (x_j + x_{j-1})/2$ and define the function

$$W(t,x) \doteq \begin{cases} U^{\sharp}_{(u;\tau,x_j)}(t-\tau, x-x_j) & \text{if } |x-x_j| \le (t-\tau)\hat{\lambda}, \\ U^{\flat}_{(u;\tau,y_j)}(t,x) & \text{if } (t,x) \in \Gamma_j. \end{cases}$$

By (9.50) and (9.52), this function $W$ provides a good approximation of $u$, for times $t = \tau + h$ with $h > 0$ small. Indeed (Fig. 9.10), recalling (9.59) we have

$$\begin{aligned}
&\limsup_{h\to 0+} \frac{1}{h} \int_{-r+h\hat{\lambda}}^{r-h\hat{\lambda}} |u(\tau+h,x) - W(\tau+h,x)|\,dx \\
&\le \sum_{j=1}^{m-1} \limsup_{h\to 0+} \frac{1}{h} \int_{x_j-h\hat{\lambda}}^{x_j+h\hat{\lambda}} \left|u(\tau+h,x) - U^{\sharp}_{(u;\tau,x_j)}(h, x-x_j)\right| dx \\
&\quad + \sum_{j=1}^{m} \limsup_{h\to 0+} \frac{1}{h} \int_{x_{j-1}+h\hat{\lambda}}^{x_j-h\hat{\lambda}} \left|u(\tau+h,x) - U^{\flat}_{(u;\tau,y_j)}(\tau+h,x)\right| dx \\
&\le 0 + \limsup_{h\to 0+} \left( \mathcal{O}(1) \cdot \sum_{j=1}^{m} \frac{\varepsilon}{h} \int_{\tau}^{\tau+h} \text{Tot. Var.}\,\{u(t); I_j\}\,dt \right) \\
&\le \limsup_{h\to 0+} \left( \mathcal{O}(1) \cdot \frac{\varepsilon}{h} \int_{\tau}^{\tau+h} \text{Tot. Var.}\,\{u(t); \mathbb{R}\}\,dt \right) \\
&= \mathcal{O}(1) \cdot \varepsilon. \qquad (9.63)
\end{aligned}$$

8. We now repeat the above computation with $u(\tau+h)$ replaced by $S_h u(\tau)$. By Theorem 9.3, the semigroup trajectory $v(\tau+h, x) = (S_h u(\tau))(x)$, being a limit of front tracking approximations, satisfies the tame oscillation condition (A4). In particular, recalling (9.58), for all $j = 1, \ldots, m$ we have

$$\text{Osc.}\,\{v; \Gamma_j\} = \mathcal{O}(1) \cdot \text{Tot. Var.}\,\{u(\tau); I_j\} = \mathcal{O}(1) \cdot \varepsilon. \tag{9.64}$$

Using (9.16), (9.52) and (9.64) we obtain

$$\begin{aligned}
&\limsup_{h\to 0+} \frac{1}{h} \int_{-r+h\hat{\lambda}}^{r-h\hat{\lambda}} |(S_h u(\tau))(x) - W(\tau+h, x)|\, dx \\
&\leq \sum_{j=1}^{m-1} \limsup_{h\to 0+} \frac{1}{h} \int_{x_j-h\hat{\lambda}}^{x_j+h\hat{\lambda}} \left|(S_h u(\tau))(x) - U^\sharp_{(u;\tau,x_j)}(h, x)\right| dx \\
&\quad + \sum_{j=1}^{m} \limsup_{h\to 0+} \frac{1}{h} \int_{x_{j-1}+h\hat{\lambda}}^{x_j-h\hat{\lambda}} \left|(S_h u(\tau))(x) - U^\flat_{(u;\tau,y_j)}(\tau+h, x)\right| dx \\
&\leq 0 + \limsup_{h\to 0+} \left( \mathcal{O}(1) \cdot \sum_{j=1}^{m} \frac{\varepsilon}{h} \int_{\tau}^{\tau+h} \text{Tot. Var.}\,\{S_{t-\tau} u(\tau); I_j\}\, dt \right) \\
&\leq \limsup_{h\to 0+} \left( \mathcal{O}(1) \cdot \frac{\varepsilon}{h} \int_{\tau}^{\tau+h} \text{Tot. Var.}\,\{S_{t-\tau} u(\tau); \mathbb{R}\}\, dt \right) \\
&= \mathcal{O}(1) \cdot \varepsilon.
\end{aligned} \tag{9.65}$$

Together, (9.63) and (9.65) imply

$$\limsup_{h\to 0+} \frac{1}{h} \int_{-r+h}^{r-h} |u(\tau+h, x) - (S_h u(\tau))(x)|\, dx = \mathcal{O}(1) \cdot \varepsilon.$$

Since $\varepsilon > 0$ can be taken arbitrarily small, this achieves a proof of (9.49). By (9.48), this establishes Theorem 9.4. □

## Problems

(1) Show that a function $u = u(t, x)$ satisfies the tame oscillation condition (A4) iff there exist constants $C, \delta > 0$ such that the following holds. For every point $x$ and every $t \geq 0$, $h > 0$ one has

$$|u(t+h, x) - u(t, x)| \leq C \cdot \text{Tot. Var.}\,\{u(t, \cdot);\ ]x - h/\delta, x + h/\delta[\}.$$

(2) Let $\mathcal{D}$ be the positively invariant domain defined at (8.1). Fix any (right continuous) $u \in \mathcal{D}$ and finitely many points $p_1 < p_2 < \cdots < p_{m+1}$. For each $j = 1, \ldots, m$ choose $x_j \in [p_j, p_{j+1}[$. Prove that the function

$$u^*(x) \doteq \begin{cases} 0 & \text{if } x < p_1 \text{ or } x \geq p_{m+1}, \\ u(x_j) & \text{if } x \in [p_j, p_{j+1}[ \end{cases}$$

lies in $\mathcal{D}$.

(3) Consider any state $\omega_0 \neq 0$ and define the function

$$u(t,x) \doteq \begin{cases} \omega_0 & \text{if } t = 1/k^2, \quad x = 1/k, \quad k = 1, 2, 3, \ldots \\ 0 & \text{otherwise.} \end{cases}$$

Show that $u$ provides an entropy weak solution to any system of conservation laws, but all conditions (A3), (A4) and (A5) fail. On the other hand, notice that (A1)–(A5) trivially hold for the right continuous version $u^*(t,x) \doteq u(t, x+) \equiv 0$.

(4) In addition to (7.1), consider a second system of conservation laws $u_t + \tilde{f}(u)_x = 0$, also satisfying the basic assumptions (♣) in Chapter 7. Call $S$, $\tilde{S}$ the corresponding Riemann semigroups. Show that, for all initial data $\bar{u} \in \mathbf{L}^1$ with sufficiently small total variation, one has

$$\|S_t\bar{u} - \tilde{S}_t\bar{u}\|_{\mathbf{L}^1} = \mathcal{O}(1) \cdot t \cdot \text{Tot. Var.}\{\bar{u}\} \cdot \|f - \tilde{f}\|_{\mathcal{C}^1(\Omega)}.$$

Hint: let $u = u(t,x)$ be a front tracking $\varepsilon$-approximate solution of (7.1). Using Problem 3 in Chapter 5, prove the estimate

$$\liminf_{h \to 0+} \frac{1}{h} \|u(t+h) - \tilde{S}_h u(t)\|_{\mathbf{L}^1} = \mathcal{O}(1) \cdot (\varepsilon + \text{Tot. Var.}\{u(t)\} \cdot \|f - \tilde{f}\|_{\mathcal{C}^1(\Omega)})$$

for almost every $t \geq 0$.

(5) Let $u$ be a solution of (7.1) obtained as the limit of front tracking approximations, and fix any point $(\tau, \xi)$ with $\tau \geq 0$. Prove that, as $\eta \to 0+$, the rescaled functions $u^\eta(t,x) \doteq u(\tau + \eta t, \xi + \eta x)$ converge in $\mathbf{L}^1_{\text{loc}}([0,\infty[\times\mathbb{R})$ to the solution of the Riemann problem (9.1) with data $u^\pm = u(\tau, \xi\pm)$.

(6) Let $S : \mathcal{D} \times [0,\infty[ \mapsto \mathcal{D}$ be the SRS generated by Burgers' equation $u_t + (u^2/2)_x = 0$ on the domain

$$\mathcal{D} \doteq \{u : \mathbb{R} \mapsto \mathbb{R};\ \|u\|_{\mathbf{L}^1} \leq 1, \text{Tot. Var.}\{u\} \leq 1\}.$$

Consider the subdomain $\mathcal{D}_1 \doteq \mathcal{D} \cap \mathcal{C}^1$ consisting of functions with continuous, uniformly bounded first derivative. Show that there exists a second semigroup $\tilde{S} : \mathcal{D} \times [0,\infty[ \mapsto \mathcal{D}$, distinct from $S$, which satisfies (9.4)–(9.5) for some constants $L, L'$ and has the following property: for every $\bar{u} \in \mathcal{D}_1$, there exists $\delta > 0$ such that $\tilde{S}_t\bar{u} = S_t\bar{u}$ for all $t \in [0, \delta]$.

Hint: let $u = h(v)$ be a smooth, non-linear change of coordinates. Then, for a suitable function $g$, every smooth solution $u$ of Burgers' equation has the form $u(t,x) = h(v(t,x))$, where $v$ is a smooth solution of $v_t + g'(v)v_x = 0$. Define the semigroup $\tilde{S}$ by setting $\tilde{S}_t\bar{u} = h(v(t,\cdot))$, where $v$ is the unique entropy solution to the Cauchy problem

$$v_t + g(v)_x = 0, \qquad v(0,x) = h^{-1}(\bar{u}(x)).$$

In connection with Theorem 9.1, this example shows that the semigroup $S$ is uniquely determined by its local action on the set $\mathcal{D}_{pc}$ of piecewise constant functions, but not by its local action on the set $\mathcal{D}_1$ of continuously differentiable functions. Observe that both $\mathcal{D}_{pc}$ and $\mathcal{D}_1$ are dense in $\mathcal{D}$, w.r.t. the $\mathbf{L}^1$ distance.

(7) With the same setting as in Theorem 9.1, let $S^* : \mathcal{D} \times [0, \infty[ \mapsto \mathcal{D}$ be a Lipschitz semigroup with the following property:

(3″) For every piecewise smooth function $\bar{u} \in \mathcal{D}$, all of whose jumps are genuinely non-linear, entropy-admissible shocks, there exists $\delta > 0$ such that the trajectory $t \mapsto S_t^*\bar{u}$ provides the unique piecewise smooth entropy solution of the Cauchy problem (7.1)–(7.2), for $t \in [0, \delta]$.

Prove that $S^*$ coincides with the unique SRS generated by (7.1).
Hint: let $u = u(t,x)$ be a front tracking $\varepsilon$-approximate solution of (7.1), defined for $t \in [0, T]$. Show that there exists times $0 = t_0 < t_1 < \cdots < t_m = T$ and a piecewise smooth function $v = v(t,x)$ with the following properties: on each open strip $]t_{i-1}, t_i[\, \times \mathbb{R}$, the function $v$ is a piecewise smooth solution of (7.1), all of whose jumps are genuinely non-linear, entropy-admissible shocks. Moreover,

$$\sum_{i=1}^{m} \|v(t_i+) - v(t_i-)\|_{\mathbf{L}^1} = \mathcal{O}(1) \cdot \varepsilon T, \qquad \sup_{t\in[0,T]} \|v(t) - u(t)\|_{\mathbf{L}^1} = \mathcal{O}(1) \cdot \varepsilon.$$

Using (3″) and the estimate (2.101), show that

$$\|v(t) - S_t^* v(0)\|_{\mathbf{L}^1} = \mathcal{O}(1) \cdot \varepsilon T \quad \text{for all } t \in [0, T].$$

(8) The following problem is concerned with solutions of Burgers' equation $u_t + (u^2/2)_x = 0$ which are right continuous, in the sense that $u(t,x) = u(t,x+)$ for all $t, x$.

(i) Construct a weak (non-entropic) solution which satisfies the bounded variation assumption (A3) but not the tame oscillation assumption (A4).
(ii) Construct a weak (non-entropic) solution which satisfies (A4) but not (A3).

Hint: consider the functions

$$u_0(x) \doteq \begin{cases} 1 & \text{if } x \geq 0, \\ -1 & \text{if } x < 0, \end{cases}$$

$$\varphi(t, x) \doteq \begin{cases} 1 & \text{if } x \in [0,\ t/2[, \\ -1 & \text{if } x \in [-t/2,\ 0[, \\ 0 & \text{if } x \in ]-\infty, -t/2, [\cup[t/2,\ \infty[. \end{cases}$$

Fix $\varepsilon > 0$ suitably small and, for $k \geq 1$, define the times $t_k \doteq \varepsilon + \varepsilon^2 + \cdots + \varepsilon^k$ so that $t_k \to \tau \doteq \varepsilon/(1-\varepsilon)$ as $k \to \infty$. For $t \in [0, \tau]$ define the function

$$u(t, x) \doteq u_0(x) + \sum_{t_k < t} \varphi(t - t_k, x - t + t_k).$$

Show that $u$ is a weak solution which satisfies (A4), but Tot. Var. $\{u(\tau, \cdot)\} = \infty$ and hence (A3) fails.

# 10
# Qualitative properties

We continue here the study of entropy weak solutions to the hyperbolic system of conservation laws

$$u_t + f(u)_x = 0 \tag{10.1}$$

with small total variation. Owing to the uniqueness results proved in the previous chapter, all these solutions can be obtained as limits of front tracking approximations. Taking advantage of this fact, our analysis will follow a natural pattern:

(i) Derive suitable a priori estimates on $\varepsilon$-approximate front tracking solutions.
(ii) Taking an appropriate limit as $\varepsilon \to 0$, recover the desired properties for exact solutions.

Since front tracking approximations are piecewise constant in the $t$–$x$ plane, step (i) mainly involves the proof of algebraic inequalities, describing how the wave strengths or the wave speeds can change at interaction times. On the other hand, as $\varepsilon \to 0$, knowing the $\mathbf{L}^1$ convergence of the approximate solutions $u_\varepsilon \to u$ is often not sufficient to obtain the desired results. Indeed, it is also essential to understand how the wave-fronts in the approximations $u_\varepsilon$ are related to the waves in the exact solution $u$.

With this in mind, in Section 10.1 we introduce the measures $\mu^i$ of $i$-waves in a $BV$ function $u : \mathbb{R} \mapsto \mathbb{R}^n$. In terms of these measures, the Glimm functionals $\mathbf{Q}$ and $\Upsilon \doteq \mathbf{V} + C_0\mathbf{Q}$, introduced in Chapter 7 for piecewise constant functions, can now be extended to general $BV$ functions. Our main result, stated in Theorem 10.1, shows that $\mathbf{Q}$ and $\Upsilon$ are both lower semicontinuous w.r.t. $\mathbf{L}^1$ convergence.

In Section 10.2 we prove an estimate of the decay of positive waves in a genuinely non-linear family. This provides an extension of the classical result proved by Oleinik in the scalar case.

Finally, Section 10.3 contains an analysis of the global structure of an entropy weak solution $u = u(t, x)$ of (10.1). Being a $BV$ function of the two variables $t, x$, the function $u$ shares the general regularity properties stated in Theorem 2.5. In addition, we will show that much stronger properties actually hold. Indeed, $u$ is continuous outside a countable family $\Gamma$ of shock curves and a countable set $\Theta$ of interaction points. Along each curve $\gamma \in \Gamma$, the solution $u$ satisfies a.e. the Rankine–Hugoniot equations and the Lax entropy conditions.

Throughout the following, we assume that the system satisfies the basic assumptions (♣) stated at the beginning of Chapter 7. We recall that, by our earlier analysis, there exists a domain $\mathcal{D}$ of the form (8.1) which is positively invariant for the flow generated by (10.1). Moreover, every function $\bar{u} \in \mathbf{L}^1(\mathbb{R}; \mathbb{R}^n)$ with suitably small total variation lies in $\mathcal{D}$. As usual, we shall assume that our $BV$ solutions are right continuous, so that $u(t, x) = u(t, x+)$ for every $t, x$.

## 10.1 Wave measures

As usual, let $A(u) = Df(u)$ be the Jacobian matrix of $f$, and call $\lambda_i(u)$, $l_i(u)$, $r_i(u)$ respectively the eigenvalues and the left and right eigenvectors of $A(u)$. We adopt here the parametrization choice (♠) introduced in Chapter 5. In particular, if the $i$-th family is genuinely non-linear we normalize the $i$-th eigenvectors so that $D\lambda_i(u) \cdot r_i(u) \equiv 1$. Let $u : \mathbb{R} \mapsto \Omega$ have bounded variation. By possibly changing the values of $u$ at countably many points, we can assume that $u$ is right continuous. The distributional derivative $\mu \doteq D_x u$ is a vector measure, which can be decomposed into a continuous and an atomic part: $\mu = \mu_c + \mu_a$. For $i = 1, \ldots, n$, we now define the scalar measures $\mu^i = \mu^i_c + \mu^i_a$ as follows. The continuous part of $\mu^i$ is the Radon measure $\mu^i_c$ such that

$$\int \phi \, d\mu^i_c = \int \phi \, l_i(u) \cdot d\mu_c \tag{10.2}$$

for every scalar continuous function $\phi$ with compact support. The atomic part of $\mu^i$ is the measure $\mu^i_a$ concentrated on the countable set $\{x_\alpha; \ \alpha = 1, 2, \ldots\}$ where $u$ has a jump, such that

$$\mu^i_a(\{x_\alpha\}) = \sigma_{\alpha,i} \doteq E_i(u(x_\alpha -), \ u(x_\alpha +)) \tag{10.3}$$

is the size of the $i$-th wave in the solution of the corresponding Riemann problem with data $u(x_\alpha \pm)$. Observe that, for every couple of states $u^-$, $u^+$, one has

$$E_i(u^-, u^+) = l_i(u^+) \cdot (u^+ - u^-) + O(1) \cdot |u^+ - u^-|^2.$$

Therefore, at every jump point $x_\alpha$ there exists a vector $\tilde{l}_i(x_\alpha)$ such that

$$|\tilde{l}_i(x_\alpha) - l_i(u(x_\alpha +))| = \mathcal{O}(1) \cdot |u(x_\alpha) - u(x_\alpha -)|, \tag{10.4}$$

$$E_i(u(x_\alpha -), u(x_\alpha +)) = \tilde{l}_i(x_\alpha) \cdot (u(x_\alpha +) - u(x_\alpha -)). \tag{10.5}$$

One can thus define the measure $\mu^i$ as

$$\int \phi \, d\mu^i = \int \phi \, \tilde{l}_i \cdot D_x u, \qquad \phi \in \mathcal{C}^0_c \tag{10.6}$$

where $\tilde{l}_i(x) = l_i(u(x))$ at points where $u$ is continuous, while $\tilde{l}_i(x)$ is some vector which satisfies (10.4)–(10.5) at points of jump. In all cases, recalling that $u$ is right continuous, the following holds:

$$|\tilde{l}_i(x) - l_i(u(x))| = \mathcal{O}(1) \cdot |u(x) - u(x-)| \quad \text{for all } x \in \mathbb{R}. \tag{10.7}$$

Call $|\mu| \doteq |D_x u|$ the measure of total variation of $u$. A useful consequence of the representation (10.7) is the following.

**Lemma 10.1.** *For every constant state $\bar{u}$ and every open interval $I$ one has the estimate*

$$\left|\int_I \varphi\, l_i(\bar{u}) \cdot d\mu - \int_I \varphi\, d\mu^i\right| = \mathcal{O}(1) \cdot \sup_{x \in I} |u(x) - \bar{u}| \cdot \int_I |\phi|\, d|\mu|, \tag{10.8}$$

*for every $\phi \in \mathcal{C}_c(I; \mathbb{R})$, $i = 1, \ldots, n$.*

Indeed, by (10.6) it follows that

$$\left|\int \phi\, l_i(\bar{u}) \cdot d\mu - \int \phi\, d\mu^i\right| \leq \int |\phi(x)|\, |l_i(\bar{u}) - \tilde{l}_i(x)|\, d|\mu|.$$

The estimate (10.8) is now a straightforward consequence of (10.7) and of the Lipschitz continuity of the maps $u \mapsto l_i(u)$. □

Intuitively, one can regard the scalar measures $\mu^i$ as the components of the vector measure $\mu = D_x u$ along the basis of eigenvectors $r_i$. Call $\mu^{i+}$, $\mu^{i-}$ the positive and negative parts of the signed measure $\mu^i$, so that

$$\mu^i = \mu^{i+} - \mu^{i-}, \qquad |\mu^i| = \mu^{i+} + \mu^{i-}.$$

The *total strength of waves* in $u$ can now be defined as

$$\mathbf{V}(u) \doteq \sum_{i=1}^{n} \mathbf{V}_i(u), \qquad \mathbf{V}_i(u) \doteq |\mu^i|(\mathbb{R}). \tag{10.9}$$

Moreover, the *interaction potential* of waves in $u$ can be defined in terms of product measures on $\mathbb{R}^2$, i.e.

$$\mathbf{Q}(u) \doteq \sum_{i<j} (|\mu^j| \times |\mu^i|)(\{(x, y); x < y\}) + \sum_{i \in \mathcal{GN}} (\mu^{i-} \times |\mu^i|)(\{(x, y); x \neq y\}). \tag{10.10}$$

By $\mathcal{GN}$ we denote here the set of genuinely non-linear families.

Observe that, in the special case where $u$ is piecewise constant, the definitions (10.9)–(10.10) coincide with (7.99). On the other hand, if $u$ is Lipschitz continuous, then its derivative $u_x(x)$ exists at almost every point $x \in \mathbb{R}$. In this case, the definitions (10.2) and (10.9) yield

$$\mathbf{V}_i(u) = \int_{-\infty}^{\infty} |l_i(u(x)) \cdot u_x(x)|\, dx.$$

The main result of this section establishes the lower semicontinuity of the functionals $\mathbf{Q}$ and $\mathbf{V} + C_0\, \mathbf{Q}$ on a domain $\mathcal{D}$ of small $BV$ functions:

$$\mathcal{D} \doteq \{u \in \mathbf{L}^1(\mathbb{R}; \mathbb{R}^n),\ \mathbf{V}(u) + C_0\, \mathbf{Q}(u) \leq \delta_0\}. \tag{10.11}$$

In the following, by $\mu^{i+}, \mu^{i-}, \mu_\nu^{i+}, \mu_\nu^{i-}$, we denote the measures of positive and negative $i$-waves in $u, u_\nu$ respectively.

**Theorem 10.1 (Lower semicontinuity of the Glimm functionals).** *With a suitable choice of the constants $C_0, \delta_0 > 0$ in (10.11), the following holds. Consider any sequence of functions $u_\nu \in \mathcal{D}$, with $u_\nu \to u$ in $\mathbf{L}^1$ as $\nu \to \infty$. Then*

$$\mathbf{Q}(u) \leq \liminf_{\nu\to\infty} \mathbf{Q}(u_\nu), \tag{10.12}$$

$$\mathbf{V}(u) + C_0 \cdot \mathbf{Q}(u) \leq \liminf_{\nu\to\infty} [\mathbf{V}(u_\nu) + C_0\, \mathbf{Q}(u_\nu)]. \tag{10.13}$$

*Moreover, for every finite union of open intervals $J = I_1 \cup \cdots \cup I_m$ the following holds:*

$$\mu^{i\pm}(J) + C_0 \cdot \mathbf{Q}(u) \leq \liminf_{\nu\to\infty} [\mu_\nu^{i\pm}(J) + C_0\, \mathbf{Q}(u_\nu)] \quad i = 1, \ldots, n. \tag{10.14}$$

**Remark 10.1.** By (10.13), the domain $\mathcal{D}$ defined at (10.11) is closed in $\mathbf{L}^1$.

It is important to observe that the lower semicontinuity of the functionals $\mathbf{Q}$ and $\mathbf{V} + C_0\,\mathbf{Q}$ holds only restricted to the domain $\mathcal{D}$. Indeed, the numerical values of the constants $\delta_0, C_0$ in (10.11) play a key role for the validity of the result.

Notice that the functional $\mathbf{V}$ is equivalent to the total variation, in the sense that

$$\frac{1}{C} \cdot \text{Tot. Var.}\,\{u\} \leq \mathbf{V}(u) \leq C \cdot \text{Tot. Var.}\,\{u\}$$

for some constant $C$ and all $u \in \mathcal{D}$. However, while the total variation is a convex and lower semicontinuous (unbounded) functional on the whole space $\mathbf{L}^1$, the total strength of waves $\mathbf{V}(u)$ may not be lower semicontinuous, even on a domain of small *BV* functions. To see this, consider a piecewise smooth solution $w = w(t, x)$ containing two shocks that interact at time $\tau$. As in Fig. 7.10, let $\sigma_i', \sigma_j'$ be the incoming shocks and let $\sigma_1, \ldots, \sigma_n$ be the waves in the Riemann problem produced by the interaction. Observe that one may well have

$$|\sigma_i'| + |\sigma_j'| < \sum_{\ell=1}^{n} |\sigma_\ell|.$$

In this case, choosing $u_\nu \doteq w(\tau - \nu^{-1})$, $u = w(\tau)$, one has the convergence $u_\nu \to u$ in $\mathbf{L}^1$, but $\mathbf{V}(u) > \lim_{\nu\to\infty} \mathbf{V}(u_\nu)$.

The proof of Theorem 10.1 will be preceded by a technical lemma. Recall that, for the sake of definiteness, we are always assuming that our *BV* functions are right continuous.

**Lemma 10.2.** *With a suitable choice of $C_0, \delta_0 > 0$ the following holds. For every $u \in \mathcal{D}$ and $a < \hat{x} < b$, the function*

$$\hat{u}(x) \doteq \begin{cases} u(x) & \text{if } x \notin [a, b[, \\ u(a) & \text{if } x \in [a, \hat{x}[, \\ u(b) & \text{if } x \in [\hat{x}, b[ \end{cases} \tag{10.15}$$

*satisfies*

$$\mathbf{Q}(\hat{u}) \leq \mathbf{Q}(u), \qquad \mathbf{V}(\hat{u}) + C_0 \cdot \mathbf{Q}(\hat{u}) \leq \mathbf{V}(u) + C_0\, \mathbf{Q}(u). \tag{10.16}$$

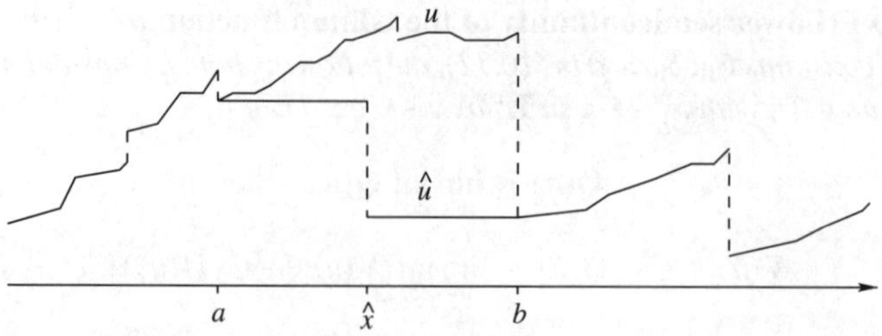


**Figure 10.1**

*Moreover, the corresponding wave measures* $\hat{\mu}^{i\pm}$ *satisfy*

$$\hat{\mu}^{i\pm}(J) + C_0\,\mathbf{Q}(\hat{u}) \le \mu^{i\pm}(J) + C_0 \cdot \mathbf{Q}(u), \qquad i = 1, \dots, n \tag{10.17}$$

*for every finite union of open intervals* $J = I_1 \cup \cdots \cup I_m$ *such that*

$$[a, b[\, \subset J \quad \text{or} \quad [a, b[\, \cap J = \emptyset. \tag{10.18}$$

Observe that the function $\hat{u}$ in (10.15) is obtained from $u$ by collapsing all wave-fronts in $]a, b]$ onto the point $\hat{x}$ (Fig. 10.1). The proof is given in three steps.

1. Consider first the case where $u$ is piecewise constant, with jumps at $x_0 < x_1 < \cdots < x_N$, say

$$u(x) = \begin{cases} 0 & \text{if } x \notin [x_0, x_N[\,, \\ u_k & \text{if } x \in [x_{k-1}, x_k[\,. \end{cases} \tag{10.19}$$

Fix any $p \in \{1, \dots, N\}$ and define the piecewise constant function $\tilde{u}$ by setting

$$\tilde{u}(x) = \begin{cases} u(x) & \text{if } x \notin [x_{p-1}, x_{p+1}[, \\ u_p & \text{if } x \in [x_{p-1}, x_{p+1}[. \end{cases}$$

Call $\sigma_i', \sigma_i''$ the waves generated by the Riemann problems $(u_p, u_{p+1})$ and $(u_{p+1}, u_{p+2})$, and denote by $\sigma_i$, $i = 1, \dots, n$, the waves in the Riemann problem $(u_p, u_{p+2})$. By Lemma 7.4, we have the estimate

$$\sum_i |\sigma_i - \sigma_i' - \sigma_i''| = \mathcal{O}(1) \cdot \sum_{(\sigma_j', \sigma_k'') \in \mathcal{A}} |\sigma_j'|\,|\sigma_k''|.$$

If the total variation of $u$ is sufficiently small, with a suitable choice of $C_0$ this implies

$$\mathbf{Q}(\tilde{u}) \le \mathbf{Q}(u), \qquad \mathbf{V}(\tilde{u}) + C_0\,\mathbf{Q}(\tilde{u}) \le \mathbf{V}(u) + C_0 \cdot \mathbf{Q}(u), \tag{10.20}$$

$$\tilde{\mu}^{i\pm}(J) + C_0 \cdot \mathbf{Q}(\tilde{u}) \le \mu^{i\pm}(J) + C_0 \cdot \mathbf{Q}(u), \qquad i = 1, \dots, n, \tag{10.21}$$

provided that either $[x_p, x_{p+1}[\subset J$ or else $[x_p, x_{p+1}[\cap J = \emptyset$. Entirely similar conclusions hold in case we define

$$\tilde{u}(x) = \begin{cases} u(x) & \text{if } x \notin [x_{p-1}, x_{p+1}[\,, \\ u_{p+1} & \text{if } x \in [x_{p-1}, x_{p+1}[\,, \end{cases}$$

and either $[x_{p-1}, x_p[\subset J$ or else $[x_{p-1}, x_p[\cap J = \emptyset$.

2. Now let $u$ be piecewise constant as in (10.19), and define $\hat{u}$ as in (10.15). By possibly inserting additional points $x_k$ we can assume $a = x_p$, $\hat{x} = x_q$, $b = x_r$. The function $\hat{u}$ can now be obtained from $u$ in a finite number of steps, replacing the values $u_k, u_{k+1}$ on two adjacent intervals $[x_{k-1}, x_k[$ and $[x_k, x_{k+1}[$ with the single value $u_k$, or with the single value $u_{k+1}$. According to (10.20)–(10.21), each of these steps does not increase the values of the functionals $\mathbf{Q}$, $\mathbf{V} + C_0\,\mathbf{Q}$ and $\mu^{i\pm}(J) + C_0\,\mathbf{Q}$. By finite induction, this establishes the bounds (10.16)–(10.17) in the case where $u$ is piecewise constant.

3. In the general case, we can approximate the *BV* function $u$ by a sequence of piecewise constant functions $u_\nu$ such that

$$\mathbf{Q}(u_\nu) \to \mathbf{Q}(u), \qquad \mathbf{V}(u_\nu) \to \mathbf{V}(u), \qquad \mu_\nu^{i\pm} \rightharpoonup \mu^{i\pm}$$

as $\nu \to \infty$. More precisely, for each $\nu \geq 1$ we construct a finite set of points $x_{\nu,0} < \cdots < x_{\nu,N_\nu}$, containing $a$, $\hat{x}$, $b$ and all the endpoints $a_\ell, b_\ell$ of the intervals $I_\ell$, $\ell = 1, \ldots, m$, and such that

$$\lim_{\nu\to\infty} x_{\nu,0} = -\infty, \qquad \lim_{\nu\to\infty} x_{\nu,N_\nu} = \infty, \qquad \lim_{\nu\to\infty} \sup_j |x_{\nu,j} - x_{\nu,j-1}| = 0.$$

We then define

$$u_\nu(x) = \begin{cases} 0 & \text{if } x \notin [x_{\nu,0},\ x_{\nu,N_\nu}[\,, \\ u(x_{\nu,j}) & \text{if } x \in [x_{\nu,j},\ x_{\nu,j+1}[,\quad 0 \leq j < N_\nu. \end{cases}$$

Calling

$$\hat{u}_\nu(x) \doteq \begin{cases} u_\nu(x) & \text{if } x \notin [a, b[\,, \\ u_\nu(a) & \text{if } x \in [a, \hat{x}[\,, \\ u_\nu(b) & \text{if } x \in [\hat{x}, b[\,, \end{cases}$$

as $\nu \to \infty$ we then have

$$\mathbf{Q}(\hat{u}_\nu) \to \mathbf{Q}(\hat{u}), \qquad \mathbf{V}(\hat{u}_\nu) \to \mathbf{V}(\hat{u}), \qquad \hat{\mu}_\nu^{i\pm}(J) \to \hat{\mu}^{i\pm}(J)\,.$$

By the previous step, the functions $\hat{u}_\nu$ and the measures $\hat{\mu}_\nu^{i\pm}$ satisfy the estimates (10.16)–(10.17). Hence the same holds for $\hat{u}$, $\hat{\mu}^{i\pm}$. □

We can now give a proof of Theorem 10.1 in several steps.

1. Let $\mu^i$ be the scalar measures in (10.6) and let $\mu_\nu^i$ be similarly defined, with $u$ replaced by $u_\nu$. By passing to a subsequence we can assume that the limits

$$\lim_{\nu\to\infty} \mathbf{V}_i(u_\nu), \qquad \lim_{\nu\to\infty} \mathbf{Q}(u_\nu)$$

exist and that the whole sequences in (10.12)–(10.14) converge to their lower limits. Moreover, we can assume that $u_\nu(x) \to u(x)$ a.e. and that the measures of total variation converge weakly in the sense of measures as $\nu \to \infty$, say

$$|\mu_\nu| \doteq |D_x u_\nu| \rightharpoonup \tilde{\mu} \tag{10.22}$$

for some positive Radon measure $\tilde{\mu}$. Observe that in this case one has $\tilde{\mu} \geq |\mu|$, in the sense that $\tilde{\mu}(J) \geq |\mu|(J)$ for every Borel set $J$.

2. Let any $\varepsilon > 0$ be given. Since the total mass of $\tilde{\mu}$ is finite, one can select finitely many points $y_1, \ldots, y_N$ such that

$$\tilde{\mu}(\{x\}) < \varepsilon, \qquad \text{for all } x \notin \{y_1, \ldots, y_N\}. \tag{10.23}$$

We now choose disjoint open intervals $I'_k \doteq ]y_k - \rho,\ y_k + \rho[$ such that

$$\tilde{\mu}(I'_k \setminus \{y_k\}) < \frac{\varepsilon}{N} \qquad k = 1, \ldots, N. \tag{10.24}$$

Moreover, we choose $R > 0$ such that

$$\bigcup_{k=1}^{N} I_k \subset [-R, R], \qquad \tilde{\mu}(]-\infty, -R] \cup [R, \infty[) < \varepsilon. \tag{10.25}$$

Because of (10.23), we can now choose points $p_0 < -R < p_1 < \cdots < R < p_r$ which are continuity points for $u$ and for every $u_\nu$, such that

$$\tilde{\mu}(\{p_h\}) = 0, \quad u_\nu(p_h) \to u(p_h) \qquad \text{for all } h = 0, \ldots, r, \tag{10.26}$$

and such that either

$$p_{h-1} < y_k < p_h, \qquad [p_{h-1}, p_h] \subset I'_k, \tag{10.27}$$

for some $k$, or else

$$|\mu|([p_{h-1}, p_h]) \leq \tilde{\mu}([p_{h-1}, p_h]) < \varepsilon. \tag{10.28}$$

Call $J_h \doteq [p_{h-1}, p_h]$. If (10.28) holds, by weak convergence for some $\nu_0$ sufficiently large one has

$$|\mu_\nu|(J_h) < \varepsilon \quad \text{for all } \nu \geq \nu_0. \tag{10.29}$$

On the other hand, if (10.27) holds, from (10.24) it follows that

$$|\mu|(J_h \setminus \{y_k\}) \leq \tilde{\mu}(J_h \setminus \{y_k\}) < \frac{\varepsilon}{N}. \tag{10.30}$$

3. If (10.28)–(10.29) hold, then the oscillation of $u, u_\nu$ on the interval $J_h$ is very small. That is, for every $x, y \in J_h$ and $\nu$ sufficiently large,

$$|u_\nu(x) - u_\nu(y)| \leq |\mu_\nu|(J_h) < \varepsilon,$$

$$|u(x) - u(y)| \leq |\mu|(J_h) \leq \tilde{\mu}(J_h) < \varepsilon.$$

Set $\bar{u}_h \doteq u(p_h)$. By the pointwise convergence $u_\nu(p_h) \to u(p_h)$ and the two estimates above it follows that

$$|u_\nu(x) - \bar{u}_h| < \varepsilon, \quad |u(x) - \bar{u}_h| < \varepsilon, \qquad \text{for all } x \in J_h, \tag{10.31}$$

for all $\nu$ sufficiently large. Using (10.8) and (10.31) we obtain

$$\begin{aligned}\left|\int_{J_h} \phi\, d\mu^i - \int_{J_h} \phi\, d\mu^i_\nu\right| &\leq \left|\int_{J_h} \phi\,(l_i(\bar{u}_h) - \tilde{l}_{\nu,i}(x)) \cdot d\mu_\nu\right| + \left|\int_{J_h} \phi\,(l_i(\bar{u}_h) - \tilde{l}_i(x)) \cdot d\mu\right| \\ &\quad + \left|\int_{J_h} \phi\, l_i(\bar{u}_h) \cdot (d\mu - d\mu_\nu)\right| \\ &= \mathcal{O}(1) \cdot \varepsilon \int_{J_h} |\phi|(d|\mu| + d|\mu_\nu|) + \left|\int_{J_h} \phi\, l_i(\bar{u}_h) \cdot (d\mu - d\mu_\nu)\right|\end{aligned} \tag{10.32}$$

for all $\phi \in C(J_h; \mathbb{R})$. By taking the supremum over all continuous functions $|\phi| \leq 1$ and using the weak convergence $\mu_\nu \rightharpoonup \mu$, we obtain

$$|\mu^i|(J_h) \leq \liminf_{\nu\to\infty} |\mu^i_\nu|(J_h) + \mathcal{O}(1) \cdot \varepsilon\tilde{\mu}(J_h). \tag{10.33}$$

Indeed, this follows from Lemma 2.8, observing that $\tilde{\mu}(\{p_{h-1}\}) = \tilde{\mu}(\{p_h\}) = 0$. On the other hand, inserting $\phi \equiv 1$ in (10.32) we obtain

$$|\mu^i(J_h) - \mu^i_\nu(J_h)| = \mathcal{O}(1) \cdot \varepsilon(|\mu|(J_h) + |\mu_\nu|(J_h)) + \left|\int_{J_h} l_i(\bar{u}_h) \cdot (d\mu - d\mu_\nu)\right|. \tag{10.34}$$

Letting $\nu \to \infty$ and using (10.33) and weak convergence, this yields

$$\begin{aligned}\mu^{i\pm}(J_h) &= \tfrac{1}{2}[|\mu^i|(J_h) \pm \mu^i(J_h)] \\ &\leq \tfrac{1}{2} \cdot \lim_{\nu\to\infty} |\mu^i_\nu|(J_h) \pm \tfrac{1}{2} \cdot \lim_{\nu\to\infty} \mu^i_\nu(J_h) + \mathcal{O}(1) \cdot \varepsilon\tilde{\mu}(J_h) \\ &= \lim_{\nu\to\infty} \mu^{i\pm}_\nu(J_h) + \mathcal{O}(1) \cdot \varepsilon\tilde{\mu}(J_h).\end{aligned} \tag{10.35}$$

4. We now take care of the intervals $J_h$ containing a point $y_k$ of large oscillation. For each $k = 1, \ldots, N$, let $h = h(k) \in \{1, \ldots, r\}$ be the index such that $y_k \in J_h \doteq [p_{h-1},\ p_h]$. For every $\nu \geq 1$ consider the function

$$\hat{u}_\nu(x) \doteq \begin{cases} u_\nu(x) & \text{if } x \notin \cup_k J_{h(k)}, \\ u_\nu(p_{h(k)-1}) & \text{if } x \in [p_{h(k)-1},\ y_k[\,, \\ u_\nu(p_h) & \text{if } x \in [y_k,\ p_{h(k)}]. \end{cases}$$

Observe that all functions $u, \hat{u}_\nu$ are continuous at every point $p_0, \ldots, p_r$ and have jumps at $y_1, \ldots, y_N$. Call $\hat{\mu}^i_\nu$, $i = 1, \ldots, n$, the corresponding measures, defined as in (10.2)–(10.3) with $u$ replaced by $\hat{u}_\nu$. Clearly $\hat{\mu}^i_\nu = \mu^i_\nu$ outside the intervals $J_{h(k)}$. By

Lemma 10.2 the following holds:

$$\mathbf{Q}(\hat{u}_\nu) \le \mathbf{Q}(u_\nu), \qquad \mathbf{V}(\hat{u}_\nu) + C_0\,\mathbf{Q}(\hat{u}_\nu) \le \mathbf{V}(u_\nu) + C_0 \cdot \mathbf{Q}(u_\nu), \tag{10.36}$$

$$\hat{\mu}_\nu^{i\pm}(J) + C_0\,\mathbf{Q}(\hat{u}_\nu) \le \mu_\nu^{i\pm}(J) + C_0\,\mathbf{Q}(u_\nu). \tag{10.37}$$

Since $u_\nu \to u$ at every point $p_h$, by (10.30) for each $k$ one has

$$\begin{aligned} |\mu^i(\{y_k\}) - \hat{\mu}_\nu^i(\{y_k\})| &= |E_i(u(y_k-), u(y_k+)) - E_i(u_\nu(p_{h(k)-1}), u_\nu(p_{h(k)}))| \\ &= \mathcal{O}(1) \cdot \{|u(y_k-) - u(p_{h(k)-1})| + |u(y_k+) - u(p_{h(k)})| \\ &\quad + |u(p_{h(k)-1}) - u_\nu(p_{h(k)-1})| + |u(p_{h(k)}) - u_\nu(p_{h(k)})|\} \\ &= \mathcal{O}(1) \cdot \frac{\varepsilon}{N} \end{aligned} \tag{10.38}$$

for each $k = 1, \ldots, N$ and all $\nu$ sufficiently large. By construction we also have

$$|\hat{\mu}_\nu^i|(J_{h(k)} \setminus \{y_k\}) = 0, \qquad |\mu^i|(J_{h(k)} \setminus \{y_k\}) = \mathcal{O}(1) \cdot \frac{\varepsilon}{N}. \tag{10.39}$$

5. Recalling the definition (10.10) and using (10.16), (10.33), (10.35), (10.38)–(10.39) and (10.25), we obtain the estimate

$$\begin{aligned} \mathbf{Q}(u_\nu) &\ge \mathbf{Q}(\hat{u}_\nu) \\ &\ge \sum_{i<j} \sum_{h<\ell} (|\hat{\mu}_\nu^j| \times |\hat{\mu}_\nu^i|)(J_h \times J_\ell) + \sum_{i\in\mathcal{GN}} \sum_{h\neq\ell} (\hat{\mu}_\nu^{i-} \times |\hat{\mu}_{\nu,i}|)(J_h \times J_\ell) \\ &\ge \sum_{i<j} \sum_{h<\ell} (|\mu^j| \times |\mu^i|)(J_h \times J_\ell) + \sum_{i} \sum_{h\neq\ell} (\mu^{i-} \times |\mu^i|)(J_h \times J_\ell) - C\varepsilon \\ &\ge \mathbf{Q}(u) - C'\varepsilon, \end{aligned} \tag{10.40}$$

for suitable constants $C, C'$ and all $\nu$ sufficiently large. Since $\varepsilon > 0$ was arbitrary, this establishes (10.12).

Using the second inequality in (10.16), an entirely similar argument yields (10.13).

To prove (10.14), we repeat the above construction including all endpoints of the intervals $I_\ell \doteq ]a_\ell, b_\ell[$ in the list of points $y_k$ considered at (10.23). In other words, we can assume that

$$\{a_1, b_1, \ldots, a_m, b_m\} \subseteq \{y_1, \ldots, y_N\}.$$

Observe that there can be at most $2m$ intervals $J_h$ which intersect $J$ but are not entirely contained in $J$. For each interval of this type, our construction yields

$$\mu^{i\pm}(J_h \cap J) = \mathcal{O}(1) \cdot \tilde{\mu}(J_h \cap J) = \mathcal{O}(1) \cdot \varepsilon.$$

Using (10.17), for all $\nu$ sufficiently large we now obtain

$$\begin{aligned} \mu_\nu^{i\pm}(J) + C_0\,\mathbf{Q}(u_\nu) &\ge \hat{\mu}_\nu^{i\pm}(J) + C_0\,\mathbf{Q}(\hat{u}_\nu) \\ &\ge \sum_{J_h\subset J} \hat{\mu}_\nu^{i\pm}(J_h) + C_0\,\mathbf{Q}(\hat{u}_\nu) \end{aligned}$$

$$\geq \sum_{J_h \subset J} \mu^{i\pm}(J_h) + C_0\,\mathbf{Q}(\hat{u}) - C\,\varepsilon$$

$$\geq \mu^{i\pm}(J) + C_0\,\mathbf{Q}(u) - C'\,m\varepsilon$$

for suitable constants $C, C'$. Since $\varepsilon > 0$ can be taken arbitrarily small, this completes the proof. □

The functionals $V, Q$ were introduced in Chapter 7 in order to control the total variation of a front tracking approximate solution. Indeed, we showed that the maps $t \mapsto Q(u(t))$ and $t \mapsto V(u(t)) + C_0\,Q(u(t))$ are non-increasing in time. By Theorem 10.1, this property can be proved also for the functionals $\mathbf{Q}$ and $\mathbf{V} + C_0\,\mathbf{Q}$, in connection with an exact solution $u = u(t, x)$. More generally, consider two horizontal segments in the $t$–$x$ plane:

$$\gamma \doteq \{(\tau, x);\ a < x < b\}, \qquad \gamma' \doteq \{(\tau', x);\ a' < x < b'\}.$$

In analogy with Definition 7.3, we say that $\gamma$ *dominates* $\gamma'$, and write $\gamma \prec \gamma'$, if $\tau < \tau'$ and

$$a + \hat{\lambda}(\tau' - \tau) \leq a' < b' \leq b - \hat{\lambda}(\tau' - \tau).$$

As usual, $\hat{\lambda}$ denotes an upper bound for all characteristic speeds. One can then define the functionals $\mathbf{V}^\gamma, \mathbf{Q}^\gamma$ as the total amount of waves and the interaction potential of $u$ restricted to the segment $\gamma$. With this notation we have

**Theorem 10.2.** *Let $u = u(t, x)$ be a solution of (10.1) obtained as the limit of front tracking approximations. For any two horizontal segments $\gamma \prec \gamma'$ one has*

$$\mathbf{Q}^{\gamma'}(u) \leq \mathbf{Q}^{\gamma}(u), \qquad \mathbf{V}^{\gamma'}(u) + C_0\,\mathbf{Q}^{\gamma'}(u) \leq \mathbf{V}^{\gamma}(u) + C_0\,\mathbf{Q}^{\gamma}(u). \tag{10.41}$$

*Proof.* Consider the domain of determinacy

$$\Gamma \doteq \{(t, x);\ t \geq \tau,\ a + \hat{\lambda}(\tau' - \tau) < x < b - \hat{\lambda}(\tau' - \tau)\}.$$

Fix a sequence $\varepsilon_\nu \to 0$. For each $\nu$, we can construct on $\Gamma$ a front tracking $\varepsilon_\nu$-approximate solution $u_\nu$ such that, as $\nu \to \infty$,

$$u_\nu \to u, \qquad V^\gamma(u_\nu) \to \mathbf{V}^\gamma(u), \qquad Q^\gamma(u_\nu) \to \mathbf{Q}^\gamma(u).$$

By Lemma 7.3 the following holds

$$Q^{\gamma'}(u_\nu) \leq Q^{\gamma}(u_\nu), \qquad V^{\gamma'}(u_\nu) + C_0\,Q^{\gamma'}(u_\nu) \leq V^{\gamma}(u_\nu) + C_0\,Q^{\gamma}(u_\nu)$$

for every $\nu \geq 1$. By Theorem 10.1, recalling (7.101) we have

$$\begin{aligned}
\mathbf{Q}^{\gamma'}(u) &\leq \liminf_{\nu\to\infty} \mathbf{Q}^{\gamma'}(u_\nu) \\
&\leq \liminf_{\nu\to\infty} (Q^{\gamma'}(u_\nu) + \mathcal{O}(1)\cdot\varepsilon_\nu) \\
&\leq \lim_{\nu\to\infty} Q^{\gamma}(u_\nu) \\
&= \mathbf{Q}^{\gamma}(u).
\end{aligned}$$

The proof of the second inequality in (10.41) is entirely similar. □

## 10.2 Decay of positive waves

Consider a scalar conservation law with a strictly convex flux, say $f''(u) \geq \kappa > 0$ for all $u \in \mathbb{R}$. Differentiating (10.1) w.r.t. $x$, we obtain

$$u_{xt} + f'(u)u_{xx} = -f''(u)u_x^2 < -\kappa u_x^2\,. \tag{10.42}$$

If $t \mapsto y(t)$ is any characteristic line satisfying $\dot{y}(t) = f'(u(t, y))$, from (10.42) we obtain

$$\frac{d}{dt}u_x(t, y(t)) = -f''(u)u_x^2(t, y(t)) \leq -\kappa u_x^2(t, y(t)).$$

By a comparison with the ODE $\dot{z} = -\kappa z^2$, one thus obtains the pointwise estimate

$$u_x(t, x) < \frac{1}{\kappa t} \qquad \tau \in ]0, T],$$

which is valid for a sufficiently regular solution defined on the strip $[0, T] \times \mathbb{R}$. In other words, for a genuinely non-linear scalar conservation law, the density of positive waves decays like $1/\kappa t$.

One cannot expect this same result to remain valid for general $n \times n$ systems, even if we assume the genuine non-linearity of all characteristic fields. Indeed, assume that two shocks interact at some time $\tau$, generating a centred $i$-rarefaction wave (Fig. 10.2). In this case, inside the rarefaction wave the gradient component $u_x^i \doteq l_i(u) \cdot u_x$ decays like $(t - \tau)^{-1}$. The norm $\|u_x^i(t, \cdot)\|_{\mathbf{L}^\infty}$ therefore can be arbitrarily large, if $\tau$ is close to $t$.

In the case of systems, the estimate on wave decay must therefore take into account the generation of new positive waves due to interactions among waves of other families.

**Theorem 10.3 (Decay of positive waves).** *Let the system (10.1) be strictly hyperbolic and let the $i$-th characteristic field be genuinely non-linear. Then there exists a constant $C''$ such that, for every $0 \leq s < t$ and every solution $u$ with small total variation obtained as the limit of wave-front tracking approximations, the measure $\mu_t^{i+}$ of $i$-waves in $u(t, \cdot)$ satisfies*

$$\mu_t^{i+}(J) \leq C'' \cdot \frac{meas(J)}{t - s} + C'' \cdot [\mathbf{Q}(s) - \mathbf{Q}(t)] \tag{10.43}$$

*for every Borel set $J \subset \mathbb{R}$.*

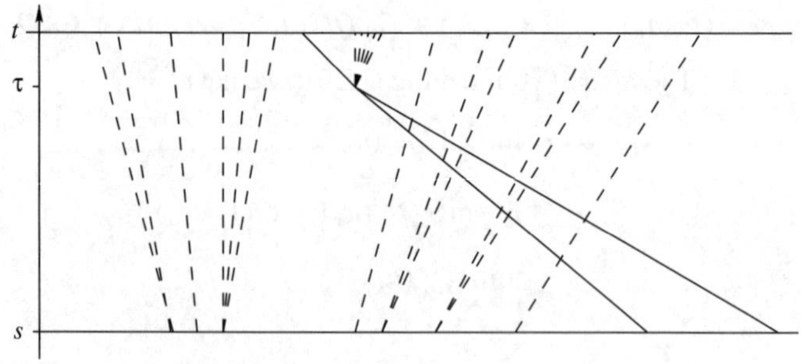


**Figure 10.2**

Intuitively, the bound (10.43) can be interpreted as follows (Fig. 10.2). The positive $i$-waves in $u(t, \cdot)$ can be split into two parts:

- The old waves, which were already present at time $s$. The density of these waves has decayed as $(t-s)^{-1}$; hence the total amount of such waves on a set $J$ is $O(1) \cdot \text{meas}(J)/(t-s)$.
- The new waves, which have been generated by interactions during the time interval $]s, t]$. The total amount of these waves is estimated by the decrease in the interaction potential, as $\mathcal{O}(1) \cdot [\mathbf{Q}(s) - \mathbf{Q}(t)]$.

The proof of Theorem 10.3 will be given in several steps.

1. Since front tracking approximations converge to semigroup trajectories depending Lipschitz-continuously on the initial data, it is not restrictive to assume $s = 0, t = T$, $u(0, x) = \bar{u}(x)$. Moreover (see Remark 7.1), we can consider a particular converging sequence $(u_\nu)_{\nu \geq 1}$ of $\varepsilon_\nu$-approximate solutions with the following additional properties:

(i) Each $i$-rarefaction front $x_\alpha$ travels with the characteristic speed of the state on the right:

$$\dot{x}_\alpha = \lambda_i(u(x_\alpha+)). \tag{10.44}$$

(ii) Each $i$-shock front $x_\alpha$ travels with a speed strictly contained between the right and the left characteristic speeds:

$$\lambda_i(u(x_\alpha+)) < \dot{x}_\alpha < \lambda_i(u(x_\alpha-)). \tag{10.45}$$

(iii) Calling $N_\nu$ the number of jumps in $\bar{u}_\nu = u_\nu(0, \cdot)$, as $\nu \to \infty$ one has

$$\varepsilon_\nu \to 0, \qquad \varepsilon_\nu N_\nu \to 0. \tag{10.46}$$

(iv) As $\nu \to \infty$, the interaction potentials satisfy

$$\mathbf{Q}(u_\nu(0, \cdot)) = Q(u_\nu(0+, \cdot)) \to \mathbf{Q}(\bar{u}). \tag{10.47}$$

2. Let $u = u(t, x)$ be a piecewise constant $\varepsilon$-approximate solution constructed by the front tracking algorithm. By a *(generalized) i-characteristic* we mean an absolutely continuous curve $x = x(t)$ such that

$$\dot{x}(t) \in [\lambda_i(u(t, x+)),\ \lambda_i(u(t, x-))] \tag{10.48}$$

for $t$ a.e. If $u$ satisfies the above properties (i)–(ii), then the $i$-characteristics are precisely the polygonal lines $x : [0, T] \mapsto \mathbb{R}$ for which the following holds. For a suitable partition $0 = t_0 < t_1 < \cdots < t_m = T$, on each subinterval $[t_{j-1}, t_j]$ either $\dot{x}(t) = \lambda_i(u(t, x))$, or else $x$ coincides with a wave-front of the $i$-th family. For a given terminal point $\bar{x}$ we shall consider the *minimal i-characteristic* through $\bar{x}$, defined as

$$y(t) = \min\{x(t);\ x \text{ is an } i\text{-characteristic}, x(T) = \bar{x}\}.$$

Observe that $y(\cdot)$ is itself an $i$-characteristic. By (10.45), it cannot coincide with an $i$-shock front of $u$ on any non-trivial time interval. All possible singularities of a minimal

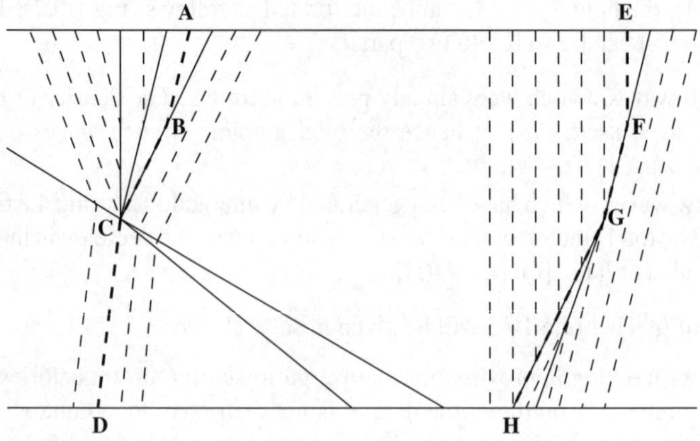


**Figure 10.3**

backward $i$-characteristic are illustrated in Fig. 10.3. Thin solid lines denote wave-fronts, while dashed lines indicate $i$-characteristics. The minimal backward $i$-characteristic starting at **A** encounters an $i$-rarefaction front at **B**. It follows this front up to the point **C** where a rarefaction fan is generated by the interaction of two $j$-shocks (with $j < i$). Along the segment **CD**, the function $u$ is again continuous, and the speed of the characteristic coincides with $\lambda_i(u)$. Another singularity occurs at point **G**, where an $i$-rarefaction impinges on the left into an $i$–shock. In the interaction the shock is completely cancelled and a weaker $i$-rarefaction emerges. In this case, the minimal backward $i$-characteristic starting at **E** follows the rarefaction front along both segments **FG** and **GH**.

3. Let $I \doteq [a, b[$ be any half-open interval. Call $t \mapsto a(t)$, $t \mapsto b(t)$ respectively the minimal backward $i$-characteristics passing through $a, b$ at time $T$, and define

$$I(t) \doteq [a(t), b(t)[\,, \qquad z(t) \doteq b(t) - a(t).$$

As usual, we call $k_\alpha$ the family of the front located at $x_\alpha$, with size $\sigma_\alpha$. We recall that, in the case of a genuinely non-linear family, the size of the jump is defined as

$$\sigma_\alpha \doteq \lambda_{k_\alpha}(u(x_\alpha+)) - \lambda_{k_\alpha}(u(x_\alpha-)). \tag{10.49}$$

By (10.49) and the Lipschitz continuity of the map $u \mapsto \lambda_i(u)$ we deduce that

$$\dot z(t) = M(t) + \mathcal{O}(1) \cdot (\varepsilon + K(t)) \tag{10.50}$$

for $t$ a.e. Here

$$M(t) \doteq \sum_{k_\alpha = i,\ x_\alpha \in I(t)} \sigma_\alpha$$

is the total amount of (signed) $i$-waves of $u(t,\cdot)$ contained in the interval $I(t) \doteq [a(t),\ b(t)[$, while

$$K(t) \doteq \sum_{k_\alpha \neq i,\ x_\alpha \in I(t)} |\sigma_\alpha|$$

denotes the total strength of wave-fronts in $u(t)$ of families $\neq i$ contained in $I(t)$.

4. To estimate the contribution of the term $K(t)$ in (10.50), we introduce the function

$$\Phi(t) \doteq \sum_{k_\alpha \neq i} \phi_{k_\alpha}(t, x_\alpha(t)) \cdot |\sigma_\alpha|$$

where

$$\phi_j(t,x) \doteq \begin{cases} 1 & \text{if } x < a(t) \\ \dfrac{b(t)-x}{z(t)} & \text{if } x \in [a(t), b(t)[ \\ 0 & \text{if } x \geq b(t) \end{cases} \quad \text{or} \quad \phi_j(t,x) \doteq \begin{cases} 0 & \text{if } x < a(t) \\ \dfrac{x-a(t)}{z(t)} & \text{if } x \in [a(t), b(t)[ \\ 1 & \text{if } x \geq b(t) \end{cases}$$

in the cases $j < i$ or $j > i$, respectively. Observe that $\Phi$ is piecewise Lipschitz continuous with a finite number of discontinuities occurring at interaction times $\tau$, where

$$\Phi(\tau+) - \Phi(\tau-) = \mathcal{O}(1) \cdot [Q(u(\tau-)) - Q(u(\tau+))]. \tag{10.51}$$

Outside the interaction times, the function $\Phi$ is non-decreasing. Indeed, due to the strict hyperbolicity of the system, we have

$$\begin{aligned} \dot{\Phi}(t) &= \sum_{k_\alpha \neq i} |\sigma_\alpha| \cdot \frac{d}{dt} \phi_{k_\alpha}(t, x_\alpha(t)) \\ &= \sum_{k_\alpha < i, x_\alpha \in I(t)} |\sigma_\alpha| \cdot \left( \frac{\dot{b} - \dot{x}_\alpha}{z} - \frac{(b - x_\alpha)\dot{z}}{z^2} \right) \\ &\quad + \sum_{k_\alpha > i,\ x_\alpha \in I(t)} |\sigma_\alpha| \cdot \left( \frac{\dot{x}_\alpha - \dot{a}}{z} - \frac{(x_\alpha - a)\dot{z}}{z^2} \right) \\ &\geq \sum_{k_\alpha \neq i} |\sigma_\alpha| \cdot \frac{c_0}{z(t)} \end{aligned}$$

for some constant $c_0 > 0$ related to the gap between distinct characteristic speeds. We are assuming here that

$$|\lambda_i(u) - \lambda_i(u')| \leq c_0, \qquad |\lambda_i(u) - \lambda_j(u')| \geq 2c_0,$$

for every couple of states $u, u'$ and every $j \neq i$. The estimate above yields the bound

$$K(t) \leq \frac{1}{c_0} \cdot \dot{\Phi}(t) \cdot z(t), \tag{10.52}$$

which is valid at all but finitely many times $t$.

5. We now provide an estimate for the term $M(t)$. Observe that $M$ can change only at times where a wave-front interaction takes place within the interval $[a(t), b(t)]$. In this

case, one has

$$\Delta M(\tau) \doteq M(\tau+) - M(\tau-) = \mathcal{O}(1) \cdot \Delta Q(\tau).$$

This yields an estimate of the form

$$M(T) - M(t) = \mathcal{O}(1) \cdot \sum_{\tau \in \mathcal{T}} \Delta Q(\tau) \tag{10.53}$$

where the summation extends over all times $\tau \in ]0, T]$ where an interaction occurs inside $[a(\tau), b(\tau)]$.

6. Inserting the estimates (10.52)–(10.53) in (10.50) we obtain

$$\dot{z}(t) + C\,\dot{\Phi}(t)\,z(t) \geq M(T) - C \cdot \left(\varepsilon + \sum_{\tau \in \mathcal{T}} |\Delta Q(\tau)|\right), \tag{10.54}$$

for some constant $C$ and $t$ a.e. We now observe that the function $\Phi$ is uniformly bounded. It can decrease only at times of interaction, where (10.51) holds. Therefore, its total variation is uniformly bounded. In particular, for some constant $K_0$ we have the estimate

$$\int_0^T \dot{\Phi}(t)\,dt \leq K_0. \tag{10.55}$$

By (10.54), two cases can occur; either

$$M(T) \leq 2C \cdot \left(\varepsilon + \sum_{\tau \in \mathcal{T}} |\Delta Q(\tau)|\right),$$

or else

$$\dot{z}(t) + C\,\dot{\Phi}(t)\,z(t) \geq \frac{M(T)}{2}.$$

In this second case, since $\dot{\Phi}(t) \geq 0$ a.e., one has

$$\frac{d}{dt}\left(\exp\left\{\int_0^t C\dot{\Phi}(s)\,ds\right\} z(t)\right) \geq \exp\left\{\int_0^t C\dot{\Phi}(s)\,ds\right\} \cdot \frac{M(T)}{2} \geq \frac{M(T)}{2}.$$

Since $z(0) \geq 0$, by (10.55) this implies

$$b - a = z(T) \geq e^{-CK_0} \cdot \frac{M(T)}{2} \cdot T.$$

In both cases the following holds:

$$M(T) \leq 2e^{CK_0} \cdot \frac{b-a}{T} + 2C\varepsilon + 2C \cdot \sum_{\tau \in \mathcal{T}} |\Delta Q(\tau)|. \tag{10.56}$$

7. Repeating the above estimates for any finite number of disjoint half-open intervals $I_\ell \doteq [a_\ell,\ b_\ell[$ we obtain

$$\sum_{\ell=1}^{m} M_\ell(T) \leq C' \cdot \left(\sum_{\ell=1}^{m} \frac{b_\ell - a_\ell}{T} + m\varepsilon + [Q(u(0)) - Q(u(T))]\right), \tag{10.57}$$

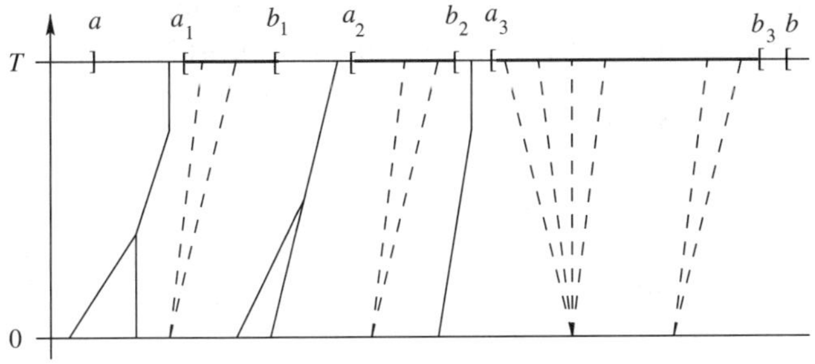


**Figure 10.4**

for some constant $C'$ independent of $m$ and of the particular $\varepsilon$-approximate solution. We use here the notation

$$M_\ell(T) \doteq \sum_{k_\alpha=i,\ x_\alpha\in[a_\ell,b_\ell[} \sigma_\alpha = \mu_T^i([a_\ell, b_\ell[)$$

to denote the sum of (signed) strengths of all $i$-waves in $u(T,\cdot)$ contained in $[a_\ell,\ b_\ell[$.

8. Now consider any open interval $]a, b[$. Let $N$ be the number of $i$-shocks of the first generation in the front tracking $\varepsilon$-approximate solution $u$, as defined in Chapter 7. We can then construct half-open intervals $I_\ell \doteq [a_\ell,\ b_\ell[,\ \ell = 1, \dots, m \le N+1$, such that the following holds (Fig. 10.4):

- Every $i$-rarefaction front in $u(T,\cdot)$ contained in $]a, b[$ falls inside one of the intervals $I_\ell$.
- No $i$-shock front of the first generation falls inside any one of the intervals $I_\ell$.

Calling $\mu_T^{i+}$ the measure of positive $i$-waves in $u(T,\cdot)$, the above two properties imply

$$\mu_T^{i+}(]a, b[) = \sum_\ell M_\ell(T) + \mathcal{O}(1)\cdot[\mathbf{Q}(u(0)) - \mathbf{Q}(u(T))] + \mathcal{O}(1)\cdot\varepsilon. \tag{10.58}$$

Indeed, the only negative $i$-waves contained in the union of the intervals $I_\ell$ must be of generation order $\ge 2$, originating from interactions during the time interval $]0, T]$. The total strength of these negative $i$-waves is bounded by the decrease in the interaction functional $Q$. The last term on the right hand side of (10.58) takes care of the difference between $Q(u)$ and $\mathbf{Q}(u)$, due to the presence of non-physical fronts. For an $\varepsilon$-approximate front tracking solution, this difference was estimated at (7.100).

Together, (10.57) and (10.58) yield

$$\mu_T^{i+}(]a, b[) \le C''\cdot\left(\frac{b-a}{T} + (N+1)\varepsilon + [\mathbf{Q}(u(0)) - \mathbf{Q}(u(T))]\right), \tag{10.59}$$

for some constant $C''$ independent of $\varepsilon$.

9. We now consider a sequence of $\varepsilon_\nu$-approximate solutions $u_\nu$ satisfying the properties (i)–(iv) stated in step 1. In (10.59) it is clearly not restrictive to assume that $C'' \ge C_0$,

the constant in Theorem 10.1. Hence (10.14) still holds with $C_0$ replaced by $C''$. Call $\mu_T^{i+}, \mu_{\nu,T}^{i+}$ the measures of positive $i$-waves respectively in $u(T,\cdot)$ and in $u_\nu(T,\cdot)$. Using (10.14), then (10.59), and finally (10.46)–(10.47) and the lower semicontinuity of the interaction functional $\mathbf{Q}$, we obtain

$$\begin{aligned}\mu_T^{i+}(]a,b[) &\leq \liminf_{\nu\to\infty}\left(\mu_{\nu,T}^{i+}(]a,b[) + C''\mathbf{Q}(u_\nu(T))\right) - C''\mathbf{Q}(u(T))\\ &\leq C''\cdot\liminf_{\nu\to\infty}\left(\frac{b-a}{T} + (N_\nu+1)\varepsilon_\nu + \mathbf{Q}(u_\nu(0))\right) - C''\mathbf{Q}(u(T))\\ &\leq C''\cdot\frac{b-a}{T} + C''[\mathbf{Q}(u(0)) - \mathbf{Q}(u(T))].\end{aligned} \tag{10.60}$$

When $s=0, t=T$, this proves (10.43) in the case where $J$ is an open interval.

The same arguments can be used in the case where $J$ is a finite collection of open intervals. Since $\mu^i$ is a bounded Radon measure, the estimate (10.43) is thus valid for every Borel set $J$. □

Given a front tracking approximate solution $u = u(t,x)$, within the above proof we introduced the concept of a (generalized) $i$-characteristic. By definition, this is an absolutely continuous function which satisfies a.e. the differential inclusion (10.48). Observe that, if the $i$-th characteristic field is genuinely non-linear, the $i$-characteristic curve passing through a given point $P=(\tau,\xi)$ may not be unique. Indeed if $P$ lies along an $i$-shock (Fig. 10.5), one can construct infinitely many backward $i$-characteristics through $P$. On the other hand, if $P$ lies along an $i$-rarefaction front (Fig. 10.6), there exist infinitely many forward $i$-characteristics starting at $P$.

The situation is completely different if the $i$-th field is linearly degenerate. In this case, the speed $\lambda_i$ is constant across $i$-contact discontinuities. It may be discontinuous only across $j$-fronts (with $j \neq i$) and across non-physical fronts. But all of these fronts are crossed transversally by $i$-characteristics. Therefore, there exists exactly one $i$-characteristic passing through any given point $P=(\tau,\xi)$. This we shall denote by $t\mapsto x_i(t;\,\tau,\xi)$. Clearly, each $i$-characteristic is a polygonal line, satisfying the equation

$$\dot{y} = \lambda_i(u(t,y)) \tag{10.61}$$

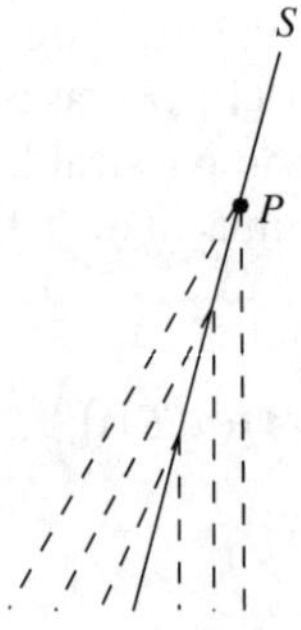


**Figure 10.5**

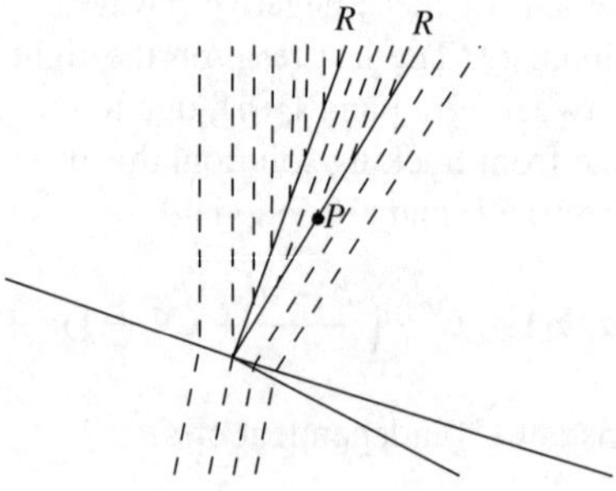


**Figure 10.6**

for all except finitely many times $t > 0$. The next lemma shows that the Cauchy problem for (10.61) not only has a unique solution, but is also uniformly well posed. This result will be used in the last section to obtain a global description of contact discontinuities.

**Lemma 10.3.** *Let the $i$-th characteristic field be linearly degenerate, and let $u = u(t, x)$ be a front tracking approximate solution of (10.1). With the above notation, one then has the estimate*

$$|x_i(t;\ \tau, \xi) - x_i(t;\ \tau', \xi')| \leq L \cdot (|\tau - \tau'| + |\xi - \xi'|), \tag{10.62}$$

*for a Lipschitz constant $L$ independent of $u$, as long as the total variation remains uniformly small.*

*Proof.* Since the $i$-th speed $\lambda_i$ remains uniformly bounded, the Lipschitz continuity w.r.t. time is clear. It thus suffices to consider the case where $\tau' = \tau$.

Let $y(\cdot)$ be a solution of (10.61) with initial data $y(\tau) = \xi$. Call $y^\theta(\cdot)$ the perturbed solution (Fig. 10.7) such that $y^\theta(\tau) = \xi + \theta$. We now estimate the quantity

$$w(t) \doteq \lim_{\theta \to 0+} \frac{y^\theta(t) - y(t)}{\theta}.$$

Observe that, for $\theta > 0$ small, all solutions $y^\theta$ will cross the same wave-fronts of $u$ in the same order. Hence, except for finitely many interaction times, the map $t \mapsto w(t)$ is well defined and piecewise constant, with jumps only at times where $y(\cdot)$ crosses a wave-front in $u$ of a family $\neq i$.

Let $t_\alpha$ be one of these crossing times, say when $y$ crosses a front of strength $\sigma_\alpha$ and speed $\dot{x}_\alpha$. An elementary computation yields

$$w(t_\alpha+) = \frac{\dot{y}(t_\alpha+) - \dot{x}_\alpha}{\dot{y}(t_\alpha-) - \dot{x}_\alpha} \cdot w(t_\alpha-).$$

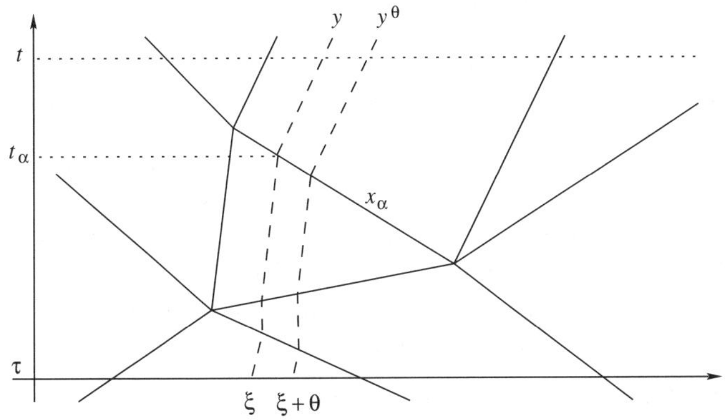


**Figure 10.7**

The Lipschitz continuity of the map $u \mapsto \lambda_i(u)$ and the strict hyperbolicity of the system imply

$$|\dot{y}(t_\alpha+) - \dot{y}(t_\alpha-)| = \mathcal{O}(1) \cdot |\sigma_\alpha|, \qquad |\dot{y}(t_\alpha-) - \dot{x}_\alpha| \geq c_0 > 0;$$

therefore

$$w(t_\alpha+) = (1 + \mathcal{O}(1) \cdot |\sigma_\alpha|) \cdot w(t_\alpha-).$$

Summing over all wave-fronts in $u$ of families $\neq i$ crossed by $y^\theta$ during a time interval $[\tau, t]$, we obtain

$$w(t) = \exp\left\{\mathcal{O}(1) \cdot \sum_{\alpha \in \mathrm{Cross}[\tau;t]} |\sigma_\alpha|\right\} \cdot w(\tau). \tag{10.63}$$

Observing that the total amount of wave-fronts of families $\neq i$ crossed by any $i$-characteristic can be estimated in terms of the total strength of waves in $u(0, \cdot)$, we deduce from (10.63) the uniform bound

$$\lim_{\theta \to 0+} \theta^{-1}(y^\theta(t) - y(t)) \leq L \tag{10.64}$$

for some constant $L$ independent of $t, \tau, \xi, u$. Using (10.64) and assuming $\xi < \xi'$ we now compute

$$\begin{aligned} x_i(t;\ \tau, \xi') - x_i(t;\ \tau, \xi) &= \int_\xi^{\xi'} \left[ \lim_{\theta \to 0+} \frac{x_i(t;\ \tau,\ x + \theta) - x_i(t;\ \tau, x)}{\theta} \right] dx \\ &\leq \int_\xi^{\xi'} L\, dx, \end{aligned}$$

proving the uniform Lipschitz continuity w.r.t. the initial data $\xi$. □

## 10.3 Global structure of solutions

The goal of this section is to prove a regularity result for entropy solutions $u = u(t, x)$ of (10.1). According to Theorem 2.6, any such solution is a *BV* function of the two variables $t, x$. Hence, it shares the regularity properties of general *BV* functions stated in Theorem 2.7. In particular, $u$ either is approximately continuous or has an approximate jump at each point $(t, x)$ with the exception of a set $\mathcal{N}$ whose one-dimensional Hausdorff measure is zero. We will show that much stronger regularity properties actually hold. Indeed, in the case of an entropy weak solution, the exceptional set $\mathcal{N}$ contains only countably many points. Moreover, $u$ is continuous (not just approximately continuous) outside $\mathcal{N}$ and outside countably many Lipschitz-continuous shock curves.

**Theorem 10.4 (Global structure of solutions).** *Let $u$ be a solution of the Cauchy problem (7.1)–(7.2) obtained as the limit of front tracking approximations. Then there exists a countable set $\Theta \doteq \{(t_\ell, x_\ell);\ \ell \geq 1\}$ of interaction points and a countable family of*

*Lipschitz-continuous curves (shocks or contact discontinuities)* $\Gamma \doteq \{x = y_m(t);\ t \in ]a_m, b_m[\ ,\ m \geq 1\}$ *such that the following holds. For each m and each* $t \in ]a_m, b_m[$ *with* $(t, y_m(t)) \notin \Theta$, *there exist the derivative* $\dot{y}_m(t)$ *and the right and left limits*

$$u^- \doteq \lim_{\substack{(s,y)\to(t,y_m(t))\\ y<y_m(s)}} u(s, y), \qquad u^+ \doteq \lim_{\substack{(s,y)\to(t,y_m(t))\\ y>y_m(s)}} u(s, y). \tag{10.65}$$

*These limits satisfy the Rankine–Hugoniot equations and the Lax entropy conditions*

$$\dot{y}_m[u^+ - u^-] = [f(u^+) - f(u^-)],$$

$$\lambda_i(u^+) \leq \dot{y}_m \leq \lambda_i(u^-) \quad \textit{for some } i \in \{1, \ldots, n\}. \tag{10.66}$$

*Moreover, u is continuous outside the set* $\Theta \cup \Gamma$.

The proof will require several steps.

1. Fix a sequence $\varepsilon_\nu \to 0$, and for each $\nu$ let $u_\nu = u_\nu(t, x)$ be an $\varepsilon_\nu$-approximate solution of the Cauchy problem (7.1)–(7.2) constructed by the front tracking algorithm. Call $\mu^{i\pm}_{\nu,0}$ the measures of positive and negative $i$-waves in the piecewise constant initial data $u_\nu(0, \cdot)$. As $\nu \to \infty$ we can assume the weak convergence

$$\mu^{i\pm}_{\nu,0} \rightharpoonup \bar{\mu}^{i\pm}, \tag{10.67}$$

where $\bar{\mu}^{i\pm}$ are the measures of $i$-waves in $\bar{u}$. To each $u_\nu$ we also associate the measure of interaction $\mu^I_\nu$ and the measure of interaction and cancellation $\mu^{IC}_\nu$, defined on the half plane $[0, \infty[\ \times \mathbb{R}$ as in (7.95)–(7.96). By possibly taking a subsequence, for some positive measures $\mu^I$, $\mu^{IC}$ we can assume the weak convergence

$$\mu^I_\nu \rightharpoonup \mu^I, \qquad \mu^{IC}_\nu \rightharpoonup \mu^{IC}. \tag{10.68}$$

We now define the countable sets $\Theta_0$, $\Theta_1$ respectively as the set containing all points of jump in the initial data $\bar{u}$, and the set of all atoms of $\mu^{IC}$:

$$\Theta_0 \doteq \{(0, x);\ \bar{u}(x+) \neq \bar{u}(x-)\}, \qquad \Theta_1 \doteq \{P \doteq (t, x);\ \bar{\mu}^{IC}(\{P\}) > 0\}. \tag{10.69}$$

2. Fix an index $i \in \{1, \ldots, n\}$ corresponding to a genuinely non-linear family. In this step we construct the countable set of $i$-shock curves in the solution $u$, by taking suitable limits of shock fronts in $u_\nu$.

Let $\varepsilon > 0$ be given. By an $\varepsilon$-shock front of the $i$-th family in $u_\nu$ we mean a polygonal line in the $t$–$x$ plane, with nodes $(t_0, x_0), (t_1, x_1), \ldots, (t_N, x_N)$, having the properties:

- (i) The points $(t_k, x_k)$ are interaction points, with $0 \leq t_0 < t_1 < \cdots < t_N$.
- (ii) Along each segment joining $(t_{k-1}, x_{k-1})$ with $(t_k, x_k)$, the function $u_\nu$ has an $i$-shock with strength $|\sigma_k| \geq \varepsilon/2$. Moreover, $|\sigma_{k^*}| \geq \varepsilon$ for at least one index $k^* \in \{1, \ldots, N\}$.
- (iii) For $k < N$, if two incoming $i$-shocks both of strength $\geq \varepsilon/2$ interact at the node $(t_k, x_k)$, then the shock coming from $(t_{k-1}, x_{k-1})$ has the larger speed, i.e. is the one coming from the left.

An $\varepsilon$-shock front which is maximal w.r.t. set theoretical inclusion will be called a *maximal* $\varepsilon$*-shock front*.

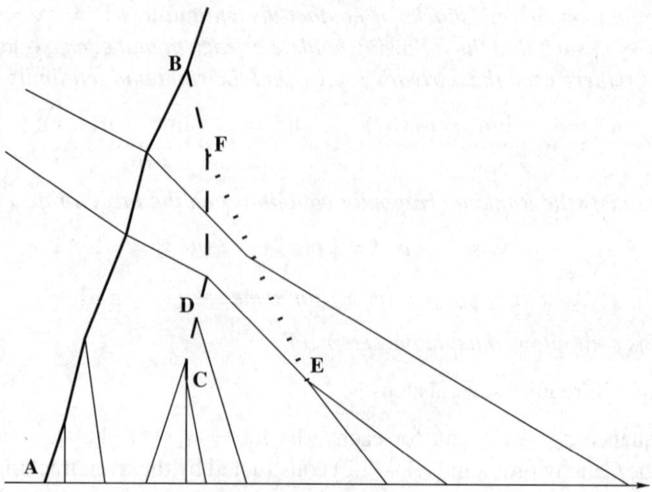


**Figure 10.8**

By (iii), two maximal $\varepsilon$-shock fronts either are disjoint or coincide. Since the total variation of $u_\nu(0, \cdot)$ is uniformly bounded, the number of maximal $\varepsilon$-shock fronts which start at time $t = 0$ is clearly $\mathcal{O}(1) \cdot \varepsilon^{-1}$. Moreover, in connection with an $\varepsilon$-shock front starting at a time $t_0 > 0$, an amount of interaction $\geq c\varepsilon^2$ must take place, for some constant $c > 0$. This is due to the fact that the strength of the front must increase from an initial value $< \varepsilon/2$ to some value $\geq \varepsilon$. Since the total amount of interaction in the solution $u_\nu$ is uniformly bounded, we obtain the estimate

$$M_\nu \doteq \#\{\text{maximal } \varepsilon\text{-shock fronts in } u_\nu\} = \mathcal{O}(1) \cdot \varepsilon^{-2}$$

which is uniformly valid as $\nu \to \infty$.

Figure 10.8 illustrates three maximal $\varepsilon$-shock fronts of the $i$-th family. The front starting at **A** has the required strength $|\sigma| \geq \varepsilon$ already at time $t = 0$. A second front (from **C** to **B**) is formed when a number of small shock fronts merge together. After **C** this front has strength $\geq \varepsilon/2$, while at **D** its strength becomes $\geq \varepsilon$. This second front terminates at **B**, when it hits the previous shock front from the right. A third front is formed at **E**, where two large $j$-shocks meet, generating a new $i$-shock of strength $\geq \varepsilon$. This maximal $\varepsilon$-shock front terminates at **F**, hitting from the right another shock front.

Call $y_{\nu,m} : [t^-_{m,\nu}, t^+_{m,\nu}] \mapsto \mathbb{R}$, $m = 1, \ldots, M_\nu$, the set of maximal $\varepsilon$-shock fronts in $u_\nu$. By possibly extracting a subsequence, we can assume that the integer $M_\nu \doteq \overline{M}$ is constant, independent of $\nu$. Moreover, we can assume that

$$y_{\nu,m}(\cdot) \to y_m(\cdot) \quad t^\pm_{m,\nu} \to t^\pm_m, \qquad m \in \{1, \ldots, \overline{M}\}, \tag{10.70}$$

for some Lipschitz-continuous paths $y_m : ]t^-_m, t^+_m[ \mapsto \mathbb{R}$.

We now repeat the above construction taking $\varepsilon = 1, 1/2, 1/3, \ldots$ . The union of all the paths thus obtained is a countable family of $i$-shock curves $y_m : ]t^-_m, t^+_m[$, $m \geq 1$. This completes our construction in the case where the $i$-th family is genuinely non-linear.

3. Now consider an index $i \in \{1, \dots, n\}$ corresponding to a linearly degenerate family. For each front tracking approximation $u_\nu$, call $t \mapsto x_{i,\nu}(t; \tau, \xi)$ the $i$-characteristic curve through the point $(\tau, \xi)$. By definition, this is the solution of the Cauchy problem

$$\dot{x} = \lambda_i(u_\nu(t, x)), \qquad x(\tau) = \xi. \tag{10.71}$$

According to Lemma 10.3, this problem is uniformly well posed. By possibly taking a subsequence, we can thus assume that

$$x_{i,\nu}(t; \tau, \xi) \to x_i(t; \tau, \xi) \quad \text{as } \nu \to \infty \tag{10.72}$$

uniformly as $t, \tau, \xi$ range within bounded sets, for some functions $x_i$ satisfying

$$|x_i(t; \tau, \xi) - x_i(t; \tau', \xi')| \le L \cdot (|\tau - \tau'| + |\xi - \xi'|). \tag{10.73}$$

Recalling (10.67)–(10.68), we now consider a positive measure $\mu^{i\star}$ on the real line, defined by setting

$$\begin{aligned} \mu^{i\star}(]a, b[) &\doteq \bar{\mu}^{i+}(]a, b[) + \bar{\mu}^{i-}(]a, b[) \\ &\quad + \bar{\mu}^{IC}\{(t, x)\,;\ t > 0,\ x_i(t; 0, a) < x < x_i(t; 0, b)\} \end{aligned} \tag{10.74}$$

for every open interval $]a, b[$. In other words, $\mu^{i\star}(]a, b[)$ measures the total amount of positive and negative $i$-waves in the initial data $u(0, \cdot)$ contained in the interval $]a, b[$, plus the total amount of interaction and cancellation occurring in the strip bounded by the two $i$-characteristics starting from $a, b$. Consider the set of atoms of $\mu^{i\star}$:

$$\mathcal{A}^{i\star} \doteq \{x \in ]a, b[;\ \mu^{i\star}(\{x\}) > 0\}.$$

Clearly, this set is at most countable, say $\mathcal{A}^{i\star} \doteq \{\xi_m;\ m \ge 1\}$. We now consider the curves

$$t \mapsto y_m(t) \doteq x_i(t; 0, \xi_m), \qquad \xi_m \in \mathcal{A}^{i\star}, \quad t > 0. \tag{10.75}$$

We will show that every $i$-contact discontinuity of $u$ is contained in one of the countably many curves (10.75).

Observe that the measure $\mu^{i\star}$ has the following property. Given any closed interval $[a, b]$ and $T > 0$, consider the strip (Fig. 10.9) bounded by the $i$-characteristics starting from $a, b$ respectively:

$$\Gamma \doteq \Big\{(t, x);\ t \in [0, T], x_i(t; 0, a) \le x \le x_i(t; 0, b)\Big\}.$$

For any space-like curve $\gamma$ contained in $\Gamma$, call $V_i^\gamma(u_\nu)$ the total strength of $i$-fronts in $u_\nu$ which cross $\gamma$. For a suitable constant $\kappa_i$ (independent of $a, b, T$) one then has the estimate

$$\limsup_{\nu\to\infty} \left\{ \sup_{\gamma\in\Gamma} V_i^\gamma(u_\nu) \right\} \le \kappa_i \cdot \mu^{i\star}([a, b]). \tag{10.76}$$

Indeed, call $A, B$ the endpoints of $\gamma$. Construct the two curves $t \mapsto a_\nu(t)$, $t \mapsto b_\nu(t)$ passing through the points $A, B$ respectively (Fig. 10.10), such that

$$\dot{a}_\nu(t) = \lambda_i(u_\nu(t, a_\nu)) + \varepsilon_\nu, \qquad \dot{b}_\nu(t) = \lambda_i(u_\nu(t, b_\nu)) - \varepsilon_\nu.$$

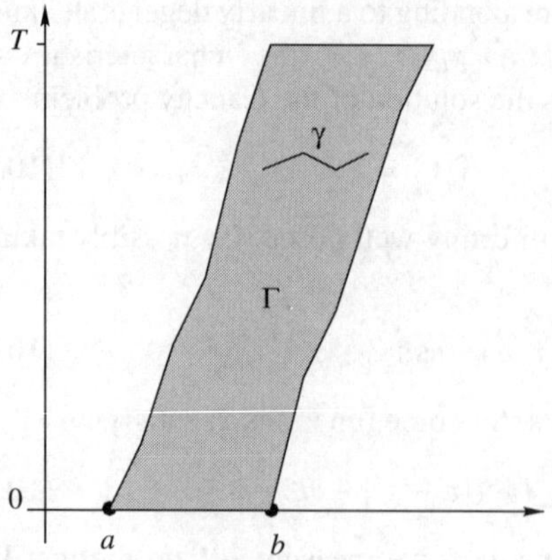


Figure 10.9

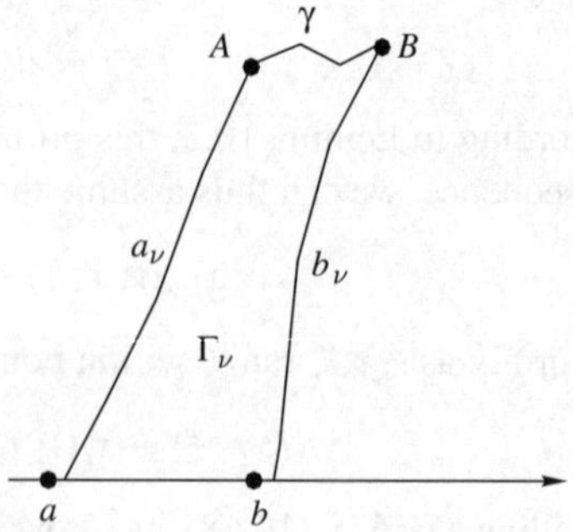


Figure 10.10

Consider the region $\Gamma_\nu$ bounded by the axis $\{t = 0\}$, by the curve $\gamma$ and by the lines $a_\nu, b_\nu$. Since no $i$-front of $u_\nu$ can enter $\Gamma_\nu$ through the sides $a_\nu, b_\nu$, by (7.98) the total strength of $i$-fronts crossing $\gamma$ is bounded in terms of the strength of $i$-waves in $u_\nu(0+,\cdot)$ contained in $[a_\nu(0),\ b_\nu(0)]$ plus the amount of interaction and cancellation occurring inside $\Gamma_\nu$. Letting $\nu \to \infty$, (10.72) and the definition of $\Gamma$ yield

$$\lim_{\nu\to\infty} a_\nu(0) \geq a, \qquad \lim_{\nu\to\infty} b_\nu(0) \leq b.$$

Hence from (10.67)–(10.68) and the definition (10.74) we deduce (10.76).

4. In the two previous steps, for each characteristic family $i \in \{1, \ldots, n\}$ we constructed a countable number of Lipschitz curves $y_m$. We now define $\Theta_2$ as the set of points where two curves $y_m, y_n$ of different characteristic families cross each other. Recalling (10.69), we consider the countable set

$$\Theta \doteq \Theta_0 \cup \Theta_1 \cup \Theta_2.$$

We claim that, with the above definitions, all properties stated in Theorem 10.4 hold.

5. Consider a point $P = (\tau, \xi) = (\tau, y_m(\tau))$ along an $i$-shock curve $y_m$ of a genuinely non-linear family. Since the function $u(\tau,\cdot)$ has bounded variation, there exist the limits

$$u^- \doteq \lim_{x\to y_m(t)-} u(\tau, x), \qquad u^+ \doteq \lim_{x\to y_m(t)+} u(\tau, x). \tag{10.77}$$

Moreover, applying the tame oscillation condition (9.39) on the triangles (Fig. 10.11)

$$\Delta^- \doteq \{(t, x);\ t \geq \tau,\ |x - (\xi - \delta)| < \delta - \hat{\lambda}(t - \tau)\},$$

$$\Delta^+ \doteq \{(t, x);\ t \geq \tau,\ |x - (\xi + \delta)| < \delta - \hat{\lambda}(t - \tau)\},$$

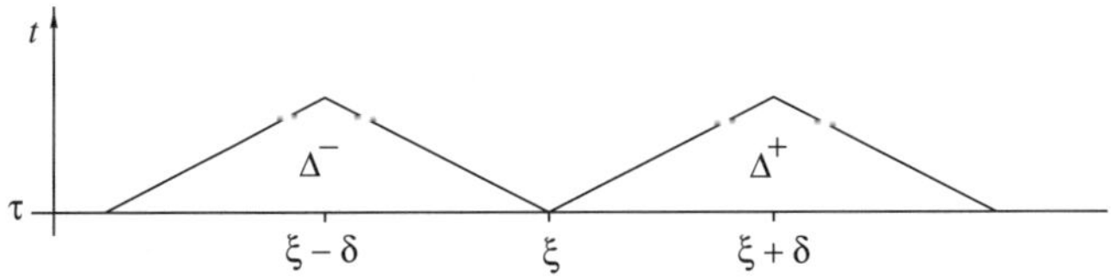


**Figure 10.11**

as $\delta \to 0+$ one obtains

$$\lim_{\substack{(t,x)\to(\tau,\xi)\\ \tau\le t<\tau+(\xi-x)/\hat{\lambda}}} u(t,x) = u^-, \qquad \lim_{\substack{(t,x)\to(\tau,\xi)\\ \tau\le t<\tau+(x-\xi)/\hat{\lambda}}} u(t,x) = u^+. \tag{10.78}$$

We claim that the limits (10.65) also exist, and hence coincide with those in (10.77)–(10.78).

By assumption, there exists a sequence of $i$-shock curves $y_{m,\nu}$ converging to $y_m$. Each $u_\nu$ has an $i$-shock of strength $\ge \delta^*$ along $y_{m,\nu}$, for some fixed $\delta^* > 0$. We claim that

$$\lim_{r\to 0+} \limsup_{\nu\to\infty} \left( \sup_{\substack{x<y_{m,\nu}(t)\\ (t,x)\in B(P,r)}} |u_\nu(t,x) - u^-| \right) = 0, \tag{10.79}$$

$$\lim_{r\to 0+} \limsup_{\nu\to\infty} \left( \sup_{\substack{x>y_{m,\nu}(t)\\ (t,x)\in B(P,r)}} |u_\nu(t,x) - u^+| \right) = 0. \tag{10.80}$$

In other words, within the ball centred at $P$ with small radius $r$, all functions $u_\nu$ take values close to $u^-$ in the region to the left of the shock front $y_{m,\nu}$, and close to $u^+$ in the region to the right of $y_{m,\nu}$.

If (10.79) did not hold, by possibly taking a subsequence we could find $\epsilon > 0$ and points $Q_\nu$ on the left of $y_{m,\nu}$ (Fig. 10.12) such that

$$Q_\nu \to P, \quad |u_\nu(Q_\nu) - u^-| \ge \epsilon \qquad \text{for all } \nu. \tag{10.81}$$

In addition, by the first limit in (10.78), since $u_\nu(t,x) \to u(t,x)$ pointwise a.e., we could also find points $P_\nu$ on the left of $y_{m,\nu}$ such that

$$P_\nu \to P, \qquad u_\nu(P_\nu) \to u^-, \tag{10.82}$$

and such that every segment $P_\nu Q_\nu$ is space-like.

By (10.81)–(10.82), for every $\nu$ large the segment $P_\nu Q_\nu$ is crossed by a uniformly positive amount of waves of $u_\nu$. The following argument shows that these waves must interact among themselves or with the $i$-shock $y_{m,\nu}$ within an arbitrarily small neighbourhood of $P$. Hence the local amount of interaction and cancellation is uniformly positive, contradicting the assumption $\mu^{IC}(\{P\}) = 0$. By possibly taking a subsequence, we are led to consider three cases:

CASE 1: In each solution $u_\nu$, the segment $P_\nu Q_\nu$ is crossed by an amount $\ge \varepsilon$ of $j$-waves in $u_\nu$, for some fixed index $j > i$ and $\varepsilon > 0$.

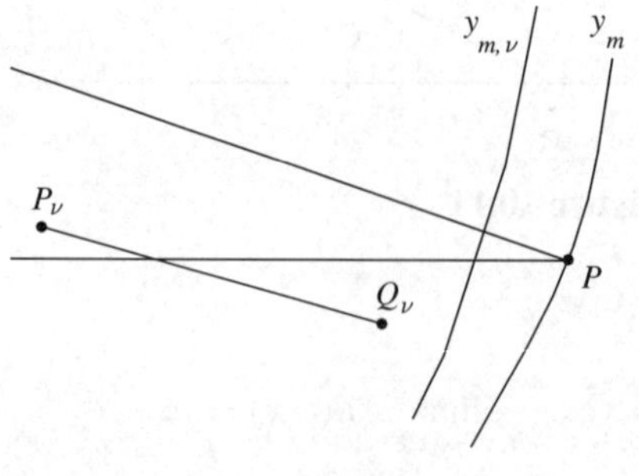


**Figure 10.12**

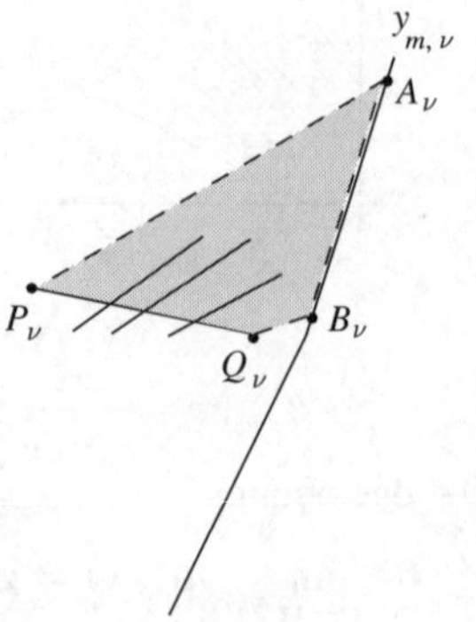


**Figure 10.13**

In this case (Fig. 10.13), for each $\nu$ we consider the open region $\Gamma_\nu$ bounded by the $i$-shock front $y_{m,\nu}$, by the segment $P_\nu Q_\nu$, by the minimal forward $j$-characteristic through the point $P_\nu$ and by the maximal forward $j$-characteristic through the point $Q_\nu$. By construction, no $j$-wave can leave $\Gamma_\nu$ through the sides $P_\nu A_\nu$ and $Q_\nu B_\nu$. Using the approximate conservation of $j$-waves (7.98) on the region $\Gamma_\nu$, we conclude that for some constant $c > 0$ the following holds. Either $\mu_\nu^{IC}(\Gamma_\nu) \geq c\varepsilon^2$, or else an amount $\geq c\varepsilon$ of $j$-waves leaves $\Gamma$ through the side $A_\nu B_\nu$. In this second case, all the outgoing $j$-fronts cross the $i$-shock $y_{m,\nu}$, determining an amount of interaction $\mu_\nu^{IC}(\overline{\Gamma}_\nu) \geq c \cdot \varepsilon\delta^*$ on the closure of $\Gamma_\nu$. By construction, the region $\Gamma_\nu$ is contained in a ball $B(P, r_\nu)$ centred at $P$ with radius $r_\nu \to 0$ as $\nu \to \infty$. This implies $\mu^{IC}(\{P\}) > 0$, against the assumptions.

CASE 2: In each solution $u_\nu$, the segment $P_\nu Q_\nu$ is crossed by an amount $\geq \varepsilon > 0$ of $j$-waves, for some $j < i$, $\varepsilon > 0$.

In this case, for each $\nu$ we consider the open triangular region $\Gamma_\nu$ bounded by the $i$-shock front $y_{m,\nu}$, by the segment $P_\nu Q_\nu$, by the minimal backward $j$-characteristic through the point $P_\nu$ and by the maximal backward $j$-characteristic through the point $Q_\nu$ (Fig. 10.14). By construction, no $j$-wave can enter $\Gamma_\nu$ through the sides $P_\nu A_\nu$ and $Q_\nu B_\nu$. Again using the approximate conservation of $j$-waves (7.98) on $\Gamma_\nu$, we conclude that the amount of interaction and cancellation $\mu_\nu^{IC}(\overline{\Gamma}_\nu)$ taking place in $u_\nu$ within the closure of $\Gamma_\nu$ remains uniformly positive as $\nu \to \infty$. As before this implies $\mu^{IC}(\{P\}) > 0$, against the assumptions.

CASE 3: In each solution $u_\nu$, the segment $P_\nu Q_\nu$ is crossed by an amount $\geq \varepsilon$ of $i$-waves, while the total amount of $j$-waves crossing $P_\nu Q_\nu$ approaches zero as $\nu \to \infty$, for all $j \neq i$.

In this case (Fig. 10.15) we construct a triangular region $\Gamma_\nu$ as follows. Consider the point $B_\nu$ where the straight line through $P_\nu Q_\nu$ intersects the curve $y_{m,\nu}$. Starting from $B_\nu$, move to the left towards $P_\nu$ until a first point $C_\nu$ is reached such that the total strength of $i$-fronts crossing the segment $C_\nu B_\nu$ is $\geq \delta^*/4$. Let $\Gamma_\nu$ be the region bounded by the segment $C_\nu B_\nu$, by the curve $y_{m,\nu}$ and by the minimal forward $i$-characteristic through $C_\nu$. Because of genuine non-linearity the slope of this minimal $i$-characteristic

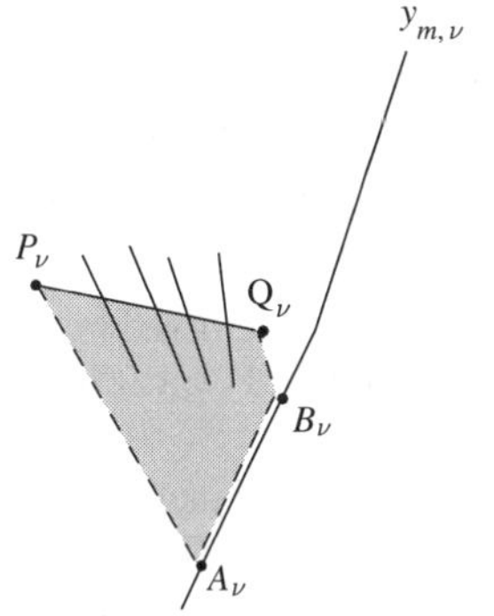


**Figure 10.14**

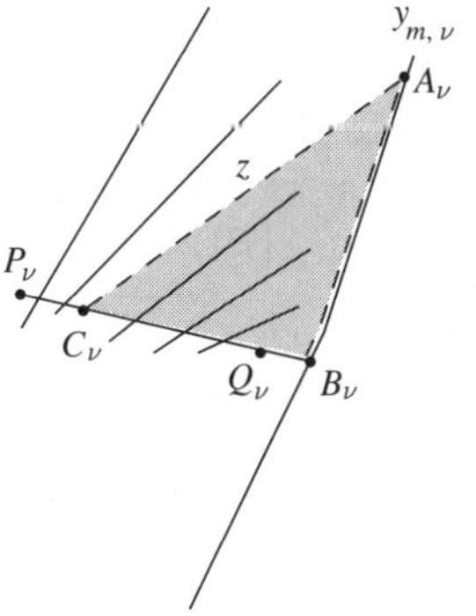


**Figure 10.15**

satisfies the estimate $\dot{x} > \dot{y}_{m,\nu} + c'\delta^*$ for some constant $c' > 0$. Therefore, if no previous interactions or cancellations occur inside $\Gamma_\nu$, this minimal $i$-characteristic will hit the $i$-shock $y_{m,\nu}$ at some point $A_\nu$, with

$$|A_\nu - P_\nu| \le \frac{|C_\nu - B_\nu|}{c'\delta^*} \to 0 \quad \text{as } \nu \to \infty.$$

Since no $i$-wave can exit from $\Gamma_\nu$ through the side $C_\nu A_\nu$, using the approximate conservation of $i$-waves (7.98) we conclude that a uniformly positive amount of interaction and cancellation takes place within the closed domain $\overline{\Gamma}_\nu$. This again implies $\bar{\mu}^{IC}(\{P\}) > 0$, against the assumptions.

In all three cases, we thus derive a contradiction. This proves (10.79) and hence the existence of the first limit in (10.65), when $y_m$ is a genuinely non-linear shock curve. The second limit is proved in the same way.

6. We now establish the limits (10.65) in the case where $y_m$ is an $i$-characteristic, for some linearly degenerate $i$-th field. Assume that, on the contrary, the first limit in (10.65) does not exist. Then there exists $\epsilon > 0$ and a sequence of points $P_k \to P$, all on the left of the curve $y_m$, such that

$$|u(P_k) - u^-| \ge \epsilon \quad \text{for all } k \ge 1. \tag{10.83}$$

For each linearly degenerate family $j \in \{1, \dots, n\}$, call $\bar{x}_j \doteq x_j(0; \tau, \xi)$ the point reached at time $t = 0$ by the $j$-characteristic through $P = (\tau, \xi)$. For any given $\varepsilon' > 0$ there exists $\delta > 0$ such that

$$\mu^{i\star}([\bar{x}_i - \delta,\ \bar{x}_i[) < \varepsilon', \quad \mu^{j\star}([\bar{x}_i - \delta,\ \bar{x}_i + \delta]) < \varepsilon' \qquad \text{if } j \ne i. \tag{10.84}$$

We can now construct a sequence of points $P_\nu, Q_\nu \to P$, both lying on the left of the curves $y_{m,\nu}$ (Fig. 10.12), satisfying (10.81)–(10.82) and such that each segment $P_\nu Q_\nu$ is space-like. From (10.84), we can also assume that the total strength of $j$-waves in $u_\nu$ crossing the segment $P_\nu Q_\nu$ approaches zero as $\nu \to \infty$, for every linearly degenerate field $j$. To derive a contradiction, by possibly taking a subsequence we are led to consider two cases.

CASE 1: There exists $\varepsilon > 0$ and two distinct indices $j \neq k$ such that each segment $P_\nu Q_\nu$ is crossed by an amount $\geq \varepsilon$ of $j$-waves and by an amount $\geq \varepsilon$ of $k$-waves in $u_\nu$.

In this case, since $P_\nu, Q_\nu \to P$, the crossing between $j$-fronts and $k$-fronts would produce an amount $\geq c\varepsilon^2$ of interaction in an arbitrarily small neighbourhood of $P$ (Fig. 10.16). Hence $\mu^{IC}(\{P\}) > 0$, against the assumptions.

CASE 2: For some $j \in \{1, \ldots, n\}$, each segment $P_\nu Q_\nu$ is crossed by an amount $\geq \bar{\varepsilon}$ of $j$-waves. In addition, for every $k \neq j$, the total strength of $k$-waves in $u_\nu$ crossing $P_\nu Q_\nu$ approaches zero as $\nu \to \infty$.

Because of (10.84), the characteristic field $j$ must be genuinely non-linear. Moreover, since $P \notin \Theta_2$, the maximum strength of $j$-fronts crossing $P_\nu Q_\nu$ approaches zero. Since the amount of waves of families $k \neq j$ crossing $P_\nu Q_\nu$ is vanishingly small, the values of each $u_\nu$ along the segment $P_\nu Q_\nu$ remain arbitrarily close to the $j$-rarefaction curve through $u^-$. In particular, this implies

$$|\lambda_j(u_\nu(P_\nu)) - \lambda_j(u_\nu(Q_\nu))| > \frac{\varepsilon}{2} \tag{10.85}$$

for all $\nu$ suitably large. Each $u_\nu$ thus has a large number of small $j$-fronts crossing $P_\nu Q_\nu$ (Figs 10.17–10.18). As $P_\nu, Q_\nu \to P$, these fronts determine a uniformly positive amount of interaction and cancellation in each $u_\nu$, within an arbitrarily small neighbourhood of $P$. Hence $\mu^{IC}(\{P\}) > 0$, against the assumptions.

In all cases, we thus derive a contradiction. This proves the existence of the first limit in (10.65), when $y_m$ is a contact discontinuity. The second limit is proved in the same way.

We now show that the states $u^-, u^+$ lie on the same $i$-rarefaction curve, i.e.

$$u^+ = R_i(\sigma)(u^-) \tag{10.86}$$

for some $\sigma \in \mathbb{R}$. Indeed, in the opposite case we could construct points $P_\nu \doteq (\tau, \xi - r_\nu)$, $Q_\nu \doteq (\tau, \xi + r_\nu)$ such that

$$r_\nu \to 0, \qquad P_\nu, Q_\nu \to P, \qquad u_\nu(P_\nu) \to u^-, \qquad u_\nu(Q_\nu) \to u^+.$$

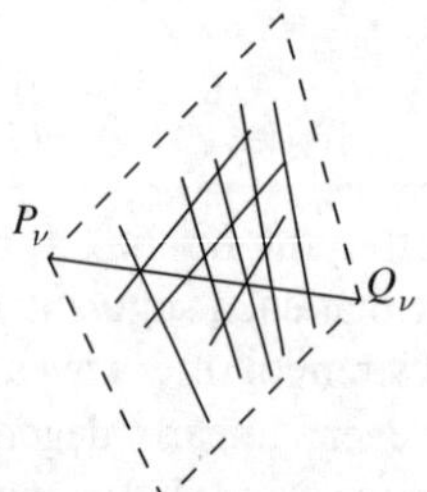


**Figure 10.16**

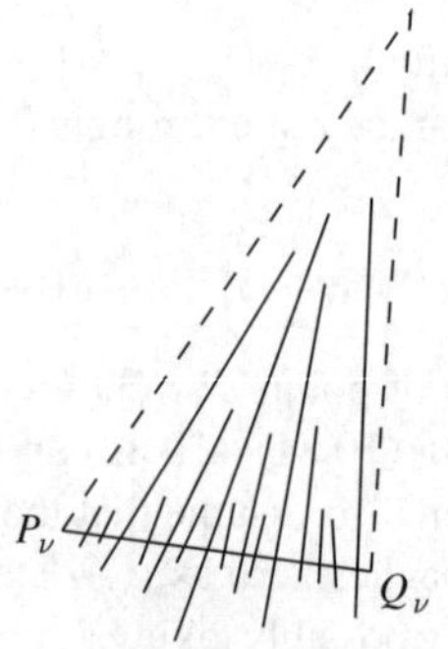


**Figure 10.17**

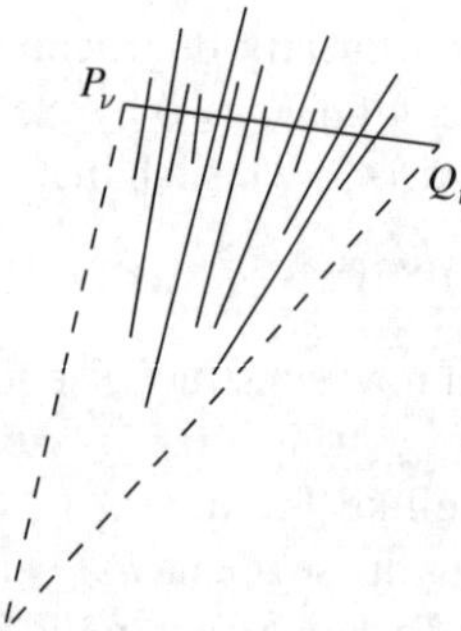


**Figure 10.18**

Each horizontal segment $P_\nu Q_\nu$ then contains a uniformly positive amount of waves of $u_\nu$, of some family $j \neq i$. By the same arguments as before, these waves determine a uniformly positive interaction and cancellation within an arbitrarily small neighbourhood of $P$. Once again this implies $\mu^{IC}(\{P\}) > 0$, against the assumptions.

7. We now establish the Rankine–Hugoniot conditions (10.66) at a given point $P = (\tau,\ y_m(\tau)) \notin \Theta$.

Consider first the case where $y_m$ is an $i$-shock. By the analysis in step 5, given any $\varepsilon > 0$ there exists a sequence of $i$-shock fronts $y_{m,\nu}$ such that $y_{m,\nu} \to y_m$ uniformly on some interval $[\tau - \delta,\ \tau + \delta]$. Call $u_\nu^\pm(t)$ the states to the right and the left of $y_{m,\nu}(t)$. According to the definition of $\varepsilon$-approximate solution, by (7.6)–(7.7) one has

$$\lambda_i(u_\nu^-(t),\ u_\nu^+(t)) \cdot [u_\nu^+(t) - u_\nu^-(t)] = [f(u_\nu^+(t)) - f(u_\nu^-(t))], \tag{10.87}$$

$$|\dot{y}_{m,\nu}(t) - \lambda_i(u_\nu^-(t),\ u_\nu^+(t))| < \varepsilon_\nu. \tag{10.88}$$

By (10.79)–(10.80) it follows that

$$\lim_{r\to 0+} \limsup_{\nu\to\infty} \Big( \sup_{|t-\tau|<r} |\dot{y}_{m,\nu}(t) - \lambda_i(u^+, u^-)| \Big) = 0. \tag{10.89}$$

By (10.89) and the convergence $y_{m,\nu} \to y_m$, from (10.87)–(10.88) we obtain (10.66).

Next, consider the case where $y_m$ is an $i$-characteristic, and the $i$-th field is linearly degenerate. In this case, by the analysis in step 6 the states $u^-, u^+$ lie on the same $i$-rarefaction curve; moreover,

$$\lim_{r\to 0+} \limsup_{\nu\to\infty} \Big( \sup_{|Q-P|<r} |\lambda_i(u_\nu(Q)) - \lambda_i(u^-)| \Big) = 0. \tag{10.90}$$

By (10.71)–(10.72) this yields

$$\dot{y}_m(\tau) = \lambda_i(u^-) = \lambda_i(u^+).$$

Together with (10.86), this implies (10.66) in the linearly degenerate case.

Having established the Rankine–Hugoniot equations, by Theorem 9.3 the Lax entropy conditions also hold.

8. Finally, consider a point $P = (\tau, \xi) \notin \Theta$ outside all jump curves $y_m$. If $u$ is not continuous at $P$, there would exist $\epsilon > 0$ and sequences $P_\nu,\ Q_\nu \to P$ such that each segment $Q_\nu P_\nu$ is space-like and

$$u_\nu(P_\nu) \to u(P), \quad |u_\nu(Q_\nu) - u(P)| \geq \epsilon \qquad \text{for all } \nu.$$

By possibly taking a subsequence, we are led to consider two cases.

CASE 1: There exists $\varepsilon > 0$ and two distinct indices $j \neq k$ such that each segment $P_\nu Q_\nu$ is crossed by an amount $\geq \varepsilon$ of $j$-waves and by an amount $\geq \varepsilon$ of $k$-waves in $u_\nu$.

This is ruled out by the same arguments used in Case 1 of step 6.

CASE 2: For some $j \in \{1, \ldots, n\}$, each segment $P_\nu Q_\nu$ is crossed by an amount $\geq \varepsilon$ of $j$-waves. In addition, for every $k \neq j$, the total strength of $k$-waves in $u_\nu$ crossing $P_\nu Q_\nu$ approaches zero as $\nu \to \infty$.

If the $j$-th family is genuinely non-linear, a contradiction is obtained as in CASE 2 of step 6.

If the $j$-th family is linearly degenerate, setting $\bar{x}_j \doteq x_j(0; \tau, \xi)$, by (10.76) this implies $\mu^{j\star}(\{\bar{x}_j\}) > 0$, against the assumptions.

In all cases, a contradiction is reached. Hence $u$ is continuous at $P$. □

## Problems

(1) Consider the domain $\mathcal{D}$ defined at (10.11). Let $\varphi : \mathbb{R} \mapsto \mathbb{R}$ be a continuous, strictly increasing function, mapping the real line onto itself. Show that, for every $u \in \mathcal{D}$, the composed function $v(x) \doteq u(\varphi(x))$ satisfies

$$\mathbf{Q}(v) = \mathbf{Q}(u), \qquad \mathbf{V}_i(v) = \mathbf{V}_i(u), \quad i = 1, \ldots, n.$$

(2) Let $\sigma_1, \ldots, \sigma_n$ be the sizes of the waves generated by the Riemann problem

$$u_t + f(u)_x = 0, \qquad u(0, x) = \begin{cases} u^- & \text{if } x < 0, \\ u^+ & \text{if } x > 0. \end{cases}$$

Prove that, for every $\varepsilon > 0$, there exists $\delta > 0$ such that the following holds. Let $u : \mathbb{R} \mapsto \mathbb{R}^n$ be any *BV* function which satisfies

$$\lim_{x \to -\infty} |u(x) - u^-| < \delta, \qquad \lim_{x \to \infty} |u(x) - u^+| < \delta.$$

Calling $\mathbf{V}_{i+}(u)$, $\mathbf{V}_{i-}(u)$ the total amount of positive and negative $i$-waves in $u$, assume that

$$\mathbf{Q}(u) < \delta, \qquad \mathbf{V}_{i+}(u) \cdot \mathbf{V}_{i-}(u) < \delta, \quad i = 1, \ldots, n.$$

Then one has

$$|\mathbf{V}_{i+}(u) - \sigma_{i+}| < \varepsilon, \qquad |\mathbf{V}_{i-}(u) - \sigma_{i-}| < \varepsilon,$$

where $\sigma_{i+} \doteq \max\{\sigma_i, 0\}$ and $\sigma_{i-} \doteq \max\{-\sigma_i, 0\}$.

(3) Let $\varepsilon_\nu \to 0$ and let $u_\nu$ be a sequence of $\varepsilon_\nu$-approximate front tracking solutions converging to $u$ in $\mathbf{L}^l_{\text{loc}}([0, \infty[ \times \mathbb{R})$. As usual, we assume that $u$ is right continuous, i.e. $u(t, x) = u(t, x+)$ for all $t, x$. Show that, by possibly taking a subsequence, the following holds. Given any space-like curve $\gamma$, the sequence $u_\nu(P)$ converges to $u(P)$ at all points $P$ on $\gamma$, except at most countably many.

Hint: by Theorem 10.4, $u$ is continuous at every point $P \in \gamma$, with the exception of at most countably many points.

# 11

# Bibliographical notes

## Chapter 1

The book by Smoller (1983) has served for many years as a basic introduction to the theory of conservation laws. In particular, it contains a clear description of the Glimm scheme and a detailed analysis of the Riemann problem for the equations of gas dynamics. More advanced material can be found in the monograph by Serre (1996), including the methods of compensated compactness and geometric optics. This book also describes initial boundary value problems, vanishing viscosity limits, and some results in several space dimensions. Rozdestvenskii and Yanenko (1983) also provide a general introduction to hyperbolic systems, with particular attention to models of gas dynamics. For various applications to mathematical physics, see Witham (1974), the classical monographs by Courant and Friedrichs (1948) or Courant and Hilbert (1962), and the recent book of Dafermos (2000). Problems in several space dimensions are discussed by Hörmander (1997) and Majda (1984). For the numerical computation of solutions to conservation laws we refer to the books of LeVeque (1990), Godlewski and Raviart (1996) or Kröner (1997).

## Chapter 2

The implicit function theorem and the contraction mapping principle are found in almost every book on advanced calculus. See for example Diéudonne (1960). For the basic theory of distributions and of functions with bounded variation we refer to the books of Rudin (1973, 1976). The main results on structure of $BV$ functions in several variables were obtained by Volpert (1967). In this direction, see also the comprehensive monographs by Evans and Gariepy (1992) and Ziemer (1989). Our proof of Rademacher's theorem follows Evans and Gariepy (1992).

In connection with the flow generated by a system of conservation laws, the elementary error estimate stated in Theorem 2.9 was first used in Bressan (1995b). The proof given here was suggested by G. Guerra.

For an introduction to measure theory and for the proof of the Riesz representation theorem we refer to Folland (1984). The weak convergence of measures is discussed in the books of Billingsley (1968) and Yosida (1980).

The basic results concerning existence and uniqueness for solutions of the ODE $\dot{x} = g(t, x)$ with continuous right hand side are classical. The case where the function $g$ is measurable w.r.t. $t$ was first studied by Carathéodory. For the basic theory of ODE's we refer to Hartman (1964). In the study of characteristics for discontinuous solutions of conservation laws, one often has to consider ODE's whose right hand side is discontinuous w.r.t. both variables $t$ and $x$. The simple example

$$x(0) = 0, \qquad \dot{x}(t) = g(t,x) \doteq \begin{cases} 1 & \text{if } x \leq 0, \\ -1 & \text{if } x > 0 \end{cases}$$

shows that the Cauchy problem in this case may have no solution. Two approaches are then possible. On the one hand, one can introduce a weaker concept of solution, so that a general existence theorem will still be available. This approach was first proposed by Filippov (1960) and further developed by Dafermos (1997, 1989) as the method of *generalized characteristics*. On the other hand, one can retain the notion of Carathéodory solutions, and impose suitable conditions on the function $g$ which guarantee the existence and uniqueness of such solutions. Results in this direction, with some applications to conservation laws, can be found in Bressan and Shen (1998a, 1998b).

## Chapter 3

The method of characteristics is the basic tool for the integration of scalar first-order PDE's, being covered in almost every textbook on the subject. The construction of a broad solution to a hyperbolic system as the fixed point of an integral operator can be found in Courant and Hilbert (1962) and Rozdestvenskii and Yanenko (1983). The local existence of continuous solutions to quasilinear hyperbolic systems was first studied in Friedrichs (1948) and Lax (1953). Positive results and counterexamples on the global existence of smooth solutions to quasilinear systems can be found in Jeffrey (1974, 1976) and John (1974), and in the more recent monograph Li (1994). Initial boundary value problems are studied in Rozdestvenskii and Yanenko (1983) and in Li and Yu (1985).

## Chapter 4

The definition of weak solution and the Rankine–Hugoniot conditions are classical. For a discussion of the various entropy-admissibility conditions, see Dafermos (1973, 1999), Lax (1971), Liu (1981) and Smoller (1983). In the scalar case, the admissibility condition (4.42) was introduced by Volpert (1967). For equations which are not in conservation form, an alternative definition of weak solutions was proposed in LeFloch (1988a). See also Cauret *et al.* (1989), Colombeau and Oberguggenberger (1990) and Oberguggenberger (1992) for the approach based on the Colombeau theory of generalized functions.

## Chapter 5

The definitions of genuinely non-linear and of linearly degenerate characteristic fields were introduced in the fundamental paper of Lax (1957), providing the general solution to the Riemann problem. The construction of a solution to the Riemann problem is here based on an application of the implicit function theorem. For general systems, this is valid only if the data $u^-, u^+$ are sufficiently close. In the case of large data, both the existence and the uniqueness of the entropy solution can fail. For a Riemann problem with multiple solutions in the large, see Sever (1989).

Systems where the shock and rarefaction curves coincide have been characterized by Temple (1983). A detailed analysis of solutions of the Riemann problem for the p-system and for the Euler equations of gas dynamics can be found in Smoller (1983). For the equations of isentropic gas dynamics, the Riemann problem with large data may have a solution when the vacuum state is present. This more general case was studied in Liu and Smoller (1980).

If the assumption of genuine non-linearity or linear degeneracy is dropped, the structure of solutions to the Riemann problem can be considerably more complicated. See Liu (1975) and Ancona and Marson (to appear) for the analysis of this more general case.

A natural conjecture is that the entropy-admissible self-similar solutions of the Riemann problem constructed in Theorem 5.3 can be obtained as limits of solutions to the parabolic system

$$u_t + A(u)u_x = \varepsilon u_{xx} \quad (A = Df), \tag{11.1}$$

as the viscosity coefficient approaches zero. For a solution containing a single shock, this was proved by Foy (1964). The general case follows from a result of Goodman and Xin (1992). A detailed analysis of general solutions to the Riemann problem with small viscosity was carried out in Yu (1999).

## Chapter 6

The construction of an entropic solution as a limit of piecewise constant approximations given here is taken from Dafermos (1972). Other constructions are possible, even for scalar conservation laws in several space dimensions: the vanishing viscosity method Kruzhkov (1970), non-linear semigroup theory Crandall (1972), or finite difference schemes Smoller (1983). In the case where the flux function is strictly convex, a variational characterization of the entropy solution is given in Lax (1957).

A proof of uniqueness of entropy weak solutions was first obtained by Volpert (1967) within a class of $BV$ functions. The fundamental paper by Kruzhkov (1970) established the uniqueness of $\mathbf{L}^\infty$ solutions. Moreover, it proved that entropy-admissible solutions form a semigroup which is contractive w.r.t. the $\mathbf{L}^1$ distance.

The decay estimate (6.42) derived in the last problem is due to Oleinik (1963).

Conservation laws where the flux function $f = f(x, u)$ also depends on $x$, possibly in a discontinuous way, were considered in Klingenberg and Risebro (1995). Measure-valued solutions to conservation laws have been studied in DiPerna (1985a) and in Poupaud and Rascle (1997).

For initial-boundary value problems, see Bardos *et al.* (1979) and LeFloch (1988b). Problems related to the boundary control of a single conservation law are discussed in Ancona and Marson (1998).

In connection with optimization problems, it is interesting to study the differentiability of the semigroup of entropic solutions w.r.t. the initial data. Indeed, one would like to describe the behaviour of a first-order perturbation to a given solution of the conservation law. In the recent literature, two distinct approaches have appeared. Bouchut and James (1999) study differentiability in terms of weak convergence of measures. On the other hand, Bressan and Guerra (1997) consider a notion of 'first-order tangent vector' based on a finer equivalence relation determined by the $\mathbf{L}^1$ distance.

## Chapter 7

The first proof of global existence for weak solutions with small total variation appeared in the fundamental paper of Glimm (1965). The striking new ideas introduced in this work have been deeply influential, providing the foundation for much of the subsequent literature on the subject. The original proof in Glimm (1965) was based on the construction of approximate solutions generated by Riemann problems, with a restarting procedure defined through a random sampling. The convergence of a deterministic version of this algorithm was later proved by Liu (1977a).

An alternative line of proof relies on front tracking approximations. A front tracking algorithm, valid for $2 \times 2$ systems, was first proposed by DiPerna (1976). To cover the general $n \times n$ case, one needs to overcome the problem of possible blow-up in the number of wave fronts. This was achieved by Risebro (1993) through a back-stepping procedure, and by the author Bressan (1992) by the introduction of non-physical fronts. Schochet (1997) proposed an alternative front tracking algorithm, obtained from Bressan (1992) by formally letting the speed $\hat{\lambda}$ of non-physical fronts tend to infinity. This has the advantage of eliminating non-physical fronts altogether. The price to pay is that now the solutions no longer have compact support, while the map $t \mapsto u(t, \cdot)$ is no longer continuous with values in $\mathbf{L}^1_{loc}$. Early applications of the front tracking method can also be found in Alber (1985) and in Wendroff (1993). The proof of Theorem 7.1 given here refines the algorithm in Bressan (1992), following the approach of Baiti and Jenssen (1998).

In all of the above works, the a priori estimates of the total variation always rely on the analysis of the wave interaction potential introduced by Glimm. By Helly's theorem, this provides the key compactness property, and hence the existence of solutions. In developing these basic techniques, a large number of related results have been obtained.

The existence of weak solutions to initial boundary value problems has been studied in Dubroca and Gallice (1990), Nishida and Smoller (1977) and Sable-Tougeron (1993) using the Glimm scheme, and in Amadori (1997) using front tracking approximations.

Non-homogeneous balance laws were considered by Dafermos and Hsiao (1982) for dissipative $n \times n$ systems and by Dafermos (1995) for the p-system with damping.

For systems which do not satisfy the assumptions of genuine non-linearity or linear degeneracy, the global existence of entropy solutions with small total variation was proved in Liu (1981) and Liu and Yang (2000). A front tracking algorithm suitable for the analysis of these more general systems was recently developed in Ancona and Marson (1999a).

By relying on the decay of positive waves due to genuine non-linearity, global existence results with initial data in $\mathbf{L}^\infty$ were obtained by Glimm and Lax (1970) and by Cheverry (1998).

For initial conditions with large total variation, the local existence of weak solutions to the Cauchy problem was proved by Schochet (1991). In general, these solutions are not globally defined, because the total variation may explode within finite time Jenssen (2000). The patterns in which blow-up may occur for hyperbolic systems are not yet understood. Some results on the amplification of the initial waves can be found in Joly *et al.* (1994) and Young (1999). For large initial data, estimates on the interval of existence can be found in Temple and Young (1996) and Cheverry (1998). In the special case of Temple class systems, a global existence theorem with large *BV* data was proved in Serre (1987). For another example of a system admitting global *BV* solutions with large data, see DiPerna (1973).

The existence of invariant regions for the flow generated by systems of conservation laws has been investigated by Hoff (1985).

For $2 \times 2$ systems, an entirely different existence proof based on the vanishing viscosity method was developed by DiPerna (1983a), using a compensated compactness argument. This applies to $\mathbf{L}^\infty$ initial data, without any restriction on the total variation. Further applications of the method of compensated compactness can be found in DiPerna (1983b, 1985a, 1985b).

It is interesting to speculate whether the global existence theorem proved by Glimm for small *BV* solutions can be extended to several space dimensions. Unfortunately, according to the analysis of Rauch (1986), in more than one space dimension there is little hope of constructing global *BV* solutions in any generality. Developing a theory of weak solutions for systems of conservation laws in several space dimensions remains an outstanding open problem. At the present time, it is not even clear in which functional space one should work, in order to prove a general existence theorem.

## Chapter 8

The existence of a Lipschitz semigroup of entropy weak solutions was conjectured by the author in Bressan (1988), and proved in a series of papers, for systems of increasing generality. The result covers the case where all characteristic fields are linearly degenerate. The existence of the semigroup for general $2 \times 2$ systems was proved in Bressan and Colombo (1995a). A construction valid in the $n \times n$ case was carried out in Bressan *et al.* (2000), based on the variational calculus for piecewise Lipschitz solutions developed in Bressan and Marson (1995) and Bressan (1995a).

All of these earlier works relied on a linearization method. In order to estimate how the distance between two solutions $u, v$ varies in time, one constructs a one-parameter family of solutions $u^\theta$ joining $u$ with $v$, as shown in Fig. 11.1. At any time $t \in [0, T]$, the distance $\|u(t) - v(t)\|_{\mathbf{L}^1}$ is thus bounded by the length of the curve $\gamma_t : \theta \mapsto u^\theta(t)$. In turn, as long as all solutions $u^\theta$ remain sufficiently regular, the length of $\gamma_t$ can be computed by integrating the norm of a generalized tangent vector $\mathbf{v}$. The advantage of this method is that tangent vectors satisfy a linearized evolution equation, which is more easily studied. From a uniform a priori estimate on the norm of these tangent vectors, one obtains a bound on the length of $\gamma_T$ and hence on the distance between $u(T)$ and $v(T)$. Unfortunately, this approach is hampered by the possible loss of regularity of the solutions $u^\theta$. In order to retain the minimal regularity (piecewise Lipschitz continuity)

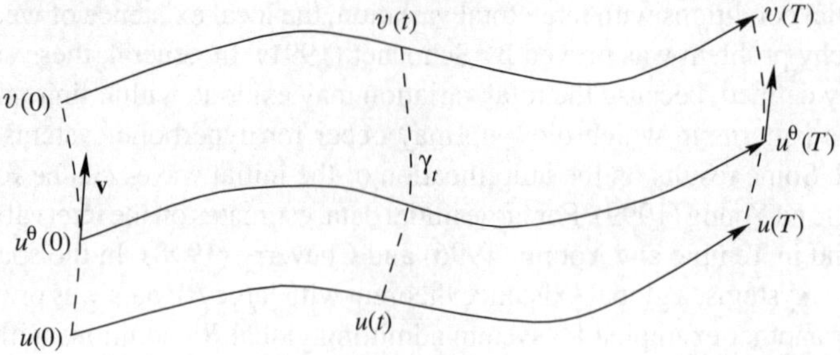


**Figure 11.1**

required for the existence of tangent vectors, in Bressan and Colombo (1995a) and Bressan *et al.* (2000) various approximation and restarting procedures had to be devised. These yield entirely rigorous proofs, but at the price of heavy technicalities.

An entirely different approach was introduced by Liu and Yang (1999b, 1999c), defining a functional $\Phi(u, v)$ which is equivalent to the $\mathbf{L}^1$ distance and decreases along couples of solutions of the hyperbolic system. In their construction, a key role is played by a new entropy functional for genuinely non-linear scalar fields, introduced in Liu and Yang (1999a). Our proof of Theorem 8.1, relying on front tracking approximations, follows Bressan *et al.* (1999) with some refinements suggested by G. Guerra. The $L^1$ stability of approximate solutions generated by the Glimm scheme is discussed in Liu and Yang (1999d).

Using the approach in Bressan and Colombo (1995a), the continuous dependence of weak solutions has been extended in several directions. The stability of perturbations of a $2 \times 2$ Riemann problem with large data was studied in Bressan and Colombo (1995b). The papers Amadori and Colombo (1997, 1998) deal with initial boundary value problems, while Crasta and Piccoli (1997) is concerned with balance laws in the presence of a source term. For some particular systems, the construction of a semigroup with domain containing arbitrarily large *BV* data was carried out in Baiti and Bressan (1997a) and Colombo and Risebro (1998). The $\mathbf{L}^1$ stability of entropy solutions for a more general class of $2 \times 2$ systems, without the assumption of genuine nonlinearity or linear degeneracy of each characteristic field, was proved by Ancona and Marson (1999b).

Furthermore, the approach developed in Bressan *et al.* (1999) was recently adapted in Amadori and Guerra (1999) to systems of balance laws with strong damping, and in Lewicka (1999) to Cauchy problems with data containing several large shocks.

When the initial data lie in $\mathbf{L}^\infty$ and have possibly unbounded variation, a counter-example in Bressan and Shen (1998a) shows that the Cauchy problem may not be well posed. For some special $2 \times 2$ systems, the continuous dependence of solutions has been established in Bressan and Shen (1998a) and Baiti and Jenssen (1997). In the case of Temple class systems where all fields are genuinely non-linear, the results in Bressan and Goatin (2000) show that $\mathbf{L}^\infty$ solutions depend Lipschitz continuously on the initial data, in the $\mathbf{L}^1$ distance.

When the system is hyperbolic, but not strictly hyperbolic, the $\mathbf{L}^1$ continuous dependence of solutions may fail. In this case, Tveito and Winther (1991) showed that the well posedness of the Cauchy problem can be obtained only w.r.t. a stronger topology on the set of initial data.

Earlier results on the well-posedness of the Cauchy problem for hyperbolic systems were obtained, in some special cases, by Liapidevskii (1974, 1975) and by Pimbley (1988).

## Chapter 9

The uniqueness of entropy weak solutions to hyperbolic systems remained an open problem for many years. Indeed, while the global existence had been established within a space of $BV$ functions, uniqueness was known only for solutions satisfying additional restrictive regularity conditions (Dafermos and Geng 1991, DiPerna 1978, 1979, Heibig 1994, LeFloch and Xin 1993, Liu 1976, Oleinik 1957, Smoller 1983). The semigroup approach, introduced in Bressan (1995b), was the first to provide a general uniqueness theorem within the same class of $BV$ functions where existence could be proved.

The definition of SRS and the characterization of semigroup trajectories by the local integral estimates in Theorem 9.2 are taken from Bressan (1995b). A uniqueness theorem for entropy weak solutions based on the tame variation condition (A5) was proved in Bressan and LeFloch (1997). The uniqueness result stated in Theorem 9.4 was obtained in Bressan and Goatin (1999) with the assumption (A4) of tame oscillation, and in Bressan and Lewicka (2000) with the assumption (A3) of bounded variation along space-like curves. The estimates proved by Trivisa (1997) imply that the 'countably regular' solutions considered by Dafermos (1989) are unique and indeed coincide with semigroup trajectories.

According to the analysis in Bressan (1995b), the $\mathbf{L}^1$ distance between an $\varepsilon$-approximate front tracking solution $u_\varepsilon$ and an exact solution $u$ can be estimated as

$$\left\| u_\varepsilon(t) - u(t) \right\|_{\mathbf{L}^1} = \mathcal{O}(1) \cdot \varepsilon(1+t).$$

This easily follows from our bounds (9.12)–(9.13). On the other hand, rigorous estimates of the difference between exact solutions and approximations generated by the Glimm scheme are more difficult to obtain. A result in this direction was proved in Hoff and Smoller (1985) for scalar equations and in Bressan and Marson (1998) for $n \times n$ systems. If $\Delta x \approx \Delta t$ is the mesh size in the Glimm approximation scheme, the analysis in Bressan and Marson (1998) shows that

$$\frac{\left\| u^{\text{approx}}(t,\cdot) - u^{\text{exact}}(t,\cdot) \right\|_{\mathbf{L}^1}}{\left| \log \Delta x \right| \cdot \sqrt{\Delta x}} \to 0 \quad \text{as } \Delta x \to 0,$$

for a suitable equidistributed sampling sequence. The proof is based on the front tracking representation of Liu (1977a) and on the error estimate stated in our Theorem 2.9.

In addition to the front tracking algorithm and to the Glimm scheme, there are other widely used methods to construct approximate solutions:

- vanishing viscosity approximations,
- relaxations,
- discrete numerical schemes.

An outstanding open problem is whether these approximations also converge to the unique semigroup solution. At present, the main difficulty stems from the lack of uniform estimates on the total variation of approximate solutions.

In some cases, mainly for $2 \times 2$ systems, the existence of a limit (for a suitable subsequence) can be proved by compensated compactness (DiPerna 1983a, 1983b, 1985b). However, it is not clear whether the limit solution is entropy admissible and if it remains in $BV$. In principle, this solution may have unbounded variation and hence take values outside the domain of our semigroup. In this case, none of the uniqueness theorems stated in Chapter 9 can be applied.

In the case of vanishing viscosity approximations, some of the available results are as follows:

- For $n \times n$ Temple class systems, as $\varepsilon \to 0+$ a proof of the convergence of the solutions $u_\varepsilon$ of the viscous system (11.1) to a weak solution of (7.1) can be found in Heibig (1994) and Schochet (1991). In this case, the a priori bound on the total variation of viscous solutions follows directly from the maximum principle.
- Assume that all characteristic fields of the system (7.1) are linearly degenerate. Then every solution with small total variation which is initially smooth remains smooth for all positive times (Bressan 1988). Clearly such solutions can be obtained as the limit of vanishing viscosity approximations. By a density argument it follows that every weak solution with sufficiently small total variation is a limit of viscous approximations.
- For a general $n \times n$ strictly hyperbolic system, let $u$ be a piecewise smooth entropic solution with jumps along a finite number of smooth curves in the $t$–$x$ plane. Using this additional regularity assumption on $u$, Goodman and Xin (1992) proved the existence of a family of viscous solutions $u_\varepsilon$ converging to $u$ in $\mathbf{L}^1_{\mathrm{loc}}$ as $\varepsilon \to 0$.
- In the case where all integral curves of the eigenvectors $r_i$ $(i = 1, \ldots, n)$ are straight lines, it was proved in Bianchini and Bressan (2000) that as $\varepsilon \to 0$ the solutions of (11.1) converge to a unique limit, depending Lipschitz-continuously on the initial data in the $\mathbf{L}^1$ norm. The result does not require the system to be in conservation form. In the standard case where $A(u) = Df(u)$, the vanishing viscosity solutions coincide with semigroup trajectories.

Relaxation problems are concerned with non-homogeneous systems of the form

$$U_t + A(U)U_x = \frac{1}{\varepsilon} g(U), \tag{11.2}$$

where $g$ is a vector field whose flow steers all points towards a manifold $\mathcal{M}$ of equilibrium states (Fig. 11.2). As $\varepsilon \to 0$, one thus expects that the limits of solutions to (11.2) can be

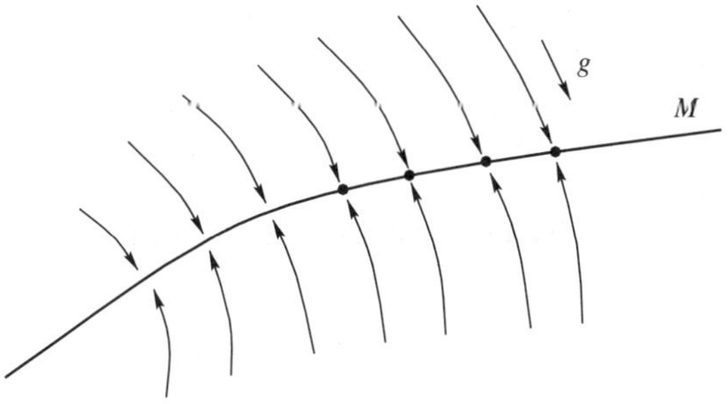


**Figure 11.2**

described by a reduced system living on the equilibrium manifold $\mathcal{M}$. There are various physical phenomena that can be described by equations of the form (11.2). Moreover, as shown by Jin and Xin (1955), the flow of the general $n \times n$ system of conservation laws (7.1) can be formally approximated in terms of the $2n \times 2n$ semilinear relaxation system

$$\begin{cases} u_t + v_x = 0, \\ v_t + \Lambda u_x = \varepsilon^{-1}(f(u) - v), \end{cases}$$

where $\Lambda$ is a suitable diagonal matrix with constant coefficients. In recent years, relaxation problems have become a topic with a rich literature. For results in this direction we refer to Chen *et al.* (1994), Chen and Liu (1993), Liu (1987), Natalini (1996), Xin (1991), Bressan and Shen (2000) and Bianchini (to appear). See also the recent survey in Natalini (1998) and references therein.

## Chapter 10

The functionals **V**, **Q** defined at (10.9)–(10.10) are straightforward extensions of those introduced by Glimm (1965). They were first used by Schatzman (1985) in connection with piecewise Lipschitz functions, and in Bressan and Colombo (1995b) in the case of general *BV* functions. The lower semicontinuity results in Theorem 10.1 are taken from Baiti and Bressan (1997b).

A general estimate of the decay of waves due to non-linearity was obtained by Liu (1981). It applies to approximate solutions generated by the Glimm scheme for $n \times n$ systems, even without the assumption of genuine non-linearity. The result stated in Theorem 10.3, on the other hand, refers to exact solutions and was proved in Bressan and Colombo (1998).

The structure of *BV* solutions constructed by the Glimm scheme is studied in DiPerna (1975b), Dafermos (1989) and Liu (1981). In the genuinely non-linear case, Theorem 10.4 was proved in Bressan and LeFloch (1999). The asymptotic structure of a small *BV* solution in a neighbourhood of a point is also described there. An interesting question is whether the wave-front structure of a solution is stable under $\mathbf{L}^1$ perturbations. In particular, one expects that the strength and the location of a large shock should not be much affected by a small perturbation in the initial data. Various results in this direction are given in Bressan and LeFloch (1999).

An important aspect of the qualitative theory of solutions, not covered in the present book, is the long-time asymptotic behaviour. Several results in this direction were derived by Glimm and Lax (1970), DiPerna (1975a), Liu (1977b) and Zumbrun (1993). For systems with dissipation or damping, a detailed analysis of solutions can be found in Hsiao (1997).

# References


H. D. Alber, A local existence theorem for quasilinear wave equations with initial values of bounded variation, Springer Lecture Notes in Mathematics **1151** (1985), 9–24.

D. Amadori, Initial-boundary value problems for nonlinear systems of conservation laws, *Nonlinear Differ. Equation Appl.* **4** (1997), 1–42.

D. Amadori and R. M. Colombo, Continuous dependence for $2 \times 2$ conservation laws with boundary, *J. Differ. Equations* **138** (1997), 229–266.

D. Amadori and R. M. Colombo, Viscosity solutions and Standard Riemann Semigroup for conservation laws with boundary, *Rend. Sem. Mat. Univ. Padova* **99** (1998), 219–245.

D. Amadori and G. Guerra, Uniqueness and continuous dependence for systems of balance laws with dissipation, Preprint (1999).

F. Ancona and A. Marson, On the attainable set for scalar nonlinear conservation laws with boundary control, *SIAM J. Control Optimization* **36** (1998), 290–312.

F. Ancona and A. Marson, A note on the Riemann problem for general $n \times n$ systems of conservation laws, *J. Math. Anal. Appl.* (to appear).

F. Ancona and A. Marson, A wave-front tracking algorithm for $N \times N$ non genuinely nonlinear conservation laws, preprint (1999a).

F. Ancona and A. Marson, Well-posedness for general $2 \times 2$ systems of conservation laws, Preprint SISSA, Trieste (1999b).

P. Baiti and A. Bressan, The semigroup generated by a Temple class system with large data, *Differ. Integral Equations* **10** (1997a), 401–418.

P. Baiti and A. Bressan, Lower semicontinuity of weighted path lengths in BV, in *Geometrical Optics and Related Topics*, F. Colombini and N. Lerner Eds., Birkhäuser, Boston (1997b), 31–58.

P. Baiti and H. K. Jenssen, Well posedness for a class of $2 \times 2$ conservation laws with $\mathbf{L}^\infty$ data, *J. Differ. Equations* **140** (1997), 161–185.

P. Baiti and H. K. Jenssen, On the front tracking algorithm, *J. Math. Anal. Appl.* **217** (1998), 395–404.

C. Bardos, A. Y. Leroux and J. C. Nedelec, First order quasilinear equations with boundary conditions, Commun. *Partial Differ. Equation* **9** (1979), 1017–1034.

S. Bianchini, On a Glimm type functional for a special Jin-Xin relaxation model. To appear in *Ann. Inst. Henri Poincare. Analyse Nonlineaire.*

S. Bianchini and A. Bressan, BV estimates for a class of viscous hyperbolic systems, *Indiana Univ. Math. J.* (2000), to appear.

P. Billingsley, *Convergence of Probability Measures*, Wiley, New York (1968).

F. Bouchut and F. James, Differentiability with respect to initial data for a scalar conservation law, in *Hyperbolic Problems: Theory, Numerics, Applications*, M. Fey and R. Jeltsch Eds., Birkäuser, Basel (1999), 113–118.

A. Bressan, Contractive metrics for nonlinear hyperbolic systems, *Indiana Univ. Math. J.* **37** (1988), 409–421.

A. Bressan, Global solutions to systems of conservation laws by wave-front tracking, *J. Math. Anal. Appl.* **170** (1992), 414–432.

A. Bressan, A locally contractive metric for systems of conservation laws, *Ann. Sc. Norm. Sup. Pisa* **IV-22** (1995a), 109–135.

A. Bressan, The unique limit of the Glimm scheme, *Arch. Ration. Mech. Anal.* **130** (1995b), 205–230.

A. Bressan and R. M. Colombo, The semigroup generated by $2 \times 2$ conservation laws, *Arch. Ration. Mech. Anal.* **133** (1995a), 1–75.

A. Bressan and R. M. Colombo, Unique solutions of $2 \times 2$ conservation laws with large data, *Indiana Univ. Math. J.* **44** (1995b), 677–725.

A. Bressan and R. M. Colombo, Decay of positive waves in nonlinear systems of conservation laws, *Ann. Sc. Norm. Sup. Pisa* **IV-26** (1998), 133–160.

A. Bressan, G. Crasta and B. Piccoli, Well posedness of the Cauchy problem for $n \times n$ conservation laws, *Am. Math. Soc. Mem.* **694** (2000).

A. Bressan and P. Goatin, Oleinik type estimates and uniqueness for $n \times n$ conservation laws, *J. Differ. Equations* **156** (1999), 26–49.

A. Bressan and P. Goatin, Stability of $L^\infty$ solutions of Temple class systems, *Differ. Integral Equations* (2000), to appear.

A. Bressan and G. Guerra, Shift differentiability of the flow generated by a conservation law, *Discrete Contin. Dyn. Syst.* **3** (1997), 35–58.

A. Bressan and P. LeFloch, Uniqueness of weak solutions to systems of conservation laws, *Arch. Ration. Mech. Anal.* **140** (1997), 301–317.

A. Bressan and P. LeFloch, Structural stability and regularity of entropy solutions to hyperbolic systems of conservation laws, *Indiana Univ. Math. J.* **48** (1999), 43–84.

A. Bressan and M. Lewicka, A uniqueness condition for hyperbolic systems of conservation laws, *Discrete Contin. Dyn. Syst.* **6** (2000), 673–682.

A. Bressan, T. P. Liu and T. Yang, $L^1$ stability estimates for $n \times n$ conservation laws, *Arch. Ration. Mech. Anal.* **149** (1999), 1–22.

A. Bressan and A. Marson, A variational calculus for shock solutions of systems of conservation laws, *Commun. Partial Differ. Equations* **20** (1995), 1491–1552.

A. Bressan and A. Marson, Error bounds for a deterministic version of the Glimm scheme, *Arch. Ration. Mech. Anal.* **142** (1998), 155–176.

A. Bressan and W. Shen, Uniqueness for discontinuous ODE and conservation laws, *Nonlinear Anal.*, TMA **34** (1998a), 637–652.

A. Bressan and W. Shen, On discontinuous differential equations, in *Differential Inclusions and Optimal Control*, J. Andres, L. Gorniewicz and P. Nistri Eds., Juliusz Schauder Center for Nonlinear Studies, Lecture Notes in Nonlinear Analysis **2** Torun (1998b), 73–87.

A. Bressan and W. Shen, BV estimates for multicomponent chromatography with relaxation, *Discrete Contin. Dyn. Syst.* **6** (2000), 21–38.

J. J. Cauret, J. F. Colombeau and A. Y. LeRoux, Discontinuous generalized solutions of nonlinear nonconservative hyperbolic equations, *J. Math. Anal. Appl.* **139** (1989), 552–573.

G. Chen, C. Levermore and T. P. Liu, Hyperbolic conservation laws with stiff relaxation terms and entropy, *Commun. Pure Appl. Math.* **47** (1994), 787–830.

G. Chen and T. P. Liu, Zero relaxation and dissipation limits for hyperbolic conservation laws, *Commun. Pure Appl. Math.* **46** (1993), 755–781.

C. Cheverry, Systèmes de lois de conservation et stabilité BV, *Mem. Soc. Math. Fr.* **75** (1998).

J. F. Colombeau and M. Oberguggenberger, On a hyperbolic system with a compatible quadratic term: generalized solutions, delta-waves and multiplication of distributions, *Commun. Partial Differ. Equations* **15** (1990), 905–938.

R. M. Colombo and N. H. Risebro, Continuous dependence in the large for some hyperbolic conservation laws, *Commun. Partial Differ. Equations* **23** (1998), 1693–1718.

R. Courant and K. O. Friedrichs, *Supersonic Flow and Shock Waves*, Wiley Interscience, New York (1948).

R. Courant and D. Hilbert, *Methods of Mathematical Physics*, Vol. II, Wiley Interscience, New York (1962).

M. Crandall, The semigroup approach to first-order quasilinear equations in several space variables, *Isr. J. Math.* **12** (1972), 108–132.

G. Crasta and B. Piccoli, Viscosity solutions and uniqueness for systems of inhomogeneous balance laws, *Discrete Contin. Dyn. Syst.* **4** (1997), 477–502.

C. Dafermos, Polygonal approximations of solutions of the initial value problem for a conservation law, *J. Math. Anal. Appl.* **38** (1972), 33–41.

C. Dafermos, The entropy rate admissibility criterion for solutions of hyperbolic conservation laws, *J. Differ. Equations* **14** (1973), 202–212.

C. Dafermos, Generalized characteristics and the structure of solutions of hyperbolic conservation laws, *Indiana Univ. Math. J.* **26** (1977), 1097–1119.

C. Dafermos, Generalized characteristics in hyperbolic systems of conservation laws, *Arch. Ration. Mech. Anal.* **107** (1989), 127–155.

C. Dafermos, A system of hyperbolic equations with frictional damping, *Z. Angew. Math. Phys.* **46** (1995), 294–307.

C. Dafermos, *Hyperbolic Conservation Laws in Continuum Physics*, Springer-Verlag, Heidelberg (2000).

C. Dafermos and X. Geng, Generalized characteristics, uniqueness and regularity of solutions in a hyperbolic system of conservation laws, *Ann. Inst. Henri Poincaré – Analyse Nonlineaire* **8** (1991), 231–269.

C. Dafermos and L. Hsiao, Hyperbolic systems of balance laws with inhomogeneity and dissipation, *Indiana Univ. Math. J.* **31** (1982), 471–491.

J. Diéudonne, *Foundations of Modern Analysis*, Academic Press, New York (1960).

R. DiPerna, Existence in the large for nonlinear hyperbolic conservation laws, *Arch. Ration. Mech. Anal.* **52** (1973), 244–257.

R. DiPerna, Decay and asymptotic behavior of solutions to nonlinear hyperbolic systems of conservation laws, *Indiana Univ. Math. J.* **24** (1975a), 1047–1071.

R. DiPerna, Singularities of solutions of nonlinear hyperbolic systems of conservation laws, *Arch. Ration. Mech. Anal.* **60** (1975b), 75–100.

R. DiPerna, Global existence of solutions to nonlinear hyperbolic systems of conservation laws, *J. Differ. Equations* **20** (1976), 187–212.

R. DiPerna, Entropy and the uniqueness of solutions to hyperbolic conservation laws, in *Nonlinear Evolution Equations*, M. Crandall Ed., Academic Press, New York (1978), 1–16.

R. DiPerna, Uniqueness of solutions to hyperbolic conservation laws, *Indiana Univ. Math. J.* **28** (1979), 137–188.

R. DiPerna, Convergence of approximate solutions to conservation laws, *Arch. Ration. Mech. Anal.* **82** (1983a), 27–70.

R. DiPerna, Convergence of the viscosity method for isentropic gas dynamics, *Commun. Math. Phys.* **91** (1983b), 1–30.

R. DiPerna, Measure-valued solutions to conservation laws, *Arch. Ration. Mech. Anal.* **88** (1985a), 223–270.

R. DiPerna, Compensated compactness and general systems of conservation laws, *Trans. Am. Math. Soc.* **292** (1985b), 383–420.

B. Dubroca and G. Gallice, Resultats d'existence et d'unicité du probleme mixte pour des systemes hyperboliques de lois de conservation monodimensionnels, *Commun. Partial Differ. Equations* **15** (1990), 59–80.

L. C. Evans and R. F. Gariepy, *Measure Theory and Fine Properties of Functions*, CRC Press, Boca Raton, FL (1992).

A. F. Filippov, Differential equations with discontinuous right hand side, *Am. Math. Soc. Transl.*, Ser. 2, **42** (1960), 199–231.

G. B. Folland, *Real Analysis*, Wiley Interscience, New York (1984).

L. Foy, Steady-state solutions of hyperbolic systems of conservation laws with viscosity terms, *Commun. Pure Appl. Math.* **17** (1964), 177–188.

K. O. Friedrichs, Nonlinear hyperbolic differential equations in two independent variables, *Am. J. Math.* **70** (1948), 555–588.

J. Glimm, Solutions in the large for nonlinear hyperbolic systems of equations, *Commun. Pure Appl. Math.* **18** (1965), 697–715.

J. Glimm and P. Lax, Decay of solutions of systems of nonlinear hyperbolic conservation laws, *Am. Math. Soc. Mem.* **101** (1970).

E. Godlewski and P.A. Raviart, *Numerical Approximation of Hyperbolic Systems of Conservation Laws*, Springer-Verlag, New York, 1996.

J. Goodman and Z. Xin, Viscous limits for piecewise smooth solutions to systems of conservation laws, *Arch. Ration. Mech. Anal.* **121** (1992), 235–265.

P. Hartman, *Ordinary Differential Equations*, Wiley Interscience, New York (1964).

A. Heibig, Existence and uniqueness of solutions for some hyperbolic systems of conservation laws, *Arch. Ration. Mech. Anal.* **126** (1994), 79–101.

D. Hoff, Invariant regions for systems of conservation laws, *Trans. Am. Math. Soc.* **289** (1985), 591–610.

D. Hoff and J. Smoller, Error bounds for Glimm difference approximations for scalar conservation laws, *Trans. Am. Math. Soc.* **289** (1985), 611–645.

L. Hörmander, Lectures on Nonlinear Hyperbolic Differential Equations, *Mathématiques & Applications* **26**, Springer-Verlag Heidelberg (1997).

L. Hsiao, *Quasilinear Hyperbolic Systems and Dissipative Mechanisms*, World Scientific, Singapore (1997).

A. Jeffrey, Breakdown of the solution to a completely exceptional system of hyperbolic equations, *J. Math. Anal. Appl.* **45** (1974), 375–358.

A. Jeffrey, *Quasilinear Hyperbolic Systems and Waves*, Pitman, London (1976).

H. K. Jenssen, Blowup for systems of conservation laws, *SIAM J. Math. Anal.* **31** (2000), 894–908.

S. Jin and Z. Xin, The relaxation schemes for systems of conservation laws in arbitrary space dimensions, *Commun. Pure Appl. Math.* **48** (1955), 235–277.

F. John, Formation of singularities in one-dimensional nonlinear wave propagation, *Commun. Pure Appl. Math.* **27** (1974), 377–405.

J. L. Joly, G. Metivier and J. Rauch, A nonlinear instability for $3 \times 3$ systems of conservation laws, *Commun. Math. Phys.* **162** (1994), 47–59.

C. Klingenberg and N. H. Risebro, Convex conservation laws with discontinuous coefficients. Existence, uniqueness and asymptotic behavior, *Commun. Partial Differ. Equations* **20** (1995), 1959–1990.

D. Kröner, Numerical Schemes for Conservation Laws, Wiley and Tauber, New York (1997).

S. Kruzhkov, First order quasilinear equations with several space variables, *Math. USSR Sb.* **10** (1970), 217–243.

P. Lax, Nonlinear hyperbolic equations, *Commun. Pure Appl. Math.* **6** (1953), 231–258.

P. Lax, Hyperbolic systems of conservation laws II, *Commun. Pure Appl. Math.* **10** (1957), 537–566.

P. Lax, Shock waves and entropy. In *Contributions to Nonlinear Functional Analysis*, E. Zarantonello Ed., Academic Press, New York (1971), 603–634.

P. G. LeFloch, Entropy weak solutions to nonlinear hyperbolic systems under nonconservative form, *Commun. Partial Differ. Equations* **13** (1988a), 669–727.

P. G. LeFloch, Explicit formula for scalar non-linear conservation laws with boundary conditions, *Math. Methods Appl. Sci.* **10** (1988b), 265–287.

P. G. LeFloch and Z. Xin, Uniqueness via the adjoint problem for systems of conservation laws, *Commun. Pure Appl. Math.* **46** (1993), 1499–1533.

R. LeVeque, *Numerical Methods for Conservation Laws*, Lecture Notes in Mathematics, Birkhäuser, Basel (1990).

M. Lewicka, $\mathbf{L}^1$ stability of patterns of non-interacting large shock waves, Preprint SISSA, Trieste (1999).

T. Li, *Global Classical Solutions for Quasilinear Hyperbolic Systems*, Wiley, New York (1994).

T. Li and W. Yu, *Boundary Value Problems for Quasilinear Hyperbolic Systems*, Duke University, Durham, NC (1985).

V. Liapidevskii, The continuous dependence on the initial conditions of the generalized solutions of the gas-dynamic system of equations, *USSR Comput. Math. Math. Phys.* **14** (1974), 158–167.

V. Liapidevskii, On correctness classes for nonlinear hyperbolic systems, *Sov. Math. Dokl.* **16** (1975), 1505–1509.

T. P. Liu, The Riemann problem for general systems of conservation laws, *J. Differ. Equations* **18** (1975), 218–234.

T. P. Liu, Uniqueness of weak solutions of the Cauchy problem for general $2 \times 2$ conservation laws, *J. Differ. Equations* **20** (1976), 369–388.

T. P. Liu, The deterministic version of the Glimm scheme, *Commun. Math. Phys.* **57** (1977a), 135–148.

T. P. Liu, Linear and nonlinear large-time behavior of solutions of general systems of hyperbolic conservation laws, *Commun. Pure Appl. Math.* **30** (1977b), 767–796.

T. P. Liu, Admissible solutions of hyperbolic conservation laws, *Am. Math. Soc. Mem.* **240** (1981).

T. P. Liu, Hyperbolic conservation laws with relaxation, *Commun. Math. Pys.* **108** (1987), 153–175.

T. P. Liu and J. Smoller, On the vacuum state for the isentropic gas dynamics equations, *Adv. Pure Appl. Math.* **1** (1980), 345–359.

T. P. Liu and T. Yang, A new entropy functional for scalar conservation laws, *Commun. Pure Appl. Math.* **52** (1999a), 1427–1442.

T. P. Liu and T. Yang, $L^1$ stability of conservation laws with coinciding Hugoniot and characteristic curves, *Indiana Univ. Math. J.* **48** (1999b), 237–247.

T. P. Liu and T. Yang, $L^1$ stability for $2 \times 2$ systems of hyperbolic conservation laws, *J. Am. Math. Soc.* **12** (1999c), 729–774.

T. P. Liu and T. Yang, Well posedness theory for hyperbolic conservation laws, *Commun. Pure Appl. Math.* **52** (1999d), 1553–1586.

T. P. Liu and T. Yang, Weak solutions of general systems of hyperbolic conservation laws preprint (2000).

A. Majda, *Compressible Fluid Flow and Systems of Conservation Laws in Several Space Variables*, Springer-Verlag, New York (1984).

R. Natalini, Convergence to equilibrium for the relaxation approximations of conservation laws, *Commun. Pure Appl. Math.*, **49** (1996), 795–824.

R. Natalini, Recent results on hyperbolic relaxation problems, in *Analysis of Systems of Conservation Laws*, H. Freisthüler Ed., Chapman & Hall/CRC, London (1998), 128–198.

T. Nishida and J. Smoller, Mixed problems for nonlinear conservation laws, *J. Differ. Equations* **23** (1977), 244–269.

M. Oberguggenberger, Case study of a nonlinear, nonconservative, nonstrictly hyperbolic system, *Nonlinear Anal.* **19** (1992), 53–79.

O. Oleinik, Discontinuous solutions of nonlinear differential equations, *Am. Math. Soc. Transl.* **26** (1963), 95–172.

O. Oleinik, On the uniqueness of the generalized solution of the Cauchy problem for a nonlinear system of equations occurring in mechanics, *Usp. Mat. Nauk.* **12** (1957), 169–176 (in Russian).

G. Pimbley, A semigroup associated with a quasi-linear system in which the coupling terms are linear – II, *Nonlinear Anal.* **12** (1988), 321–340.

F. Poupaud and M. Rascle, Measure solutions to the linear multidimensional transport equations with nonsmooth coefficients, *Commun. Partial Differ. Equations* **22** (1997), 337–358.

J. Rauch, BV estimates fail for most quasilinear hyperbolic systems in dimension greater than one, *Commun. Math. Phys.* **106** (1986), 481–484.

N. H. Risebro, A front-tracking alternative to the random choice method, *Proc. Am. Math. Soc.* **117** (1993), 1125–1139.

B. L. Rozdestvenskii and N. Yanenko, *Systems of Quasilinear Equations and Their Applications to Gas Dynamics*. AMS Translations of Mathematical Monographs **55**, Providence, RI (1983).

W. Rudin, *Functional Analysis*, McGraw-Hill, New York (1973).

W. Rudin, *Real and Complex Analysis*, McGraw-Hill, New York (1976).

M. Sable-Tougeron, Méthode de Glimm et problème mixte, *Ann. Inst. Henri Poincaré – Analyse Nonlineaire* **10** (1993), 423–443.

M. Schatzman, Continuous Glimm functionals and uniqueness of solutions of the Riemann problem, *Indiana Univ. Math. J.* **34** (1985), 533–589.

S. Schochet, Sufficient conditions for local existence via Glimm's scheme for large BV data, *J. Differ. Equations* **89** (1991), 317–354.

S. Schochet, The essence of Glimm's scheme, in *Nonlinear Evolutionary Partial Differential Equations*, X. Ding and T. P. Liu Eds., American Mathematical Society – International Press, Providence, RI (1997), 355–362.

D. Serre, Solutions á variations bornées pour certains systèmes hyperboliques de lois de conservation, *J. Differ. Equations* **68** (1987), 137–168.

D. Serre, *Systemes de Lois de Conservation I, II*, Diderot Editeur, Paris (1996).

M. Sever, Uniqueness failure for entropy solutions of hyperbolic systems of conservation laws, *Commun. Pure Appl. Math.* **42** (1989), 173–183.

J. Smoller, *Shock Waves and Reaction-Diffusion Equations*, Springer-Verlag, New York (1983).

B. Temple, Systems of conservation laws with invariant submanifolds, *Trans. Am. Math. Soc.* **280** (1983), 781–795.

B. Temple and R. Young, The large time stability of sound waves, *Commun. Math. Phys.* **179** (1996), 417–466.

K. Trivisa, A priori estimates in hyperbolic systems of conservation laws via generalized characteristics, *Commun. Partial Differ. Equations* **22** (1997), 235–268.

A. Tveito and R. Winther, Existence, uniqueness and continuous dependence for a system of hyperbolic conservation laws modeling polymer flooding, *SIAM J. Math. Anal.* **22** (1991), 905–933.

A. I. Volpert, The spaces BV and quasilinear equations, *Math. USSR Sb.* **2** (1967), 225–267.

B. Wendroff, An analysis of front tracking for chromatography, *Acta Appl. Math.* **30** (1993), 265–285.

G. B. Witham, *Linear and Nonlinear Waves*, Wiley, New York (1974).

Z. Xin, The fluid-dynamic limit of the Broadwell model of the nonlinear Boltzmann equation in the presence of shocks, *Commun. Pure Appl. Math.* **44** (1991), 679–713.

K. Yosida, *Functional Analysis* (6th edn), Springer-Verlag, Berlin (1980).

R. Young, Exact solutions to degenerate conservation laws, *SIAM J. Math. Anal.* **30** (1999), 537–558.

S. H. Yu, Zero-dissipation limit of solutions with shocks for systems of hyperbolic conservation laws, *Arch. Ration. Mech. Anal.* **146** (1999), 275–370.

W. P. Ziemer, *Weakly Differentiable Functions*, Springer-Verlag, New York (1989).

K. Zumbrun, *N*-waves in elasticity, *Commun. Pure Appl. Math.* **46** (1993), 75–95.

# Index